A TEXTBOOK OF SPECTROSCOPY

A TEXTBOOK
OF
SPECTROSCOPY

[For B.Sc. (Hons.) and M.Sc. Students]

By

M.S. YADAV

M.Sc., Ph.D.

ANMOL PUBLICATIONS PVT. LTD.

NEW DELHI - 110 002 (INDIA)

ANMOL PUBLICATIONS PVT. LTD.
H.O.: 4374/4B, Ansari Road, Darya Ganj,
New Delhi-110 002 (India)
Ph.: 23278000, 23261597
B.O.: No. 1015, Ist Main Road, BSK IIIrd Stage
IIIrd Phase, IIIrd Block,
Bangalore - 560 085 (India)
Visit us at: www.anmolpublications.com

A Textbook of Spectroscopy

Second Revised Edition, 2003
Reprint, 2008

PRINTED IN INDIA

Printed at Mehra Offset Press, Delhi.

Contents

Preface

In recent years, there has been no dearth of elegant works dealing with the subject of spectroscopy at the undergraduate and postgraduate levels. There are, however, very few books which hold the interest of the inquisitive students. The present book has been an attempt to hold the interest of the inquisitive students. The order of the topics covered in this book may not be able to satisfy every one. However, the order of the topics covered in this book can do some justification for Indian students. Each topic covered in this book has been self-sufficient in itself and has been explained with a large number of figures. Every topic has been described in the light of modern developments in a simple language and elegant style. An attempt has been made to include the latest development in the instrumentation of various spectroscopic techniques.

We would like to thank several of my immediate colleagues and friends for making useful suggestions during the preparation of manuscript.

Suggestions or criticisms towards the further improvement of the book shall be gratefully acknowledged.

—**The Author**

Preface

1

Introduction to Spectroscopy

Historical Development

Towards the end of the Napoleonic War in Europe, Joseph Von Fraunhoffer carried out one of the first spectroscopic experiments. He simply focussed the direct light from the sun on to narrow slit, dispersed the light from this image with a prism, and looked carefully at what he saw. (This was effectively the same experiment which Issac Newton had carried out some 150 years earlier but with much higher dispersion and some ability to measure the wavelengths). In addition to the familiar rainbow pattern formed by spreading out sunlight into its component wavelengths, he saw several black lines super imposed which correspond to the image of the slit at particular wavelengths. Not knowing the explanation of these features, he simply produced a list of the lines, labelling the major ones alphabetically A, B, C.....H.

It was many years later (1859) that the correct explanation for these lines was suggested by G. R. Kirchoff, namely that the light from the sun was being absorbed at these particular wavelengths by chemical species which are present in the solar atmosphere. The double line in the yellow, labelled D by Fraunhoffer, was assigned to the sodium atom because the wavelength corresponded exactly with the bright lines seen when salt was introduced into the flame of Bunsen burner. By working systematically through all conceivable possibilities, Kirchoff was able to assign all Fraunhoffer lines to a variety of different elements, such as Fe and Ca. This early achievement demonstrates the power of spectroscopy to identify specific atoms and molecules remotely because their spectra are different and uniquely characteristic. This ability is still exploited today, being used to identify and monitor molecules in remote astrophysical sources like the interstellar clouds, in the further research of our own atmosphere, and even the exhaust gases of cars as they speed past on the high way.

We get information about the structure of matter by studying its physical and chemical properties. Study of absorption and emission spectra etc. Generally more precise information can be obtained from spectroscopy. The spectroscopic properties of atoms and molecules results from interaction with electromagnetic waves.

Atoms of molecules in their normal state are in the most stable electronic, vibrational and rotational state. In the stable state the atom or molecule possesses a particular minimum energy E_0. By interaction with radiations, atoms or molecules may undergo transitions into energy enriched, excited states. These then possess a higher energy written as E*. The energy $\Delta E = E^* - E_0$ is needed to bring about this transitions from a stable ground state to excited state. This transition is known as *excitation* and so the energy needed is called *excitation energy*. This energy is supplied by one quantum of radiation and is absorbed by an atom or molecule. When energy associated with one quantum is smaller than excitation energy (ΔE) no excitation takes place. Two or more photons with a lower energy than that required for excitation can not combine their energies to produce the excitation energy ΔE.

Photons with the energy hv striking atoms or molecules which required the energy $\Delta E = hv$ to get excited are absorbed by these atoms or molecules. Thus the intensity of radiations decreases after passing through. The measurement of radiation intensity during irradiation of a substance as a function of wavelengths, wave number or radiation energy results in the absorption spectrum of that substance.

In an atom, electrons occupy only particular energy levels known as atomic orbitals. Thus atoms can be excited only by those photons which supply the energy difference between these energy levels. Absorption can occur only at particular energies or wave-lengths. Therefore absorption spectra are line spectra.

X-ray spectroscopy can also be used to find out the structure of organic molecules. For spectroscopic analysis of organic molecules ultraviolet, infra red, ESR, microwave and *nmr* spectroscopy are found to be useful. They provide rapid and exact information with instruments which can be operated easily.

The various branches of spectroscopy involve measurement of two important experimental parameters namely the energy of the radiation absorbed or emitted by the system and the intensity of spectral lines.

Most of the spectroscopic studies are done under two important heads *i.e.* Atomic spectroscopy and molecular spectroscopy.

(i) Atomic Spectroscopy

It is concerned with the interaction of electromagnetic radiations with atoms which are most commonly in their lowest energy state known as *ground state*. In this case transitions of electrons occur from one electronic energy level to another.

(ii) Molecular Spectroscopy

It is concerned with the interaction of electromagnetic radiations with molecules. In this case transitions occur between rotational and vibrational energy levels in addition to electronic transitions. The spectra obtained are therefore quite complicated in this case than those of atomic spectroscopy.

Molecular spectra extend from visible to microwave region and it can provide information about molecular vibrations and rotations to reveal a great deal about molecular structure.

It is possible to get a lot of information about molecular structure from atomic and molecular spectra.

Newton was the first to carry out spectroscopic investigations in seventeenth century. The introduction of quantum mechanics made possible the quantitative treatment of spectroscopy.

The major advantages of using spectroscopic methods are :

(i) These take a very little time to provide the required informations.

(ii) Only very small quantities (milligram or micrograms) of the substance is required.

(iii) The desired observation on the substance does not affect any chemical change in the material. It is therefore, recorded unchanged

(iv) These are found to be highly reliable in establishing identity of compounds.

(v) The information collected is in the form of permanent record.

(vi) Study of reaction rates can be easily carried out by these methods by continuous analysis.

(vii) These methods provide for more information about molecular structure.

(viii) It is possible to detect almost all groups and other structural features of a compound by using spectroscopic methods.

Interaction of Radiant Energy and Matter

There are a number of processes by which radiation may react with matter. Some of these are later used to explain the chemical and physical reactions within the system being measured, and some are required to explain the components of the instruments. Specific instruments have been developed to measure each interactive property. A requisite process in one instrument may contribute to an error in another. For easy explanation the properties have been paired :

1. Absorption and transmission
2. Emission and fluorescence.

It is not to be inferred that these are only reactions of energy with matter. The listing includes only those which are most prominent in instrumental analysis.

Absorption and Transmission

The electrons in chemical particles such as atoms, ions, or molecules possess a limited number of discrete energy levels which they can occupy in the particle. When electromagnetic energy like light is impinged upon these particles the electrons may be promoted from the lower energy **(ground)** state to one of a higher energy **(excited)** state provided the energy of the impinging radiation corresponds to one of the energy differences between the ground and excited states. During the process the energy of the incoming radiation is transferred to the particles at wavelengths corresponding to these discrete energy differences and the electromagnetic radiation at these wavelengths is decreased or **absorbed.** If the intensity of the incoming radiation is in excess of that needed at any particular wave-length ($E = hc/\lambda$) the remainder is **transmitted.**

Due to the limited number of permissible atomic energy levels, atoms can absorb at only a few selected frequencies. The plot of absorbance vs. wavelength shows a **line absorption spectrum** where each line corresponds to a permitted transition. This type of spectrum is utilized for atomic analysis in atomic absorption and x-ray absorption instruments.

On the other hand, the number of possible energy transitions in molecules is much greater than in atoms. More electrons are present in molecules and the number of possible *electronic* transitions is vastly increased. In addition, energy is also needed for *vibration* of the atoms within the molecule and *rotation* of the molecule around its gravitational center, the total energy of the molecule being.

$$E_{total} = E_{electronic} + E_{rotational} + E_{vibrational}$$

Because of the number of permissible combinations of the three components and because the energies of these various combinations are frequently close to each other, they cannot be resolved by the instrument and the plot of absorbance vs. wavelength appears as a **band absorption spectrum.** See Figure 1.1.

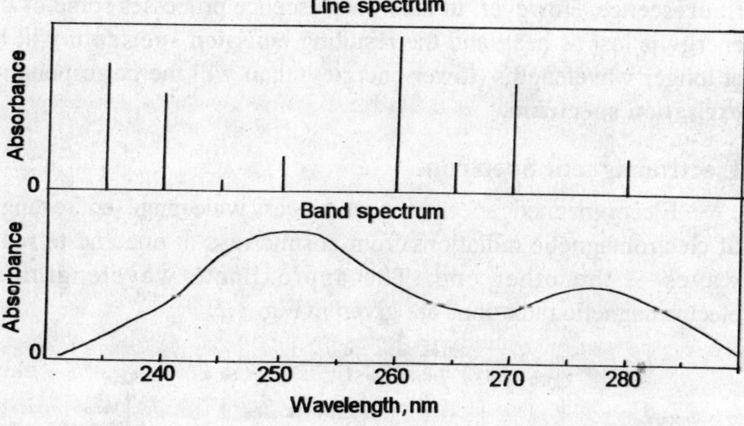

Fig. 1.1: Absorption spectra. The top one is a typical line or atomic spectrum. and the bottom one is a typical band or molecular spectrum.

Emission and Fluorescence

After electrons in the chemical entities are excited, the energy of excitation will subsequently be released and the electrons will be returned to their respective ground or stable states. This excess energy release is usually dissipated as heat when the particles collide, but it might also be released in photochemical reactions or as radiant energy. It is not uncommon for several of these processes to occur simultaneously.

If the excited particles are gaseous atoms, a limited number of discrete energy transitions which correspond to specific wavelengths may occur and a discontinuous **emission line spectrum** is obtained. Emission line spectra are used as qualitative and quantitative tools in flame photometry, emission spectroscopy, and X-ray spectroscopy. The differences in the three will be discussed latter.

An emission band spectrum is obtained if the energy states of the sample are so close to each other and so numerous, as in an excited molecule, that the energies of the various transitions to the ground state cannot be separated. This principle is used in molecular fluorescence and phosphorescence instruments. Molecular fluorescence may be roughly considered the reverse process of molecular absorption. If the emitting energy is quantitatively equal to the excitation energy, the emission process is designated as **resonance fluorescence.** However, in most fluorescence processes some of the energy is lost as heat, and the resulting **emission spectrum** will be at longer wavelengths (lower energies) than will the corresponding **excitation spectrum.**

Electromagnetic Spectrum

Electromagnetic spectrum covers a very wide range (entire range) of electromagnetic radiations from cosmic rays at one end to radio waves at the other end. The approximate wavelengths of electromagnetic radiations are given in Fig. 1.2.

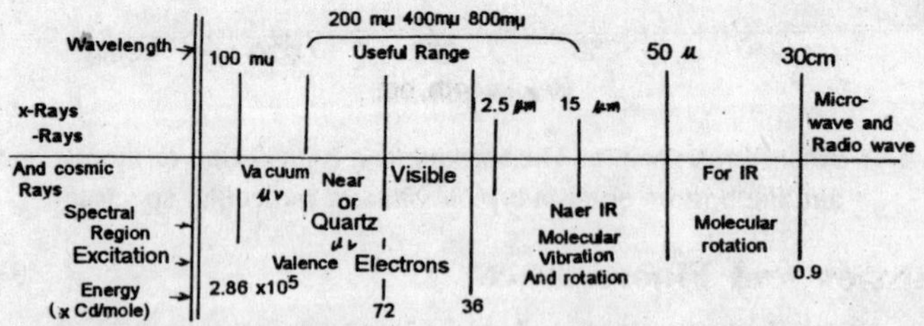

Fig. 1.2: Electromagnetic spectrum

The major characteristics of various spectral regions are as under:

(i) γ-ray region (10^8 – 10^{20} Hz or 100pm – 1pm)

It consists of cosmic rays and γ-rays and lies between 0 2 to 1° A (or .1nm). The γ-rays are shortest waves emitted by atomic nuclei. The energy changes involve the rearrangement of nuclear particles, having energies of 10^2 to 10^{11} J/gram atom.

(ii) X-ray region (10^{16} 10^{18} H_2 or 10nm to 100 pm)

It lies between 1 to 10A°. X-ray are emitted by or absorbed by movement of electrons close to nuclei of relatively heavy atoms. These transitions involve energy changes involving the inner electrons of an atom or a molecule, could be of the order of 10,000 kJ/mol.

(iii) Visible and ultraviolet regions ($10^{14.5}$ – 10^{16} H_2 1μm – 10 nm)

U.V. region is between 200 to 400 nm and the visible region is between 400 – 740 nm. The ultraviolet and visible region are adjacent and the boundry is somewhat arbitrary since it depends upon human perception of light. The commonly used unit of wavelength in this regions is **nanometer** (1nm = 10^{-9} m).

The ultra violet region can further be extended into **vacuum ultra violet,** then (1 – 180 nm) and ultra violet (180 – 400 nm). In the ultra violet region air starts absorbing. The quartz also starts absorbing below 180 nm, quartz is used for constructing cells and windows. Due to this the measurement of radiations below 180 nm becomes a bit more difficult and complicated. However, now instruments are available for use in vacuum ultra violet region.

To describe radiations some times their frequency is expressed. Longer the wavelengths smaller is the frequency. They are related to each other as

$$= \frac{c}{\lambda} \quad \text{or} \quad c = v.\lambda$$

where

c = 3 × 10^{10} cm/sec (velocity of hight)

λ = wavelength

v = frequency.

All the waves possess the same velocity (3×10^{10} cm/sec)

I.R. Region (Infra red region) ($10^{12} - 14^{14.5}$ H_2 or 10µm – /µm)

It lies between 0.7 – 200 neutron (µ) and can be furthur divided into following sub-regions near infra red – 0.7 – 2.5 µ

infra red – 2.5 – 15 µ

For infra red – 15 – 200 µ

($\mu = 10^{-4}$ cm)

In I. R. region the electromagnetic spectrum is associated with the changes in the vibration of the molecules *i.e.,* molecular vibrations. It is the most important region for an organic chemist. The Energy Changes are of the order or 10^4 J/mole.

Microwave region (0.1 mm to 1 cm wavelength)
($10^{10} - 10^{12}$Hz or 1 cm – 1.00 µm)

Thus region corresponds to changes in rotation of the molecules. Separation between rotational levels of molecules are of the order of hundreds of joules per mole.

Radio frequency region (10 cm – 1 cm wavelength)
($10^6 - 10^{10}$ H_2)

The energy changes involved in this region are due to the reversal of a spin of nucleus or electrons. This is of the order of 0.001 – 10 Joules/mole.

Different types of Molecular Energies

The internal energy (E) possessed by a molecule consists of energy due to translational motion (which is not of concern in molecular spectroscopy), vibrational energy, rotational energy and electronic energy etc. In spectroscopy we are mainly interested in $E_{vib.}$, $E_{rat.}$, $E_{elec.}$ and it has been found that $E_{elec.} > E_{vib.} > E_{rot.}$. These three components of molecular energy are also governed by laws of quantum theory and they can have only discrete and quantised values.

Translational Energy (E_t)

Because of very small difference between discrete energy levels the kinetic energy of translation for all practical purposes may be considered as continuous though it has discrete energy levels. It is given by

$$E_t = \tfrac{1}{2} mv^2 = \frac{n^2h^2}{8ma^2}$$

thus allowed values of E_t (K. E.) depend upon the square of translational quantum number n and it satisfies the conditions $n.\ \lambda/2 = a$. Where a = amplitude.

The above formula is for a particle moving in one dimension box and the wave aspect of particle vanishes at $x = 0$ and $x = a$, because particle can not penetrate into regions beyond these points. Such a wave is also called a *standing wave.* The corresponding formula for a particle confined with in a rectangular box of sides, a, b, c is given by

$$E_t = \frac{h^2}{8m}\left(\frac{n_1^2}{a^2} + \frac{n_2^2}{b^2} + \frac{n_3^2}{c^2}\right)$$

where n_1, n_2, n_3 are three integral quantum numbers corresponding to three degrees of freedom of the particle and they specify the allowed energy levels. When $a = b = c$ the container becomes a cube and then the allowed energy levels are

$$E_t = \frac{h^2}{8mn^2}\left(n_1^2 + n_2^2 + n_3^2\right)$$

the volume of cube is a^3, thus

$$E_t = \frac{h^2}{8mv^{2/3}}\left(n_1^2 + n_2^2 + n_3^2\right)$$

From this *we observe that as the volume increases, the spacings of energy levels decreases* and as the volume becomes infinity the spacings of energy levels becomes zero and so the energy varies continuously and not in quantum jumps.

The lowest allowed energy state in a three dimensional box is designated as (111) and in this state $n_1 = n_2 = n_3 = 1$, the next state is (211) but there are three states of the same energy for a cubical box *i.e.*, (211), (121), (112). This energy level is said to possess *statistical weight or degeneracy* g = 3. *It is to be noted that levels but not states may have degeneracy.*

Since the number of translational energy levels is very large and the difference between them is very small so continuous range of translational energies are available. The concept of translational energy is not of much use in molecular spectroscopy.

Quantised translational energy can be observed by infrared spectroscopy under special circumstances. From the equation of E_t in three dimensional box it can be seen that large separations between energy levels occur when both particle mass and available volume are small. For example a small volume can be created by trapping a hydrogen molecule in a solvent cavity in a liquid at low temperature. Under these conditions it is possible to observe excitation of H_2 from one translational level to another by absorption of light in for infra red region. Fig. 1.3 shows the absorption spectrum of isolated H_2 molecules trapped in liquid organ solvent cages at 90°K.

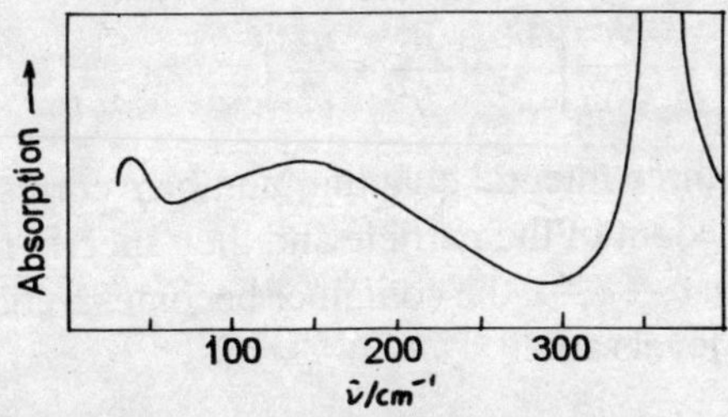

Fig. 1.3: Absorption spectrum of isolated H molecules trapped in liquid argon at 90°K. The absorption band is due to uptake of radiation into translational motion of the H_2

Rotational Energy

It is the energy associated with the system due to rotation of the molecules considering atoms as fixed point masses. The value of E_{rot} is given by

$$E_{rot} = \frac{1}{2} I\omega^2$$

where I is the moment of Inertia and ω is the angular velocity of a rotating molecule.

The rotational energy of a simple linear molecule is given by

$$E_{rot} = J(J+1)\frac{b^2}{8\pi^2 1}$$

vhere J is the rotational quantum number. It can have values J = 0,

1, 2........ . Each rotational energy level possesses a (2J + 1) fold degeneracy.

The pattern of rotational energy levels is shown in Fig. 1.4

J	E
3	$12h^2/8\pi^2 I$
2	$6h^2/8\pi^2 I$
1	$2h^2/8\pi^2 I$
0	0

Fig. 1.4: Energy levels of a rigid linear rotor

Because the energy depends as J(J + 1) the spacing between levels increases as J increases. So the size of quantum of rotational energy graves as J increases.

The equation for E_{rot} is not applicable in liquid phase where molecular collisions are frequent. Only polar molecules can rotate in electromagnetic field.

There is only one rotational degree of freedom in case of monoatomic molecules and three rotational degrees of freedom for a non-linear tri- or poly-atomic molecules but a linear tri- or poly-atomic molecule can have only two degrees of freedom because of the restriction imposed on its rotation about the axis passing through the nuclei of bonded atoms.

Vibrational Energy

This type of energy arises due to oscillation of atoms (considered as point masses) about equilibrium position. It is given by

$$E_{vib} = hv\left(v + \tfrac{1}{2}\right)$$

Here V = frequency of vibration and V is the vibrational quantum number with allowed values V = 0, 1, 2.......

Making calculation from the above equation it can be easily seen that the value of E_{vib} may not be zero for the ground state. In this case the ground state (V = 0) has an energy equal to $hv/2$ and this called *zero point energy.*

Even in the limit of absolute zero when all translational and rotational motion ceases in a crystal, the molecules would still be oscillating internally about their equilibrium inter nuclear distance R_e. The zero point energy is by no means small or negligible.

For a diatomic molecule vibration frequency is given by the equation.

$$V = \frac{1}{2\pi}\sqrt{\frac{K}{\mu}}$$

where K is a **force constant** and μ is the **reduced mass.**

The vibrational degrees of freedom for tri-, poly-atomic non-linear molecules is $(3n - 6)$ and for such linear molecules is $(3n - 5)$.

The vibrational energies are so large that quantum effects are dominant. When a molecule absorbs energy, the increase in vibration motion of the molecule is generally accompanied by increased rotation of the same molecule. This is the basis of infra red absorption spectroscopy.

Electronic Energy

It is the energy associated with the system due to the motion of the electrons about the nuclei of atoms of a molecule. Any increase in electronic energy of a molecule is because of an increase in the kinetic energy required to take the electron from ground state to excited state and it varies with the type of molecular bond in which the electron exists. When we speak of a hydrogen atom with its electron in 1*s*, 2*p*, 3*s* etc., orbital, in each case the atom is associated with different *electronic energy* and is in different *electronic energy level*. The separation between electronic energy levels is generally large and is 10 to 100 times larger than typical separations between vibrational levels. These levels can be studied by adsorption or emission spectroscopy. Visible or ultraviolet light can be used for the excitations of electron from one electronic energy level to the other.

There is no general formula for electronic energy levels except for simple hydrogen like atoms.

With few exceptions molecules are found in their lowest electronic energy levels at temperatures below 1000 K.

Interaction of Radiations with Matter

As the radiation passes from a vacuum through the surface of a portion of matter, the electrical vector of radiation interacts with the atoms and molecules of the medium. The nature of the interaction depends upon the properties of matter and may lead to transmission, absorption or scattering of radiation.

Some of the phenomenon that are likely to occur when electromagnetic radiation passes through matter are those of dispersion reflection, refraction, scattering, absorption, polarisation etc. Some of these are as follows :

(i) When the energy of the photons of radiations correspond to appropriate values they may be absorbed by matter for electronic, vibrational or rotational changes. Some times a combination of these changes may occur. When atoms or molecules absorb energy they acquire the excited state which is unstable and so they give out energy quickly and lose the energy in the form of heat energy or in the form of radiations of suitable wavelength.

(ii) The radiations passing through matter are not essentially completely absorbed and it is possible that some of them may pass into matter and undergo scattering or reflection etc. Some of them may be remitted at the same wavelength or different wavelengths.

(iii) Some times the radiations are neither absorbed nor scattered in such cases they only undergo changes in orientation or polarisation.

(iv) In some cases the excited molecules formed due to absorption do not lose energy quickly but lose it after some time. In such cases the energy is remitted in the form of electromagnetic radiations mostly of larger wavelength then the wavelength of radiations absorbed. This happens in case of fluorescence and phosphorescence.

Absorption and Emission Spectra

In spectroscopy we are mainly concerned with the interaction of electromagnetic radiations with matter. This interaction of radiation and matter may result in variation in intensity of electromagnetic radiation with frequency (or wavelength). These variations can be observed in two ways, firstly in which the sample itself emits radiations and secondly where the sample absorbs radiations from a continuous source. The spectrum obtained in first case is called **emission spectrum** and in second case as **absorption spectrum.**

Atomic Absorption of Electromagnetic Radiations

The absorption of electromagnetic radiations by an atom results in increase in their electronic energy. The absorption occurs only if the energy of attacking photon is equal to the energy difference (ΔE) between two quantised levels of absorbing atom. Due to this increase in electronic energy electron jumps from an inner orbit to any outer orbit and so the atom gets excited if it was previously in the ground state. This excited atom when reverts to ground state, it loses energy either in one straight jump or in a number of steps in the form of electromagnetic radiations. This effect appears as *line spectrum.* Such explanation explains the spectrum of hydrogen or hydrogen like atoms. In case of multi electron atoms the energy absorbed may bring about electronic transitions from $1s \rightarrow 2s$ or $2s \rightarrow 2p$ or $3d \rightarrow 4p$ etc. These electronic transitions correspond to X-ray radiations, u.v., radiations and visible radiations respectively.

These lines spectra are generally obtained by using light sources like mercury lamp, sodium lamp, neon discharge tube etc.

Some important points about line spectrum are :

(i) The spectral lines are regularly spaced.

(ii) The wavelengths of lines emitted are the characteristics of the element under consideration and are independent of the compound in which this element is present.

(iii) When these lines are viewed through an instrument having high resolving power. These lines can be seen as regularly spaced but they differ in their intensities.

Molecular Absorption of Electromagnetic Radiation

When electromagnetic radiations are absorbed by a molecule it gets excited and on returning to its ground state it emits radiations which appear as **band spectrum.** These are characteristic of the molecule under consideration. If observed with an instrument having high resolving power each band appears as composed of a large number of very fine lines. If these lines are observed along the wavelength edge they become closer and closer and finally they coincide. This edge is known as **band head.** These spectra are due to molecules and these bonds disappear when emitting substance is heated to a temperature at which the molecule dissociates into atoms.

For rotational transitions only a small amount of energy is needed because the rotational energy levels of a molecule are very closely spaced so such transitions can be observed in 'far infra red' and 'microwave' regions of spectrum. This information may be quite useful about molecular structure but is of little importance in analytical chemistry.

Vibrational transitions need larger amounts of energy than required for rotational transitions and so the vibrational transition occur in infra red region of spectrum. The occurrence of pure vibrational transitions is not possible because they are super imposed by rotational transitions. Hence in a typical vibrational spectrum we find complex bands and not single lines.

The electronic transitions occur only when the molecule absorb photons corresponding to visible light or *u.v.* radiations. The electronic transition from a lower orbital to a higher orbital is always accompanied by a vibrational and a rotational change. These change introduce 'fine structure' into the spectrum and so the molecular structure involves a band of wavelengths and not a single line. The spectrum obtained for electronic transitions is a broad absorption spectrum.

Spectroscopic Displacement Law

W. Kossel and A. Sommerfeld in 1919 gave the law known as *spectroscopic law*, it states, "The spark spectrum of an element is similar to the first spark spectrum of the element one place higher in the periodic table or to the second spark spectrum of the element two places higher in the periodic table and so on." The law is found to be generally applicable to fine structures of spectral lines.

Techniques of Spectroscopy

Of the various types of spectroscopic techniques known, the most useful are (i) infra red spectroscopy (IR spectroscopy), (ii) ultra violet and visible spectroscopy (*u.v.* spectroscopy) and (iii) Nuclear magnetic resonance spectroscopy (NMR spectroscopy).

The other methods being

(i) Microwave spectroscopy
(ii) Raman spectroscopy
(iii) NQR spectroscopy
(iv) Electron spin resonance (ESR) spectroscopy
(v) X-ray spectroscopy
(vi) Electronic spectroscopy

(vii) Atomic absorption spectroscopy or
Absorption flame photometry

(viii) Flame Emission spectroscopy or Flame photometry

(ix) Emission spectroscopy

(x) Molecular luminescence fluorometry

(xi) Mossbauer spectroscopy

(xii) Neutron diffraction spectroscopy

(xiii) Man spectroscopy

(xiv) C–13 NMR spectroscopy (CMR spectroscopy)

(xv) Inductively coupled plasma atomic emission spectroscopy.

2

Absorption Spectroscopy

Principle of Absorption Spectroscopy

When a beam of electromagnetic radiations is passed through a substance, the radiation can either be absorbed or transmitted depending upon the frequency of radiations and molecular structure of the substance under examination. If the molecule has higher energy molecular orbitals available, usually an anti-bonding orbital, it can absorbs light energy of a particular wavelength to excite its electrons to pass from one of its quantised levels to a higher energy level, by absorbing an energy equal to ΔE corresponding to photo energy ($E = hv$). There are two types of jumps when a molecule absorbs energy:

(a) **Electronic Excitation.** These can be achieved by electromagnetic radiations of wavelength ranging from 200 – 800 $m\mu$ *i.e.*, ultra violet (200 – 400 $m\mu$) and visible absorption spectra (400 – 800 $m\mu$).

(b) **Nuclear vibration.** These can be achieved by electromagnetic radiations following with in the range of I.R. (2.5 – 1.5 μ) or radio wave region (10 m–1 cm wavelength).

In either of these spectral ranges, the action of light absorption at a particular wavelength is a direct function of and indication of a molecular structural feature possessing the capability of absorbing the specific discrete energy packet.

The spectroscopic examination consists in passing radiations of continuously changing wavelength through a sample of the compound and measuring the light intensity before and after passing through the sample. A record of amount of light absorbed by a sample as a function of wavelength of radiations is called **Absorption spectrum.** In principle all **absorption spectra** should consist of *lines* but actually we get *absorption bands*. The ultraviolet and visible absorption spectrum is recorded as maxima or vertical peaks and the infrared absorption spectrum is recorded as minima or inverted peaks. In the case of infra-red, it is the transmission of X-rays *i.e.*, opposite of absorption which is recorded.

Nowadays accurate spectroscopic analysis of complex organic and inorganic compounds has become a standardised technique and the spectral features noted are a sure test for the establishment of molecular structure.

Absorption Spectroscopy

Since infra red and ultraviolet spectroscopy depend upon absorption or radiations by molecules they are collectively known as *Absorption spectroscopic analysis technique*. The molecules gain energy on absorption of radiation and it depends upon the wavelength/frequency of the radiation and the structure of the molecules. The gain in energy by the molecule is measured in absorption spectroscopy.

The spectrum of the compound in *u.v.* and I.R. spectroscopy is a plot which shows how much radiation is absorbed (or transmitted) at each frequency. It can be highly characteristic of the structure of the compound.

The intensity of absorption (or transmission) is repeated along the *ordinate* and the wavelength or frequency is repeated along the *abscissa*.

In *u.v.* spectroscopy the intensity of absorption is represented by the Greek symbol epsilon (ε) along the ordinate and is known as *molecular extinction coefficient*.

Atomic Absorption Spectroscopy

Since the mid 1950's when Alan Walsh introduced atomic absorption spectroscopy as an instrumental technique for quantitative determination of trace metals in liquids, the growth of analytical atomic absorption spectroscopy has been phenomenal. About 60-70 elements including rare earth metals has been determined making use of this technique.

Direct application of technique is limited to metals with the exception of B, Si, As, Se and Te. A number of non-metals have been estimated using indirect methods.

In atomic absorption spectroscopy sample preparation is not essential, it is an ideal tool for non-chemist *e.g.*, the biologist, engineer etc.

Principle

The absorption of energy by ground state atom in the gaseous state is the basis of atomic absorption spectroscopy.

When a solution containing metallic species is introduced into a flame, we get the vapour of the metal in which some of the metal atoms may be raised to an energy level sufficiently high to emit the characteristic radiation of the metal. This phenomenon is generally used in *emission flame photometry.*

However, a large number of metal atoms remain in ground state and these are receptive of light radiations of their own specific resonance wavelength (*i.e.*, the same wavelength as they would emit if excited). Thus when a light of this wavelength is allowed to pass through a flame having atoms of the metallic species, a part of the light is absorbed which will be proportional to the density of atoms in flame. So in absorption spectroscopy the amount of light absorbed is determined knowing the value of absorption it is possible to find out the concentration of metallic element, since absorption is proportional to the density of metal atoms in flame. This can be expressed mathematically as under.

At any frequency v total amount of light absorbed is equal to

$$\frac{\pi e^2}{mc} Nf.$$

where e = charge on electron of mass m.

c = velocity of light.

N = Total number of atoms which can absorb at frequency v in the light path.

f = oscillator strength.

Oscillator strength is the ability for each atom to absorb at frequency, v_o. Since π, e, m and c are constants so the above equation can be simplified as

Total amount of light absorbed = constant × N × f

From this simplified equation we can see that

(i) It contains no term involving the wavelength (or frequency) of absorption other than the indication of the actual absorption wavelength.

(ii) It contains no term involving temperature.

So we conclude that absorption by atom is independent of the wavelength of absorption and temperature of atoms. This gives a distinctive advantage to absorption spectroscopy over flame emission spectroscopy.

The other important facts about absorption spectroscopy are

(i) In atomic absorption spectroscopy the signal is obtained from the difference between the intensity of source in the absence of metallic present in the solution and the decreased intensity obtained when metallic elements are present in the optical path.

(ii) In atomic absorption spectroscopy, the absorption depends upon the number of unexcited atoms and the absorption intensity is independent of temperature of the flame.

(iii) In atomic absorption spectroscopy the relation between absorbance and concentration is nearly linear and so the Beer's law is obeyed over a wide range of concentration.

Instruments

A host of instruments are available commercially for absorbance measurements in the ultraviolet and visible spectral regions. The simplest of these are colorimeters, in which the eye serves as transducer. The most complex are double beam, recording instruments which cover the entire spectral range from about 185 to 3000 nm. Some typical instruments are given here.

Fig. 2.1: Shows an atomic absorption spectro photometer.

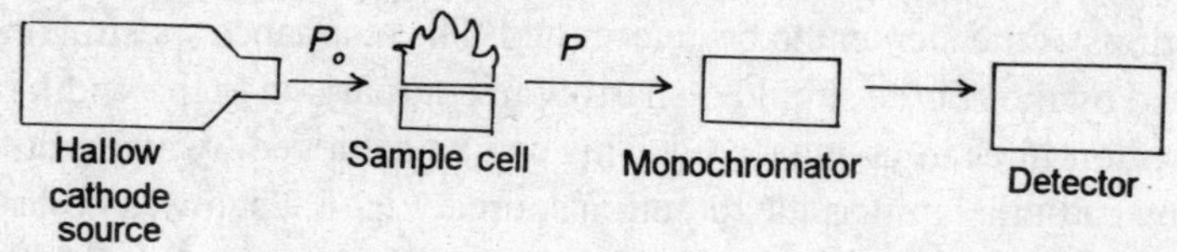

Fig. 2.1.

Light of a particular wavelength (obtained by using a special kind of lamp) that can emit the spectral lines corresponding to the energy required for an electronic transition from ground state to an excited state is allowed to pass through the flame. In the mean time the sample solution is aspirated into the flame. The solution gets dispersed into a mist consisting of very small droplets before it enters the flame and evaporates in the flame to form first the dry salt and then the salt vapours. A part of the vapours get dissociated into atoms of element to be measured. Hence, free unexcited atoms are present in the flame and these can absorb some radiations from the external source if the radiations correspond exactly to the energy needed for a transition of element from ground electronic state to some excited state. The unabsorbed radiation from flame is allowed to pass through a monochromator which isolates the exciting spectral lines of light source. From the monochromator these unabsorbed radiations are led into the detector which is then registered by a photodetector. The output of photodetector is amplified and measured on a recorder. Absorption is measured by the difference in transmitted signal in the presence and absence of test element.

Any type of atomic absorption spectrometer may be used but it must contain the following components:

(i) The radiation source
(ii) Chopper
(iii) Atomic vapour producer
(iv) An assembly for *Nebulisation*
(v) Monochromator
(vi) Detector
(vii) Amplifier
(viii) Read-out device.

1. Radiation Source. The source should emit stable, intense radiations of the element to be determined. The resonance spectral line emitted by the source should be narrow in comparison to the width of absorption lines to be measured. This can be achieved by adopting a hollow cathode lamp as the radiation source. Fig. 2.2 shows a hollow cathode lamp.

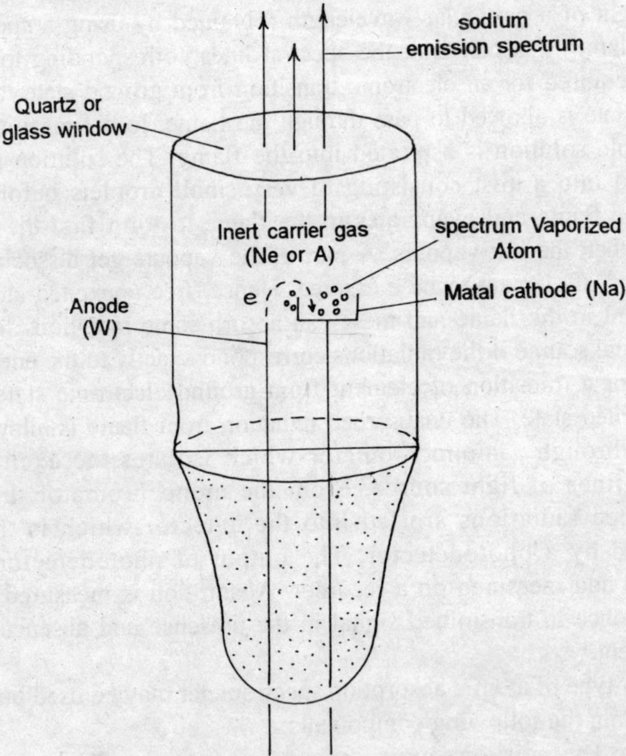

Fig. 2.2: Schematic diagram of a hollow cathode source lamp for atomic absorption of sodium

It consists of anode (W) of tungsten wire and metal cathode of element whose spectrum is desired which terminate in a hollow cup. The cathode may be constructed from an inert element into which the desired element or salt of the desired element is kept. The hollow cathode is generally filled with neon or argon gas at low pressure. The lamp envelope is made from the glass and there is a window of desired properties.

The emission spectrum produced by a hollow cathode lamp is a sharp line spectrum of the cathode material and the gas filled in it. The hollow cathode lamp is generally filled with helium or neon gas. The filled gas performs following functions:

(i) It acts as the main source of current carrying capacity in the hollow cathode.
(ii) It dislodges atoms from the surface of cathode.
(iii) It is mainly responsible for the excitation of ground state metal atoms

To work with the apparatus a d.c. voltage of 300-500 V is used and a pressure of 1-5 torr is maintained in the hollow cathode lamp. At higher pressure the discharge tends to be unstable and at lower pressures there is an increase in vapourisation of hollow cathode metal which also increases the operating temperature.

The spectral lines produced by hollow cathode lamp are quite narrow and they can be completely absorbed by the atoms. Thus, atomic absorption can be easily detected and measured.

Each hollow cathode lamp emits the spectrum which is characteristic of the metal used in cathode. A copper cathode emits copper spectrum and a zinc cathode emits zinc spectrum. The spectral lines emitted by any specific cathode are only absorbed by atoms of that very element present in the sample under analysis. It is therefore essential to use a different hollow cathode lamp for each element to be analysed by absorption spectroscopy.

Arc lamps (or gaseous discharge lamps) containing metals or metal salts are also used in atomic absorption spectroscopy. Such lamps are found to be useful for the alkali metals, zinc, cadmium and mercury.

2. Chopper. It is a rotating wheel that is interposed between the hollow cathode lamp and flame. Its function is to break the steady light from the lamp into an intermittent or pulsating light which gives a pulsating current in a photo cell. A steady current also exists because such a current is caused by light which is emitted by a flame. However, only the pulsating current is amplified and recorded. It is therefore possible to measure the absorption of light without interference from the light emitted by the flame.

3. Atomic Vapour Producer. To make absorption by atoms possible it is essential to reduce the sample to the atomic state and it can be conveniently achieved by using a flame to convert the liquid sample into vapour. It can also convert molecular entities into atomic vapours. Commonly used burners for this purpose are : (i) the total consumption burner and (ii) the premixed burner.

Fig. 2.3: shows a total consumption burner.

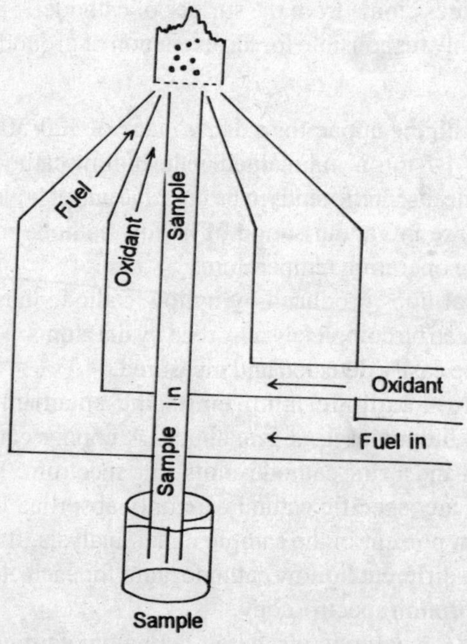

Fig. 2.3 : Schematic diagram of a total consumption burner. The fuel, oxidant and sample all mix at the flame.

In this burner the sample solution, the fuel and oxidising gases are passed through seperate passages to meet at the opening of the base of the flame. The sample being in liquid form, it is broken by the flame into droplets which are then evaporated or burnt. The residue left behind is reduced to atoms. Such burners do not used oxygen, with hydrogen or acetylene. They give very hot flames. Such a burner has a low efficiency, is noisy and hard to use.

Fig. 2.4: shows a premixed burner.

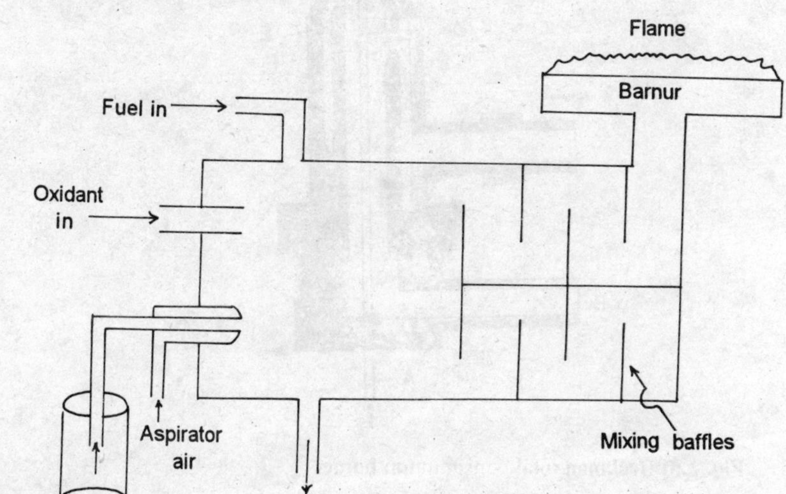

Fig. 2.4: Schematic diagram of a premix burner. The fuel, oxidant, and sample are mixed prior to their introduction into the flame.

A mixture of liquid sample and premixed gases ($C_2H_2 + O_2$) is allowed to enter the base M from where these gases enter the region A. From region A the unburnt hydrocarbons and liquid droplets enter the region B. Region B is a region of free heating and has a thickness of about 1 mm. In this region liquid evaporates and a residue is left behind. To carry out the heating in this region heat is obtained from region C by conduction and convection and by diffusion of radicals into it which initiate the combustion. When the residue is burnt in regions C and D it produces atom. The production of atoms starts in region C and gets completed in region D.

Such burners are found to be useful in the study of atomic absorption of metals belonging to groups IB, IIB, IA etc.

4. Nebulisation. *Nebulisation* is the process of formation of small droplets from the liquid sample. This process is completed before the liquid sample enters the burner. This is achieved by *pneumatic nebulisation* which makes use of a gas moving at high velocity. A Beckman total consumption burner shown in Fig. 2.5 is used for this purpose.

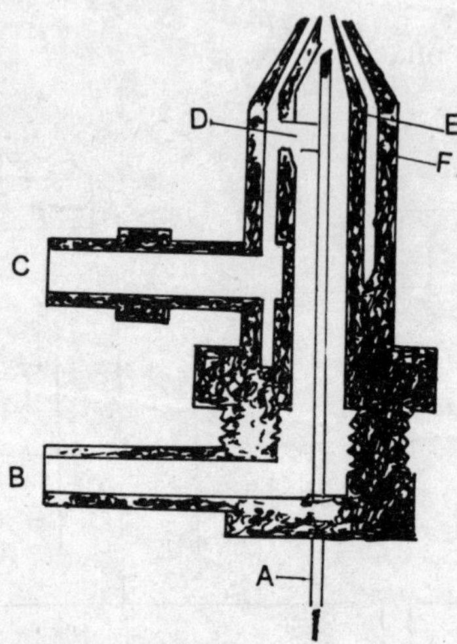

Fig. 2.5: Beckman total consumption burner

Various part of the burner are:

(i) A, It is a solution capillary.
(ii) B, It is an aspirating gas inlet.
(iii) C, It is a fuel gas inlet.
(iv) D, It is the centring screw.
(v) E, It is the gas inlet.
(vi) F, It is a Jacket.

A back pressure of 250 torr occurs at the tip of burner in such a burner. This high pressure is due to high velocity of aspirating gas as it comes out of the orifice. As the liquid is drawn up the capillary it is broken up into droplets by high velocity gas stream.

5. Monochromators. The function of monochromator is to select a given absorbing line from spectral lines emitted from the hollow cathode. The required wavelength is usually preferred by a prism or a grating and passed through a narrow slit. By carefully rotating the grating or the prism it is possible to change the wavelength continuously and thus total assembly is known as **monochromator.** It is so designed to give light of one wavelength accurately.

Commercially packaged atomic absorption instruments are generally provided with a monochromator of about 0.5 m focal length having a linear reciprocal dispersion in the range 16-35 A°/mm. For many elements high dispersion is not necessary and in such cases a resolution of the order of 0.5 A° is desirable. However, when cathode of the hollow cathode lamp is made up of transition metals a high dispersion is needed because of the complicated emission spectrum from such a cathode.

6. Detectors. For detection generally a photomultiplier tube is used. It is found to have a good stability when used with a stable power supply.

A photomultiplier tube (Fig. 2.6) has an evacuated envelope which contains a photo cathode and a series of electrodes called **dynodes** and an anode. It is fixed to the terminals of the power supply.

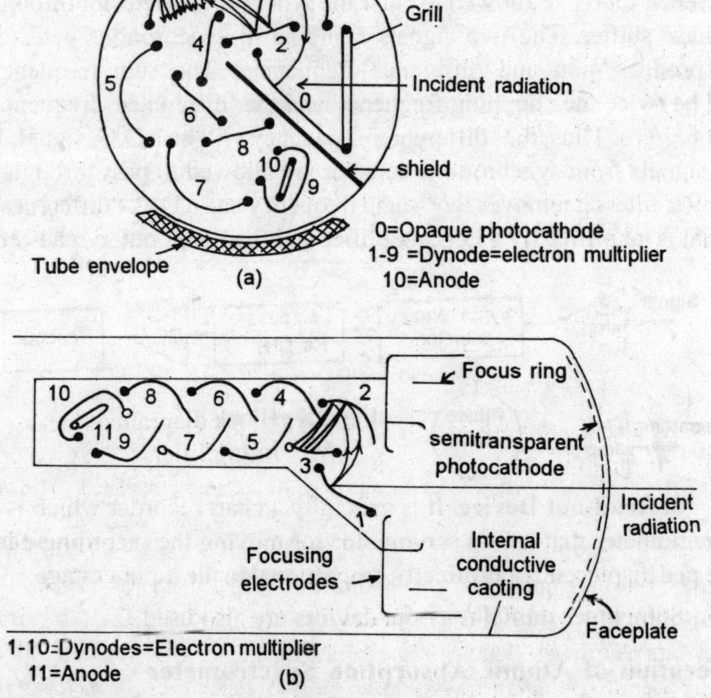

Fig. 2.6: Photomultiplier design. (a) The circular-cage multiplier structure in a "side-on" tube and (b) the linear-multiplier structure in a "head-on" tube.

A photon on striking the photo cathode dislodges an electron and the photon itself gets accelerated to dynode 1 thus, liberating two or more electrons from this dynode. In the same way electrons from dynode are accelerated to dynode 2 where more electrons are liberated and in this way current gets multiplied at each dynode and the resultant electron current is received by the anode to produce an EMF across RL which goes to the external amplifier and read out system.

7. Amplifier. The current from detector is fed to amplifier whose function is to amplify the electric current many times 'Lock in' amplifiers are preferred as they provide a very narrow frequency band pass and help to achieve an excellent signal to noise ratio.

Fig. 2.7 shows a "lock in" amplifier. The signal from detector is led to an AC amplifier to modulate the energy source. It then enters the synchronous detector and simultaneously a signal from the reference source is allowed to enter the synchronous detector through a phase shifter. The two signals combine in synchronous detector and produce 'sum' and 'difference' frequencies. The 'sum' frequency will be twice the chopping frequency and the 'difference' frequency will be zero. Thus, the 'difference' frequency will be a D.C. signal. If the signals from synchronous detector are allowed to pass through a low RC filters it removes the "sum" frequency and a D.C. "difference" signal is amplified by a D.C. amplifier and then goes out to read-out.

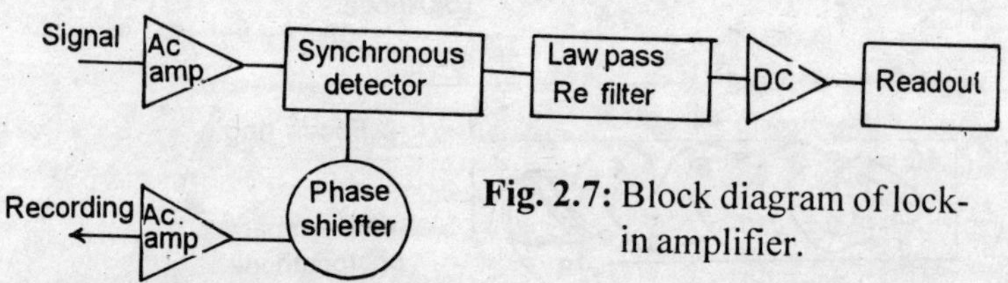

Fig. 2.7: Block diagram of lock-in amplifier.

8. Read-out Device. It is generally a chart recorder which is a potentiometer that uses a servomotor for moving the recording pen. The pen displacement is directly proportional to the input voltage.

Sometimes digital read-out devices are also used.

Operation of Atomic Absorption Spectrometer

To start with the spectrometer is adjusted at zero absorbance when a blank solution is sprayed in the film and the unobstructed light from hollow cathode lamp passes on to the photomultiplier tube when a solution containing the absorbing species is introduced a part of it is absorbed which decreases intensity of light falling as the photomultiplier and produces a deflection in spectrometer needle.

Standard solutions of the elements to be determined are used to obtain a calibration curve by which the contents in the test solutions can be measured. For quantitative analysis the sample is atomised and the absorbance is measured under exactly the same conditions under which the calibration curve was obtained.

Applications of Atomic Absorption Spectroscopy

This type of spectroscopy is widely used in analytical chemistry, ceramics, mineralogy, biochemistry, metallurgy, soil analysis etc.

1. Qualitative analysis

As already discussed a different hollow cathode is used for each element to be tested so an element which is used in construction of cathode of a hollow cathode lamp can be detected only. Since this analysis involves checking for a particular element at a time so it is a laborious process and is seldom used.

2. Quantitative Analysis

Absorption spectroscopy is one of the most useful tools available to the chemist for quantitative analysis. It is based as the amount of radiation absorbed by the sample which is directly proportional to $N \times f$. So it can be used to calculate N (the number of absorbing atoms in the light path). Since the efficiency of producing atoms from a sample can not be known so N cannot be used to calculate the concentration of element in the sample. Mostly the quantitative analysis is based on calibration curves.

Atomic absorption spectroscopy can be used for the determination of Al, Ba, Ca, Cu, Ph, Ni, K, Na, Sn, Zn, Sb, Be, Cr, Au, Mg, Ph, Si, Sr, U, As, Bi, Co, Fe, Hg, Pt, Ag, and Ti.

It is a very sensitive method and quite accurate. For better accuracy the parameters that influence quantitative determination can be controlled as under :

(i) *Flame system*. The flame height can be adjusted for optimum and uniform flame by setting the oxidising gas to a fixed flow rate and regulating the flow of fuel so as to get peak absorption when suitable standard solution is sprayed into the flame.

(ii) *Wavelength*. The wavelength is so selected that it limits the selection of most absorbing lines.

(iii) *Concentration of solutions*. The concentration of test solution and concentrated solutions can be adjusted for better sensitivity.

3. Simultaneous Multicomponent System Analysis

It is possible to carry out multicomponent analysis simultaneously if a multielement emission source is available. Before 1973 such determinations could not be carried out for lack of such a multielement emission source. In 1973 Mitchell described a multielement absorption system making use of a multielement hollow cathode source and a vidicon, detection system. Mitchell detected eight elements (Zn, Cd, Ni, Co, Fe, Mn, Cu and Ag) simultaneously in the spectral region 2320 to 3280 A°.

4. Determination of Metallic Elements in Biological Materials

The atomic absorption spectroscopy is being increasingly used for detection of trace elements in biological materials. The recently decomposition of carbonaceous material, is the use of 50% H_2O_2. Sometimes it has been used in combination with conc, sulphuric acid and nitric acid. This procedure gives good results for blood, liver and meat products.

The quantities of hydrogen peroxide and the acid needed depend on the material under investigation and on the nature of trace metal, *e.g.*, sulphuric acid will become unsuitable if lead is being determined because lead sulphate formed is insoluble.

5. Determination of Metallic Elements in Food Industry

The interest of a food analyst is to detect the presence of toxic elements in foodstuff. Most common toxic elements present being Cu, Zn, Ni. To detect their presence the metal is extracted from the solid foodstuff by digestion with dil H_2SO_4 or HNO_3 or with 50% H_2O_2. This method is used for determination of Nickel in vegetable oils, and of copper in beers.

6. Determination of Ca, Mg, Na and K in Blood Serum.

These elements are determined in blood serum for the diagnosis of many pathological conditions, diabetes and primary aldosterionism. The method of atomic absorption is considered to be most suitable for determination of Ca, Mg, Na. K in blood serum.

To carry out this determination the sample solution of serum is diluted 10, 20 or 50 time in presence of Lanthanum chloride. The test solution is then aspirated and absorbance measured which is then compared with absorbance of standard solutions. The presence of Lanthanum chloride overcomes the possible under estimation of calcium due to phosphate suppression. Calcium and magnesium can be estimated in presence of Lanthanum chloride.

7. Determination of Lead in Petrol

Lead compounds tetraethyl lead [Ph $(C_2H_5)_4$] and tetra methyl lead [Ph$(CH_3)_4$] are added to petrol and anti knocking agents. Before analysing the sample of petrol for lead it is necessary to know which of the two compounds or the compositions of the mixture of these two compounds have been used as antiknocking agent in the sample under considerations. In the absence of such an information the results may not be very reliable and in such cases indirect methods are used.

(i) *Direct Method.* If we know the nature of antiknocking agent added then the standard samples are prepared which contain tetraethyl lead or tetramethyl lead or the mixture of these two in the same ratio as used as an additive. The solvent used is cyclohexanone. These standards are then aspirated to the atomic absorption spectro photometer at 2833A° using air-acetylene and the response curves are drawn between absorbance and μg/ml. These should cover the range 0–5 μg/ml in case of tetraethyl lead and 0–50 μg/ml in case of tetramethyl lead. The typical response curves for both of these are shown in Fig. 2.8.

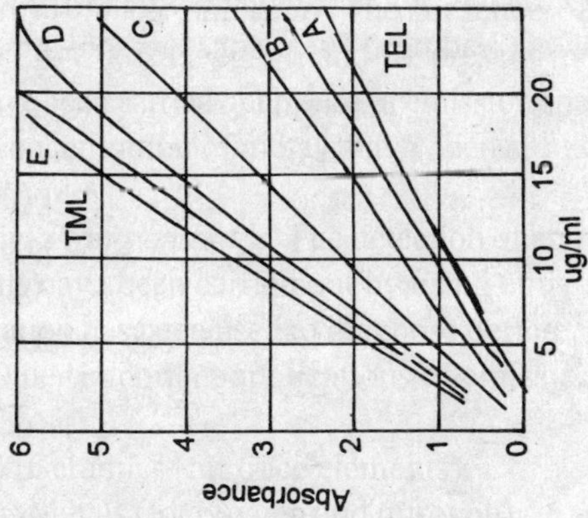

Fig. 2.8: Typical response curves for TEL/TML

The given sample of petrol is now diluted with cyclohexanone and is then aspirated at 2833A° and the value of absorption noted. To obtain the lead content it is then compared with standard curve.

(ii) *Indirect Method.* In the absence of any information about the nature of lead additive a standard sample of lead containing 50, 25, 10, 5 μg/ml are prepared and aspirated at 2833A° and the absorbance is noted in each case.

Now the petrol sample to be analysed is treated with bromine water, which converts lead to lead bromide, and extracted with dilute nitric acid. The extract is aspirated and the absorbance noted. The lead content is then calculated as under:

Ph as μg/ml (*wt/wt*) present in petrol

$$= \frac{s_2 - s_1}{s_0 - s_2} \times c \times \frac{1}{10 \times \text{sp.gr}}$$

where s_0 = scale reading for blank 1% HNO_3

s_1 = scale reading for sample extract

s_2 = scale reading for standard

c = conc. of standard Ph μg/ml (*wt/wt*)

sp.gr. = specific gravity of original sample

Problems

1. For the analysis of cement samples, a series of standards were prepared and the emission intensity for sodium and potassium was measured at 590 and 768 nm, respectively. Each standard solution contained 6300 μg/ml of calcium as CaO to compensate for the influence of calcium upon the alkali readings. The results are shown below:

Concentration	Emission	Reading
μg/ml	Na_2O	K_2O
100	100	100
75	87	80
50	69	58
25	46	33
10	22	15
0	3	0
Cement A	28	69
Cement B	58	51
Cement C	42	63

For each cement sample 1,000 g was dissolved in acid and diluted to exactly 100 ml. Calculate the per cent of Na_2O and K_2O.

2. In Problem 1, what contributed to the emission reading of the blank at the analytical wavelength for sodium, but did not for potassium? A small quartz spectrometer was employed to obtain the results would the blank reading be larger, smaller, or the same if a filter photometer equipped with glass absorption filters had been employed.

3. Boron gives a series of fluctuation bands due to the radical BO_2 that lie in the green portion of the spectrum. Although the overlapping band systems present a problem in the measurement of the flame background, the minimum between adjacent band heads can be used. These results were obtained :

Boron Present	Emission Reading	
μg/ml	518-nm Peak	505-nm Minimum
0	36	33
50	44	36
100	52	39
150	60.5	42.5
200	68.5	45.5

What are the concentrations of boron in these unknowns?

A	45	36.5
B	85	65
C	66	50

4. A calibration curve for strontium, taken at 460.7 nm, was obtained in the presence of 1000 μg/ml of calcium as CaO and also in the absence of added calcium. These results were

Strontium Present	Emission Reading	
μg/ml	No Calcium	Calcium Added
0	0	13
0.25	2	18.5
0.5	6	24
1.0	16	36
2.5	44	70
5.0	94	125
7.5	150	181
10.0	200	238

(a) Graph the calibration curve on rectilinear graph paper and also on log-log paper. (b) What might be the cause of the upward curvature in the region of low concentrations on the rectilinear graph when calcium is absent ? (c) Why does the addition of calcium straighten the calibration curve and increase the net emission reading for strontium?

5. Calculate the iron content in a diethyldithiocarbamate extract using the following data:

Absorbance Blank	Units Sample	Iron Added, μg/200 ml
0.0020	0.0009	None
0.0018	0.0284	2.00
0.0022	0.0484	4.00
0.0020	0.0677	6.00

6. A sample of mineral ash gave a meter reading of 37. Solutions B and C, containing the same quantity of unknown solution plus 40 and 80 μg/ml of added potassium, respectively, gave net meter readings of 65 and 93. Calculate the quantity of unknown potassium in the original solution.

7. A metal naphthenate sample, ashed and diluted to a fixed volume, gave a reading of 29. Solutions B and C, containing the same quantity of unknown solution plus 25 and 50 μg/ml of barium, gave readings of 53 and 78, respectively. Calculate the quantity of barium in the original solution.

8. To illustrate the effect of aqueous-organic solvents upon droplet size, calculate the mean droplet diameter for (a) water, (b) 50% methanol-water, and (c) 40% glycerol-water. Pertinent data follow :

System	Surface Tension, dyn/cm	Viscosity, dyn/cm²	Density, g/cm³	Velocity of Aspirating Gas, m/sec	Q_{air}/Q_{liq}
Ethanol, 50%	28	0.029	0.934		
Glycerol, 40%	68.6	0.039	1.102	279	2540
Methanol, 50%	30.60	0.027	0.946	198	9540
Methyl isobutyl ketone	24.6	0.0051	0.801		
Water	73	0.010	1.00	198	6400

9. For (a) water, (b) 50% (v/v) ethanol-water, and (c) methyl isobutyl ketone as solvents, plot the droplet diameters for solution flowrates ranging from 0.1 to 5 ml/min. A values for a typical nebulizer,

assume the velocity of the aspirating gas to be 333 m/sec and Q_{gas} to be 8.5 liter/min. Other data are give in Problem 8.

10. Calculate the fraction of cesium atoms ionized in a flame at 2000°K when the total cesium concentration in the flame gases is (a) 10^{-4} atm, (b) 10^{-6} M, (c) 10^{-7} atm.

11. Calculate the fraction of lithium atoms ionized in a premixed laminar flame at 3000°K when the concentration sprayed into the flame is (a) 10^{-2} M, (b) 10^{-3} M, (c) 10^{-4} M.

12. What individual amounts of (a) cesium, (b) rubidium, (c) potassium, or (d) lithium should be added individually to a flame at 2500°K and at 2800°K to suppress the ionization of a solution containing 0.23 μg/ml of sodium?

	K_2, atm	
Element	2500°K	2800°K
Li	1.48×10^{-9}	2.63×10^{-8}
Na	4.8×10^{-9}	7.40×10^{-8}
K	1.8×10^{-7}	2.08×10^{-6}
Rb	3.9×10^{-7}	3.98×10^{-6}
Cs	1.45×10^{-6}	1.32×10^{-5}

13. If lithium (670.8 mm) is used as an internal standard for the determination of sodium (589.0 nm) and potassium (766.5 nm), what is the maximum permissible temperature variation of the flame if one desires to maintain deviations in intensity ratios less than 1%? Assume a flame temperature of 2000°K.

14. From the absorbance traces shown for arsenic by atomic absorption, estimate (a) the signal-to-noise ratio, (b) the sensitivity of the method, and (c) the detection limit.

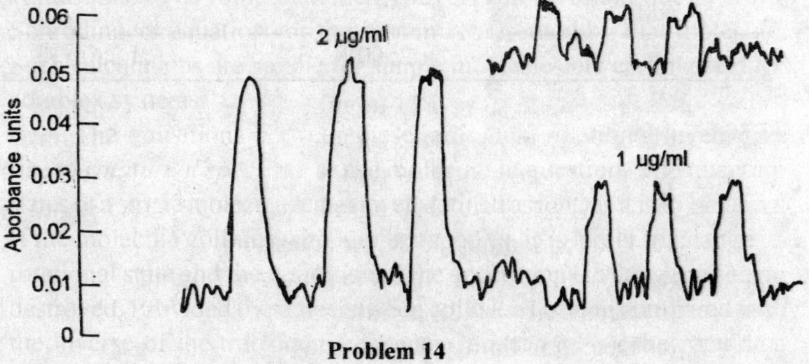

Problem 14

Answers to Problems

1. Cement A; 0.14% Na_2O, 0.62% K_2O
   Cement B; 0.37% Na_2O, 0.43% K_2O
   Cement C; 0.22% Na_2O, 0.55% K_2O
2. Emission from molecular band system of CaO and CaOH Blank would be larger because reading would increase approximately with the square of the band pass of filter or monochromator.
3. (a) 55 μg/ml (b) 170 μg/ml (c) 130 μg/ml
4. (b) Ionisation of strontium atoms in flame (c) Ionisation in (b) is repressed by large excess of calcium atoms. Addition of calcium also contributes a significant background emission due to CaOH.
5. 0.715 μg/200 ml
6. 52.8 μg/ml
7. 29.7 μg/ml
8. Water : 27.0 μm, 50% methanol-water : 18.7 μm 40% glycerol-water : 29.4 μm.
9.

Flow Rate, ml/min	Droplet Diameter, μm Water	5% Et OH–H_2O	MIBK
1.0	16.1	12.0	10.9
2.0	18.3	16.3	13.0
3.0	21.0	21.8	15.7
5.0	28.0	35.9	22.5

10. (a) 0.0094 (b) 0.094 (c) 0.29
11. (a) 0.35 (b) 0.75 (c) 0.97
12. To suppress the ionisation of 0.23 μg/ml of sodium to 0.01, add:

Element	at 2500 K	at 2800 K
Cs	0.44 μg/ml	0.74 μg/ml
Rb	1.05 μg/ml	1.6 μg/ml
K	1.04 μg/ml	1.4 μg/ml
Li	22.50 μg/ml	19.5 μg/ml

13. 13.4°K for sodium, 15°K for potassium.
14. (a) S/N = 13.3 (or 6.7 peak to peak) for 2 μg/ml
    (b) Sensitivity = 0.22 μg/ml
    (c) Detection limit = 0.30 μg/ml.

3

Emission Spectroscopy

Introduction

Atomic emission spectroscopy (also known as flame emission spectroscopy or flame photometry) finds wide spread application in elemental analysis. The most important use of this technique is in the determination of Na, K, Li, Ca specially in biological fluids and tissues. For reasons of convenience, speed and relative freedom from interferences emission spectroscopy is preferred for determination of these elements. This method is used for determination of at least half the elements in the periodic table and so it can be considered to be the most important tool of analysis.

In this type of spectroscopy the radiation emitted by the excited sample are studied. It is related to the study of atoms and is used as a method of elemental analysis. It is possible to identify and quantitatively determine almost all metallic elements even in very low concentrations. Either solids, liquids or gases can be identified but generally solids or evaporated solutions are used. In emission spectroscopy, the sample is excited by absorbing thermal or electrical energy and then the radiations emitted by the excited sample are studied for both qualitative and quantitative analysis.

THEORY

Emission Spectra. When a substance is heated to a high temperature either by thermal or electrical method it gets excited and excited substance then emits certain radiations when it comes to the ground state. If this emitted light is allowed to pass through to prism or a grating and then examined with a spectroscope, then the spectrum obtained is known as emission spectrum.

Emission spectrum can be classified as *continuous spectrum* if it is spread over a considerable wavelength region uninterrupted. In such a spectrum there are no sharp lines or discrete bands. Such a spectrum is generally obtained by heating bulk of a substance. Such a spectra depends on the temperature and not on the chemical composition and so is not quite useful for spectrochemical analysis. *To avoid* this the *sample is volatilised before the excitation.*

Band spectra are characteristic of molecules so they are some times referred to as molecular spectrum. Such a spectrum consists of a group of lines so close to each other that they ordinarily appear as continuous bands. To avoid this type of spectra the molecules are provided sufficient energy so as to break them into atoms.

Line spectrum. Consisting of discrete irregularly spaced lines is obtained by the emission of radiations from a substance in the atomic state. It is called *atomic spectrum*. It depends upon the type of atom and is quite useful in emission spectroscopy.

Emission spectroscopy is based upon the line spectra produced by excited atoms Excitation of the electron to higher orbitals can be brought about by the heat of a flame or an electric arc or spark etc.

Arc, spark and plasma source are generally richer in lines because of their greater energy. The lines from neutral atoms predominate in an arc spectrum whereas in a spectra excited by a spark and plasma contain lines that arise from ions. This is because of lower energy provided by arc source as compared to spark or plasma sources.

Thus the central problem in emission spectroscopy is the large influence of the source upon both the pattern and the intensity of lines produced by a species. Another problem is the sensitivity which is inversely proportional to the complexity of the spectra.

Arc and Spark Sources

In arc and spark sources, sample excitation occurs in the gap between a pair of electrodes. Passage of electricity through the electrodes provides the necessary energy to atomise the sample and excite the resulting atoms to higher electronic states.

The usual arc source for a spectrochemical analysis is formed with a pair of graphite or metal electrodes spaced about 1 to 20 mm apart. The arc is initially ignited by a low current spark which causes momentary formation of ions for current production in the gap once the arc is struck thermal ionisation maintains the current. Alternatively the arc can be started by bringing the electrodes together to provide heat for ionisation, then the electrodes are separated to the desired distance.

In the typical arc current lie in the range 1 to 30V *d.c.* source usually has an open circuit voltage of about 200V. *a.c.* source voltage range from 2200 to 4400V.

Fig. 3.1 shows the essential components of a circuit for *d.c.* arc current.

The solid or liquid sample is put on the lower electrode in the arc gap. The variable resistance in the circuit is used to adjust current and the inductance cool for stabilizing the flow of the current. The direct current arc has following advantages:

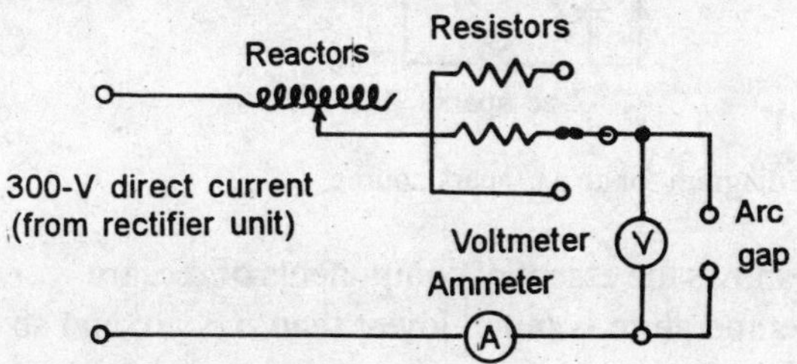

Fig. 3.1: Circuit for a dc. arc

(i) Because of high temperature generated it acts as a sensitive excitation source.
(ii) The energy provided by such a source is mainly thermal which is sufficient to excite almost all the metal elements.
(iii) The emission species are neutral atoms and not ions.

Following difficulties may arise while working with arc source:

(i) A fall in resistance occurs rapidly with increase in current.
(ii) There is a variation in temperature along the length of the arc. It is maximum near the cathode.
(iii) There is a fluctuation in emission intensities because of current

fluctuations resulting from the changing arc length and temperature. It could be reduced by adjusting KL Values.

For efficient excitation a stallwood jet is put across the arc. In this jet a mixture of argon and oxygen in the ratio 30 : 20 is sprayed around the electrodes which cools the electrodes as well as the outer layer of plasma and also excludes atmospheric gases. It increases the sensitivity by a factor of 2 or more. It also eliminates the formation of cyanogen (C_2N_2) whose intense bands often provide a heavy background for emission lines of interest.

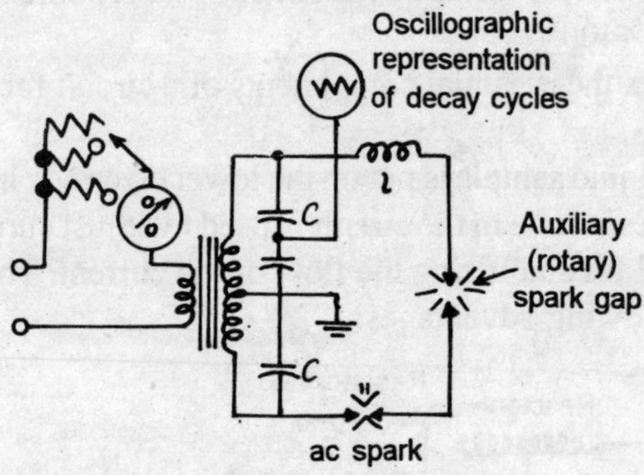

Fig. 3.2: Circuit diagram for an a.c. spark source.

Fig. 3.2 shows the essential components of *a.c.* arc.

In this temperature is much lower than *d.c.* arc and so it has a lower sensitivity.

Spark Source

Fig. 3.3 shows a typical electrical circuit for driving an *a.c.* across a spark gap.

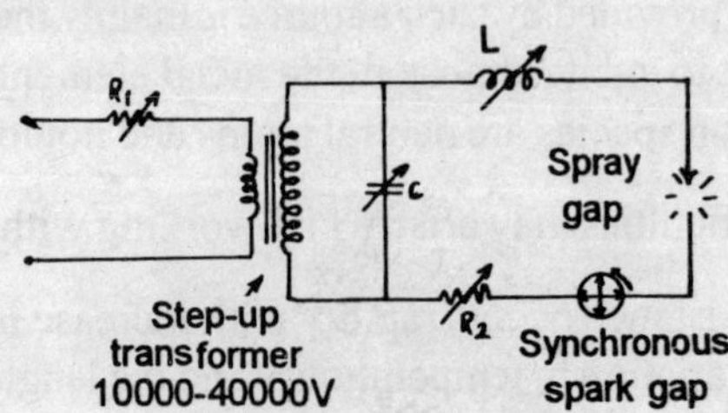

Fig. 3.3:

In the primary circuit a voltage of 110 – 220V is maintained. The transformer converts line power to 15000 – 40000V, which then charges the capacitor. When the potential becomes large enough to break down the air gaps shown, a series of oscillating discharge follows. The voltage then drops until the capacitor is in capable of forcing furthur current across the gaps, following which the cycle is repeated.

During the initial phase of the discharge the *instantaneous* current may exceed 1000A and the temperature is estimated to be as great as 40,000K. Thus while the average electrode temperature of a spark source is much lower than that of an arc, the energy in the small volume of the streamer may be several times greater. Thus ionic spectra are more pronounced in a high voltage spark than in an arc.

Argon Plasma Source

Plasma is a gaseous mixture containing a significant fraction of atomic or molecular species in the form of ions. The plasma used for emission analysis is generally a mixture of argon ions and atoms. When a sample is injected into this medium atomisation occurs as a consequence of high temperature 10,000K. Argon ions once formed in a plasma are capable of absorbing sufficient power from an external source to maintain a sufficiently high temperature to continue the ionisation process and thus sustain the plasma indefinitely. Three power sources that have been used in argon *plasma spectroscopy* are d.c. electrical source, the second is powerful radio frequency source and the third is powerful microwave frequency generator. Of the three the radio frequency source appears to offer the greatest advantage in terms of sensitivity and freedom from interference. There arc also referred to as *inductivity coupled* plasma sources (ICP).

Electrodes

Two types of electrodes are commonly used in emission spectroscopy *i.e.*

(i) Self electrodes.

(ii) Graphite electrodes.

(i) Self Electrodes

When the material to be investigated can withstand high temperature and is also a good conductor, then it is used as such for electrodes. This type of electrodes are called self electrodes *e.g.* alloys or metal powders which can be used as self electrodes after being pressed into solid discs or cylinders.

(ii) Graphite Electrodes

The electrodes of graphite (a good conductor) are used in those cases when the substance to be investigated is not capable of withstanding high temperature or if it is not a good conductor. In case of graphite electrodes the substance under investigation is placed in a small cavity in the lower graphite electrode and the upper graphite electrode is brought to a point in a pencil sharpner. Generally such electrodes have centre posts or are watched to give a narrow neck or both. The function of centre post is to decrease wandering of arc source and to improve reproducibility and the function of narrow neck is to increase sensitivity.

Sample Handling

Sample for arc and spark source spectroscopy may be solid or liquid. These must be distributed regularly upon the surface of at least one of the electrodes that serve as the source.

If the sample is a metal or an alloy, one or both electrodes can be formed from the sample by milling by turning or by casting the molten metal in mould. Care must be taken to avoid contamination of the surface. In case of non-metallic samples they are supported on an electrode whose spectrum will not interfere with the sample analysis. Carbon (Graphite) is an ideal electrode material because it can be obtained in pure form, is a good conductor, has good heat resistance and can be readily shaped. Such electrodes are available in many shapes, sizes and forms. Silver or copper electrodes are also used when these elements are not of analytical interest. The surfaces of these electrodes can be cleared and reshaped after each analysis.

For handling liquid samples, sample holders are used. The porous base of cup container slowly passes the sample into the discharge, that can be sustained till the base of the holder is burnt away.

Various types of these sample holders shown in the diagram below (Fig. 3.4) can be used for aqueous and also for non-aqueous solvents. The errors in spectra are caused by ignition of the organic solvents in the discharge.

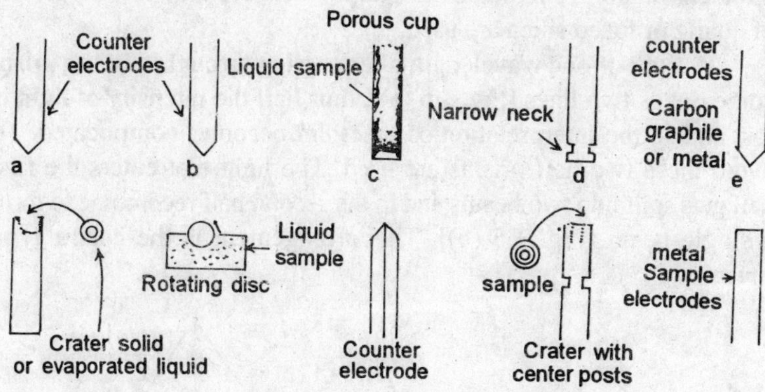

Fig. 3.4: Schematic diagrams of various electrodes: (a) crater for solid or evaporated liquid samples, (b) rotating disc for liquid sample, (c) porous cup for liquid sample, (d) crater with center posts, (e) metal electrodes and/or samples.

Liquids may be (a) evaporated after being placed in the cup of a graphite electrode, or may be (b) mixed with a powdered graphite, placed in a cup, and dried, or (c) a porous graphite electrode may be immersed in the solution and dried prior to excitation. A rotating disc may also be used. The surface of the rotating disc electrode is continually wetted from the liquid sample in a container and the sample is continually entered into the excitation area. Because of the volatile and flammable properties of organic solvents, most liquid sample spectroscopic procedures are limited to aqueous systems.

Solid samples may be ground and placed in the electrode cup directly or placed on top of a flat sample holder. Frequently solid samples will be ground with graphite in order to increase conductivity properties, or if the sample is conducting it may be machined and used as the electrode.

Monochromators

The dispersing element may be either a prism or a grating. High dispersion prism instruments employ a Littrow-mount monochromator, a cornu-type often used in lower dispersion spectrographs.

Prism Monochromators. When prism is to be used as a monochromator it should be transparent to u.v. radiations thus prisms of quartz or fused silica are used.

A light of one wavelength after passing through a quartz prism come out as two lines [Fig. 3.5 (a)], thus half the intensity of light is lost and so the interpretation of spectrum becomes complicated. To avoid these two half prisms are used. The light that enters the first half gets split into two beams and in the second half recombine to form a single beam [Fig. 3.5 (b)]. This arrangement is the **cornu type prism.**

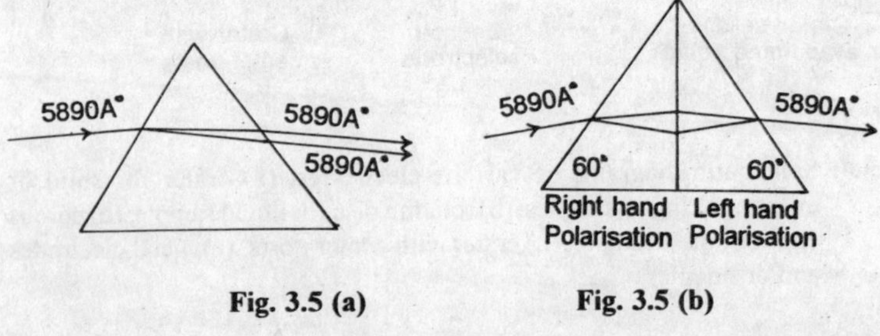

Fig. 3.5 (a) **Fig. 3.5 (b)**

The other difficulty that generally arises in use of prism as a monochromator is that the dispersion is not constant over a wide wavelength and so the identification of emission lines or unknown wavelengths is not possible simply by measuring their dispersion. In such a case identification is done by comparing these wavelengths with emission lines of known wavelength.

Grating Monochromators

These are used in more sophisticated instruments having high resolving power. These are costlier but they have replaced prism monochromators as they give linear dispersion and it makes the dispersion independent of wavelength and the resolution is constant. The identification of spectral lines becomes simple.

A variety of geometric configurations is available. Most grating instruments utilize an arrangement of mounting the grating on the Rowland circle, a circle the radius of which is one-half the radius of the grating. The images of the slit will be focused somewhere on the circle circumference when the slit and concave grating both lie on the Rowland circle. The schematic diagrams of several mountings which utilize this principle are shown in Fig. 3.6(a).

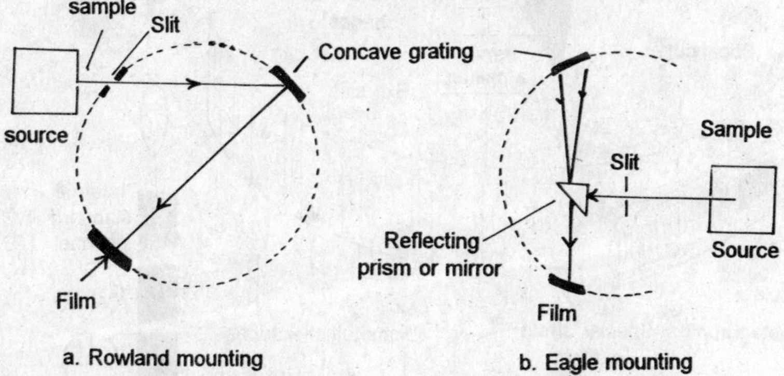

Fig. 3.6 (a): Schematic diagrams of emission spectrographs which utilize the Rowland circle : (a) Rowland mounting; (b) Eagle mounting.

The main disadvantage in such type of monochromator is the overlapping of high order wavelengths. However it can be avoided by using filters or by use of such detectors which are not sensitive to high orders.

Detectors

A photographic emulsion can serve as a detector, an amplifier and an integrated device for evaluating the intensity time integral for a spectral line. Such types of detectors are referred to as photographic plates or films.

The other type of detector generally used are photomultipliers. For routine emission analysis, photoelectric detection of integrated line intensities can be employed. Instruments for this purpose are equipped with as many as 48 photomultiplier tubes, which are located behind fixed slits along the focal curve of the spectrometer [Fig. 3.6 (b)]. The position of each slit must correspond exactly to the position of a line of interest.

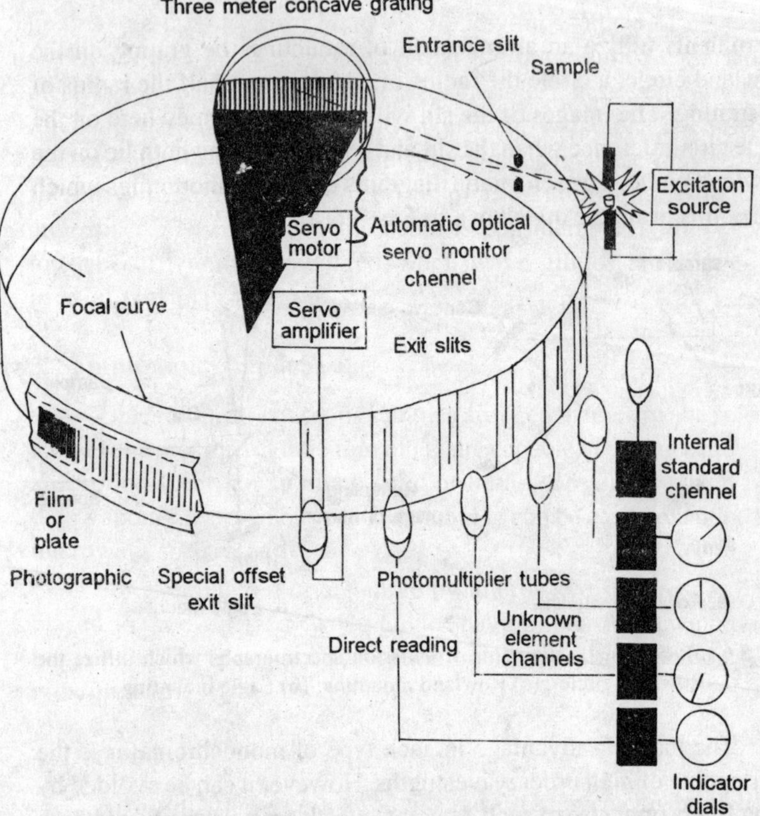

Fig. 3.6 (b): Optical system of a direct-reading grating spectrometer.

Photoelectric multielement instruments are highly automated, complex, expensive and permit measurement of only certain elements, which are chosen at the time of instrument manufacture. These also have better precision and have a reproducibility of the order of 1% relative of the amount present.

However we get certain advantages when we use photographic plates as detectors. Some of these are :

(i) It is possible to record a large number of spectral lines simultaneously.

(ii) It provides a permanent record of spectrum.

(iii) The emission intensity can be integrated by a photographic emulsion over a period of time.

(iv) Photographic emulsion are highly sensitive in *u.v.* and visible regions which are of utmost importance in emission spectroscopy.

(v) The cost of such detectors is much less in comparison to photomultipliers.

Photographic Detection

The intensity of spectral line is registered on a photographic emulsion which is developed using a suitable developer in the recommended developing time at a specified temperature.

The density of a film is measured by a densitometer. The darkness of the exposed area is a function of exposure E which is defined by

$$E = I_\lambda . t$$

where I_λ is the intensity of radiation and t is exposure time. In obtaining spectra for quantitative purposes t remains constant for both sample and standards and thus the exposure is directly proportional to the line intensity and thus to the concentration of the emitting species.

The blackness of a line on a photographic plate is expressed in terms of its optical density D, which is defined as

$$D = \log \frac{I_0}{I}$$

where I_0 is the intensity of beam of radiation after it has passed through an unexposed portion of emulsion and I is its intensity after being attentuated by the line.

To convert the optical density of a line to its original intensity (or exposure), it is necessary to obtain empirically a plate calibration curve, which is a plot of the optical density as a function of the logarithm of relative exposure. Such a plot is obtained experimentally by exposing portions of a plate to radiation of constant intensity for various lengths of time. Generally the absolute intensity of the source will not be known so that only relative exposure can be calculated.

Fig. 3.7 shows typical calibration curve for an emulsion obtained at two wavelengths. With the aid of such curves, the measured optical density of various optical lines can be related to their relative intensities.

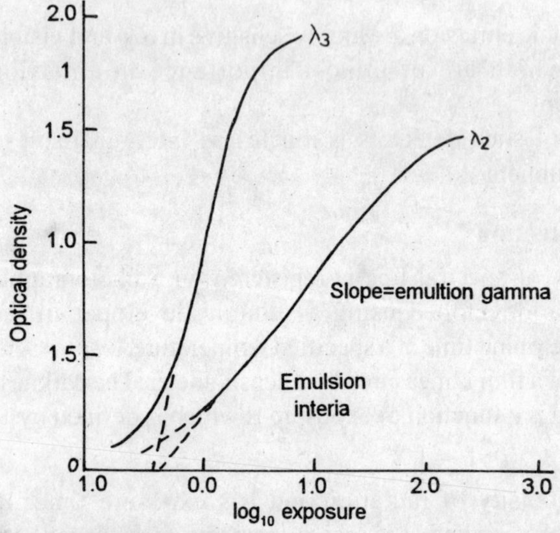

Fig. 3.7: Typical plate calibration curves at two wavelengths

Several calibration curves are needed if a spectrum covers a large wavelength range because of dependence of slope on wavelength region.

To make it clear let us consider the curve shown in Fig. 3.8.

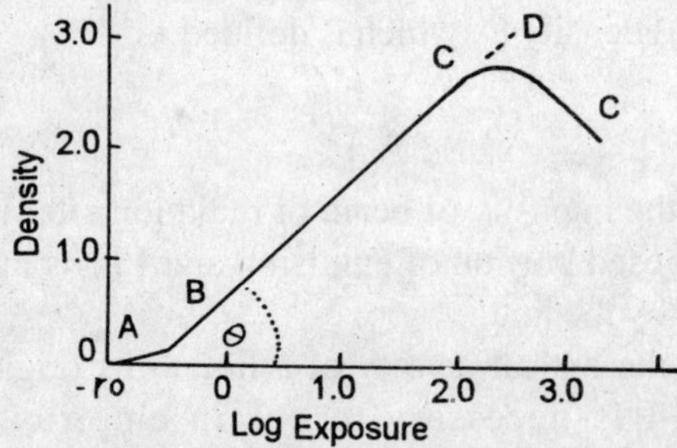

Fig. 3.8:

In this curve in region BC density is directly proportional to the logarithm of intensity, this region is the useful range and the slope of this region is termed γ of the emulsion of photographic plate.

$$\gamma = \tan \theta$$

If γ has a high value it indicates high degree of contrast and a low value of γ indicates a low degree of contrast.

A photographic emulsion consists of light sensitive crystals of silver halides suspended in gelatin. Such an emulsion is formed to be sensitive to blue, violet and shorter wavelengths. By adding suitable dyes to silver salts photographic emulsions sensitive to red, green and I.R. region of spectrum can be obtained.

Spectrographs

The *spectrometer* is a usual device where as a spectrograph is a permanent record. The important types of spectrographs are distinguished by whether they use a prism or grating as dispersing medium.

Prism Instruments. These are either *Cornu type prism* (having two half prism) or Littrow type prism. These are shown in Fig. 3.9 and Fig. 3.10 respectively.

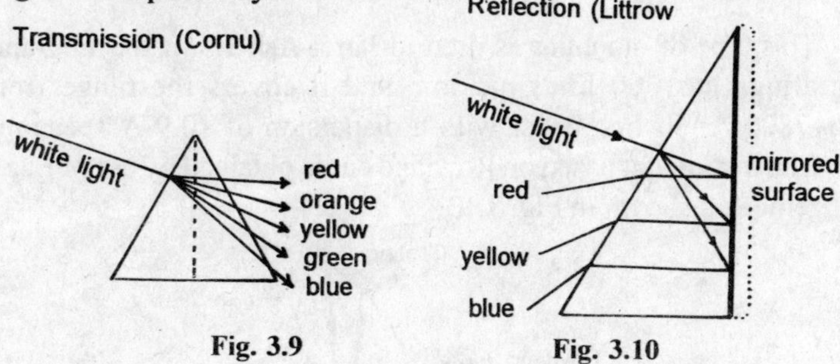

Fig. 3.9 Fig. 3.10

Grating Instruments

It is possible to substitute the prism by a grating keeping the rest of the arrangement as in above cases.

A variety of commercial grating spectrographs are marketed. A few of than are those containing.

(i) A Rowland mounting for a concave grating (Fig. 3.11).

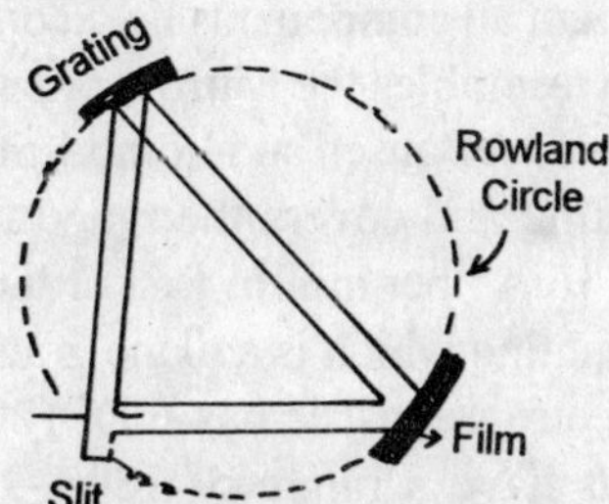

Fig. 3.11:

(ii) Wads worth mounting which uses a concave mirror along the concave grating (Fig. 3.12).

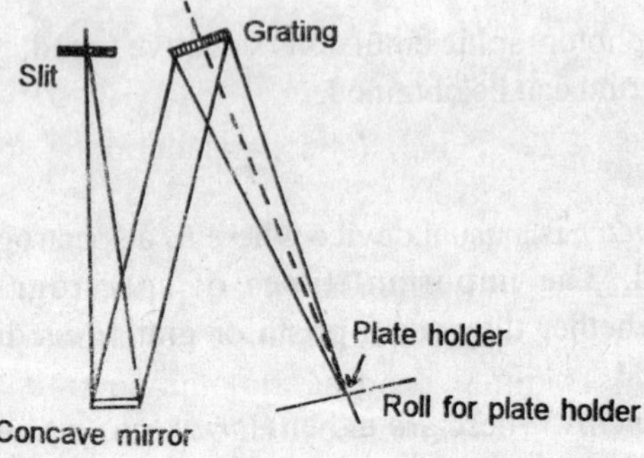

Fig. 3.12: Wadsworth mounting for a concave grating

This type of mounting is used in Jarrel-Ash 1.5 m spectrograph. Its gratings has 600 lines per mm and it covers the range from 2200–7800 A° in first order with a dispersion of 10.9 A° per mm.

The instrument based on Rowland circle obtained by use of Eagle arrangement is shown in Fig. 3.13.

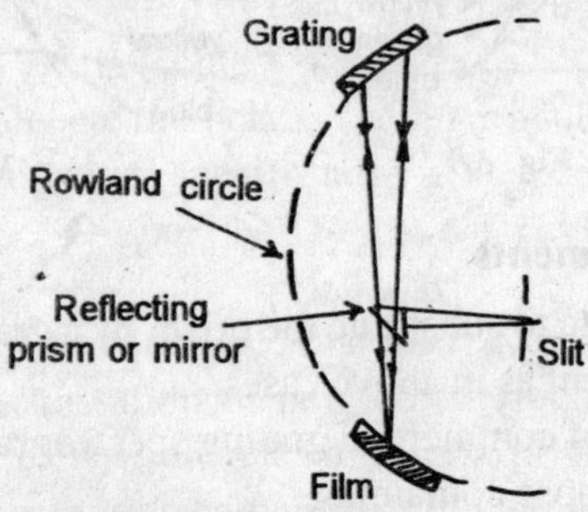

Fig. 3.13: Eagle mounting for a concave grating.

In this instrument all components lie on one side of the Rowland circle and the set up resembles the Littrow prism arrangement.

Another instrument Bausch and Lomb 1.5m spectrograph makes use of Eagle mounting and covers the range from 225–6250 A°. It has a dispersion of 16 A° per mm in first order.

(iii) Ebert mounting which is a plane grating (Fig. 3.14) is used mostly in large spectrographs. It has 600–1200 lines per mm and these spectrographs have a range of 1800–30,000 A°. The main advantages of such a spectrograph are :

(a) their considerably greater wavelength range and (b) capability to efficiently observe high order visible and *u.v.* spectra.

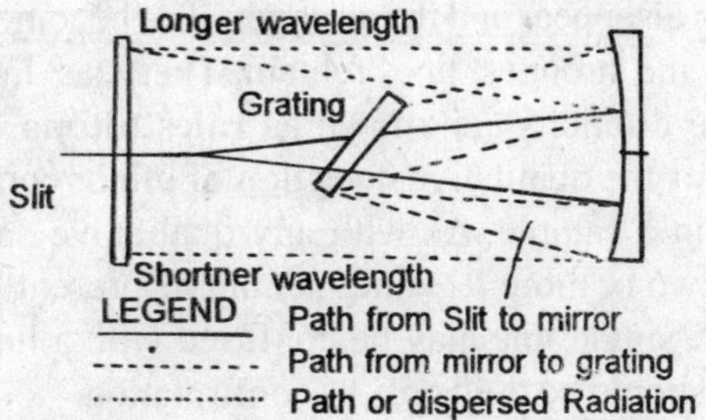

Fig. 3.14:

Qualitative Procedures

Emission spectroscopy is a destructive technique which offers the possibility of qualitatively identifying and quantitatively determining most of the metals. With special adaptation, a few of the non-metals like carbon, phosphorus, and sulfur may also be determined. It is for elemental analysis only and provides no information about the bonding in either molecules or alloys.

Comparison of Spectra and Lines

Qualitative identification of the elements can be done by comparison of the sample spectrum with the spectrum produced by pure known standard samples, or the wavelengths of the spectrum of the sample may be measured and compared to values in standard tables.

Due to the complexity of the spectrum of a single element, and also the possible close proximity of the lines of one element to those of another, care should be taken in obtaining and interpreting these measurements. Instrument settings must be duplicated and precise methods of measuring line positions must be used if a complex spectrum of two or more elements is to be analysed.

Generally a minimum of two or three lincs is necessary for the positivc identification of an element. Once an element in a mixture is identified, the lines of that element are eliminated from the lines of the mixture. Each successive identification becomes more straightforward.

Last Lines, RU Lines

As the concentration of a metallic element in a powder mixture is made successively more and more dilute, the number and intensities of the lines in the resulting spectra will change. The more concentrated the element, the more complex will be the spectrum. Upon dilution the weaker lines disappear and the stronger lines become less intense. Ultimately only the strongest lines remain. These last lines to remain upon successive dilutions are known as **raies ultima** or **RU lines.** They are used for the qualitative detection of minor constituents and trace elements in a sample. As with any qualitative spectrographic determination, two or more RU lines should be present for a positive identification. A single line may be confused with a line of a major constituent and should be used only as confirmation.

Procedure for Qualitative Analysis

Photographic emission spectrography permits the detection of some seventy elements by brief excitation of a few milligrams of the sample. The concentration of an element with an accuracy of very high order can be rapidly estimated by subjective judgment of line blackening. In this application the photographic emission spectrography is unsurpassed.

The sensitivity of the method depends upon the nature and amount of sample the type of excitation and the instrument used. Table 3.1 lists some representative data of lower limits of detection that can be achieved with d.c. are excitation.

Table 3.1: Detection Limit with a d.c. Arc

Element	*Wavelength*	*Lower limit of detection*	
	A°	*per cent*	*micrograms*
Ag	3280.68	0.0001	0.01
As	3288.12	0.002	0.2
B	2497.73	0.0004	0.04
Ca	3933.66	0.0001	0.01
Cd	2288.01	0.001	0.1
Cu	3247.54	0.00008	0.008
K	3446.72	0.3	30
Mg	2852.12	0.00004	0.004
Na	5895.92	0.0001	0.01
P	2535.65	0.002	0.2
Pb	4057.82	0.0003	0.03
Si	2516.12	0.0002	0.02
Sr	3464.45	0.02	2
Ti	3372.80	0.001	0.1
Zn	3345.02	0.003	0.3

Excitation Technique

For excitation a d.c. arc is commonly used. The current and exposure times are adjusted so that the sample volatilises completely. Generally currents 5 to 30 A for 20 to 100 sec are used. A weight of about 2 to 50 mg of sample are packed into the cavity of the graphite electrode or evaporated upon its surface. The other electrode of graphite is used to produce an arc. Sample electrode is generally made the anode because of its higher temperature.

Identification of Elements

For qualitative analysis a standard iron spectrum is usually recorded as a reference on each plate or film. Identification of unknown lines is greatly fecilitated with a projector that allows the simultaneous comparison of this plate with a master iron spectrum upon which the wavelengths of the more intense iron lines have been identified. The master spectrum also indicates the positions of several of most sensitive lines for each of the elements. The analytical plate and comparison plate are then aligned so that the two iron spectra coincide exactly. Identification of three or four lines of a particular element constitutes nearly certain evidence for its presence in the sample.

In case of complex spectrum the process of detection can be simplified. If the spectrum contains many lines some strong and some weak thus the concentration of metal is decreased when weaker lines will disappear. It is further diluted when less strong lines also disappear and only a few lines are left. The lines that are left are known as **raies ultima** of **RU lines.**

For an unknown sample, emission spectrum containing **RU lines** is recorded and by comparing these with **RU lines** of known sample the metal can be identified.

To carry out determination of wavelengths of unknown lines, linear inter-polation is done between the lines of known wavelength. In the case of prismatic dispersion the interpolation may be done linearally only over a short-range and only in those regions where the dispersion is quite high. In case shown in Fig. 3.15 interpolation between known wavelength λ_1 and λ_2 will locate λ_x. It may be written as

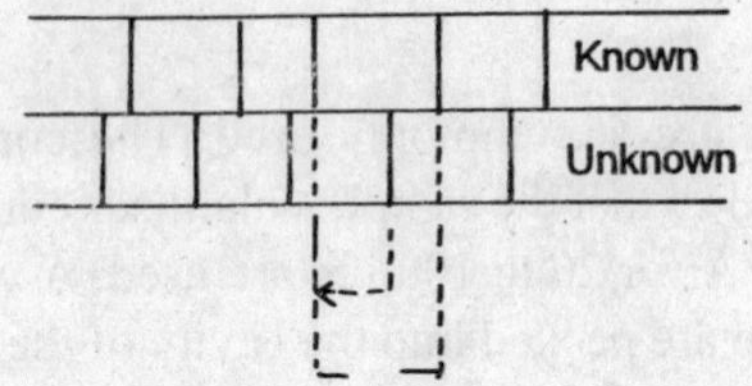

Fig. 3.15:

$$\frac{d_1}{d_2} = \frac{\lambda_x - \lambda_1}{\lambda_2 - \lambda_1}$$

or $$\lambda_x = (\lambda_2 - \lambda_1)\frac{d_1}{d_2} + \lambda_1$$

Hartman Method is used for exact calculations, but this method is tedious than interpolation method. Hartman equation is as under

$$\lambda_x = \lambda_0 + \frac{C}{d_x - d_0}$$

where λ_0, C and d_0 are constants and have to be evaluated experimentally. The process can be carried out by substituting the wavelengths λ_x and distance d_x of three unknown lines from an arbitrary point and then solving the three equations simultaneously.

Quantitative Procedures

Most of the quantitative determinations depend upon the measurement of the intensity of one or more lines and a method of correlation of these intensities or intensity ratios with concentrations. Refined repetition of conditions for both samples and a series of standard is necessary so as to duplicate quantitative results. Care should be taken to maintain the highest purity of the samples and standards in any prior handling, for example, in concentrating or asking pretreatments.

The more intense lines may be detected even at very low concentrations. However, upon increasing concentration the fraction of vaporized unexcited atoms increases and the fraction of excited atoms decreases. The presence of vaporized atoms leads to a decrease in expected intensity because of the following two reasons :

1. As they are not excited they do not directly contribute to the measured intensity.

2. The vaporized atoms will absorb the emitted radiation from the excited atoms. This phenomenon is used in the atomic absorption.

The intensity of a line in a series of successively more concentrated mixtures will increase up to a point, will reach a maximum, and will then decrease with an increase in concentration. This line **reversal** prohibits the use of direct intensity—concentration correlations at higher concentrations. The lines may, be used for measuring lower concentration levels where reversal does not occur.

Emission intensity is affected by the matrix, this effect arising from differences in the rate of vaporization of the sought element in different matrices. All the samples and standards should be reduced to a common matrix. For example, a metallic element is an ore sample could be quantitatively compared to the same element in a soil sample only if both samples, as well as the series of standards, were mixed with the same inert matrix material.

In summary, the conditions of sample preparation, excitation conditions, films exposure and film development must be meticulously controlled and duplicated.

Generally the following methods are used for quantitative analysis:

(i) Comparison Sample Method

To analyse a sample quantitatively by this method, the spectra of the sample under consideration and various standards are photographed alternately on the same plate under similar conditions. The plates are then developed and the concentration of the desired constituent is found by the comparison of darkness of the lines of this constituent with the same lines on the standards. For comparison either visual or photometric method is used.

This method is found to give quite accurate results if constant excitation and exposure conditions are maintained. However, even if the experimental conditions are rigidly controlled, it is not possible to determine the concentration of the element from only the measurement of a given lines intensity because factors like packing and particle size cause the quantity of sample vapourised into the discharge to vary from one sample to the next. It can be controlled to a large extent if one use internal standard.

(ii) Internal Standard Method

In quantitative emission analysis the major problem is the very large member of variables that affect the blackness of the image of a spectral line on a photographic plate or the intensity of the line reaching the photoelectric detector. Most variables that are associated with the excitation and the photographic processes are difficult or impossible to control completely. To compensate for their effects, an internal standard is generally used.

An *Internal standard* is an element incorporated in a fixed concentration into each sample and each calibration standard. The relative intensity of a line of internal standard is then determined for each sample or standard at the same time the relative intensities of the elements of interest are measured. A ratio of the relative intensity of the analyte to that of standard is then employed for concentration determinations. Experimentally a direct proportionality often exists between this ratio and concentration, some times the relationship is non-linear, in such cases the resulting curve is used for determination of concentration.

For a better understanding of this method let us consider the measurement of manganese in steel. For this determination iron is selected as internal standard and the emission spectrum of carbon steel having iron have been recorded on photographic plate. The intensities of the selected lines of manganese and iron are determined on a microphotometer. The ratio of manganese-to-iron intensities will be calculated from the photographic calibration curve.

As the iron concentration is not changing appreciably, different percentages of manganese in a number of standard samples will give different line intensities and so different ratios. Then the ratios of manganese-to-iron line intensities are plotted against per cent manganese on logarithmic graph paper. Such a plot generally obtained is a straight line.

From the calibration curve we can calculate the percentage of manganese in carbon steel if we know the value of log intensity ratio of manganese-to-iron.

In this method the *ideal standard* should have the following properties :

(i) Its concentration in samples and standards is always same.

(ii) Its chemical and physical properties should be as similar as possible to those of the element under investigation. It is essential because only under such conditions the internal standard provide adequate compensation for the variables associated with volatilizations.

(iv) The ionisation energies of the internal standard and the element of interest should be similar to assume that both have the same distribution ratio of atoms to ions in the source.

(v) The lines of the standard and the analyte should be similar in intensity and should be in the same spectral region so as to provide adequate compensation for emulsion variables.

Example

The following data were obtained (after prior treatment) for the determination of boron in citrus leaves. The 2498 A° boron line and the 4323 A° line of the internal standard, palladium, were used. Pd was added in the same concentration to all standards and unknown.

Data				*Calculated values*		
Sample standard	*Concentration of boron, ppm*	*log Intensity of B line*	*Intensity of Pd line*	*Ratio* I_b/I_{pd}	*Log Conc. boron*	*Log of* I_B/I_{pd}
1	4	7.9	1.80	0.44	0.60	– 0.36
2	6	9.6	16.5	0.58	0.78	– 0.24
3	10	12.1	17.3	0.70	1.00	– 0.15
4	14	12.0	15.0	0.80	1.15	– 0.10
5	18	15.6	17.9	0.87	1.26	– 0.06
6	22	17.4	18.6	0.94	1.34	– 0.03
Unknown	...	11.8	18.2	0.65	...	...

(a) Confirm the calculated values in the three right-hand columns. (b) Plot the ratio I_B/I_{Pd} against concentration. (c) Plot the ratio I_B/I_{Pd} against the log of the concentration. (d) Plot the log of the ratio against the log of the concentration. (e) Compare the three graphs (Fig. given) and confirm that the concentration of the unknown is 8.5 pp boron.

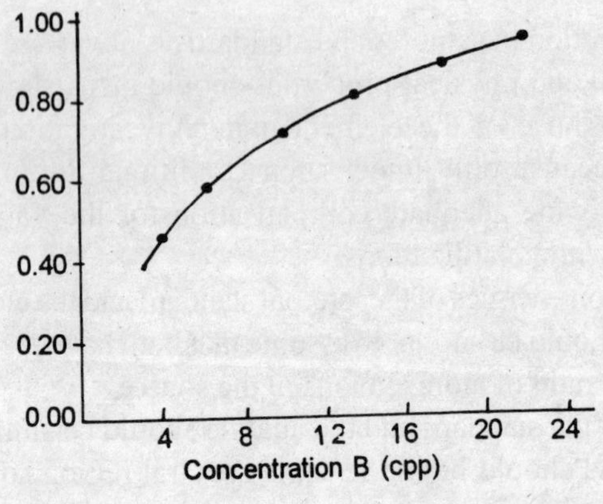

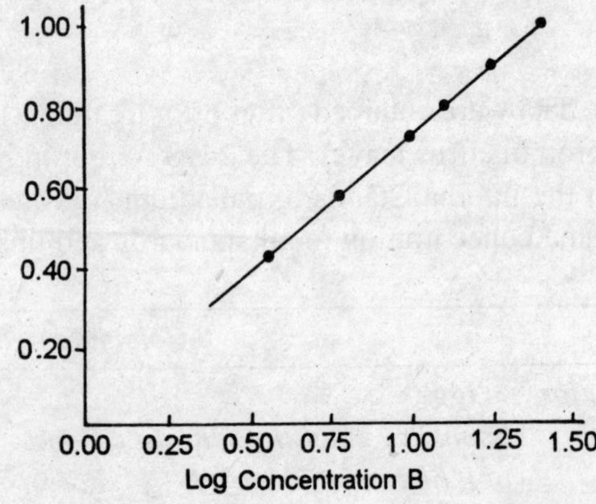

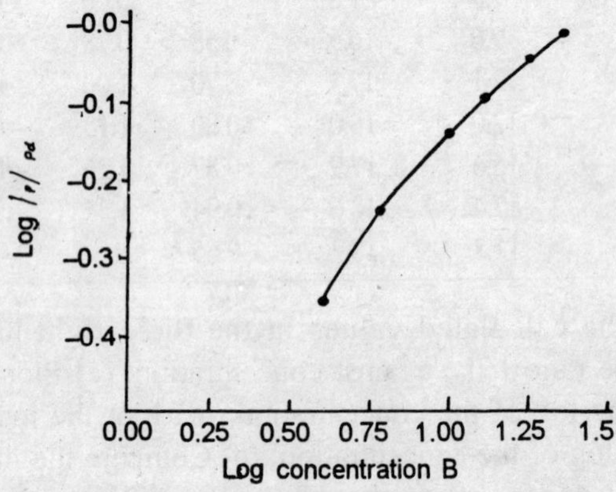

Examples of working curves of emission spectroscopy. (See previous Example)

(iii) Densitometer

When the spectra are recorded on photographic film, the relative intensities of two or more lines may be determined by a **densitometer** (Fig. 3.16). It has a light source, such as a tungsten lamp, a film holder, a slit, and a detector and readout system. The exposed film is placed between the sources lamp and detection system. The intensity of the light which passes through the clear portion and that which passes through the dark lines is successively determined and compared and the logarithm of the ratio of these two transmission intensities is calculated. Ideally, a plot of the log of this ratio against concentration will yield a straight line.

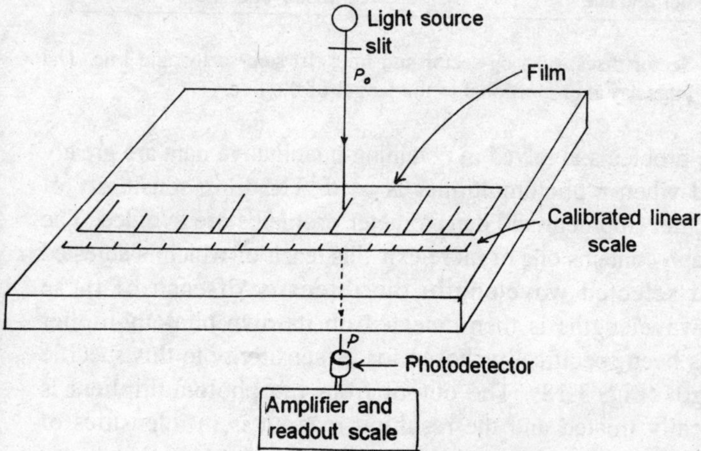

Fig. 3.16: Schematic diagram of an optical densitometer

(iv) Sector Discs

As an alternate to a densitometer, rotating **log-sector** or **step-sector** (Fig. 3.17) discs are placed in front of the slit. As the log-sector disc rotates, the slit becomes continuously narrower and less intense radiation passes to the detector. The width, density and length of the resulting lines become less. The same principle is applicable with a step-sector. In this case, however, definite increments instead of a continuous change will be recorded. In both cases, the strongest lines appear as the longest lines and the length of the line is proportional to the intensity, which, in turn, is proportional to the log of the concentration. If an internal standard is used, the difference in the lengths of the selected unknown and internal standard lines is plotted against the log of the concentration of the unknown.

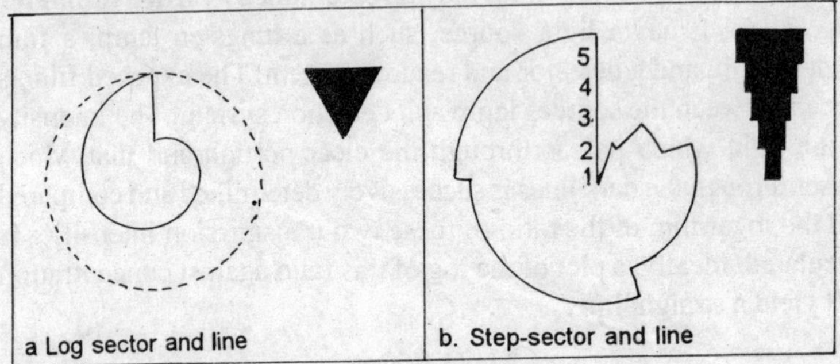

Fig. 3.17: Sector discs: (a) Log-sector and line; (b) step-sector and line. The intensity is proportional to the length of the line.

The problems involved in obtaining quantitative data are greatly simplified when a photomultiplier is used. The film, sensitivity of emulsion, development and densitometer problems are avoided. The spectrograph contains one or more exit slits, each of which is adjusted to pass a selected wavelength; the intensity of each of these selected wavelengths is then detected on its own photomultiplier which has been specifically chosen for its sensitivity to this specific wavelength (Fig. 3.18). The output from the photomultipliers is electronically treated and the resulting readout is in intensities of specific lines, ratios of intensities, or per cent concentration.

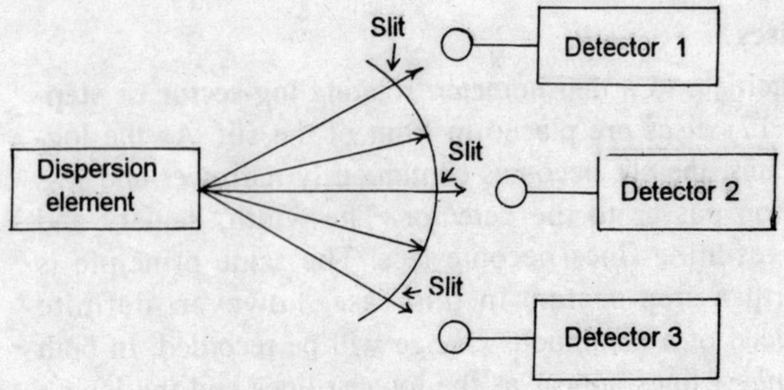

Fig. 3.18: Schematic diagram of a multidetector system

(v) Visual Spectroscopes

In visual spectroscopes the emission spectra are projected to a viewing eyepiece. Also projected simultaneously onto the same eyepiece is a linear scale which is calibrated in wavelength. Separate lines and, hopefully, the element from which these lines originated may be identified. An internal standard with a readily identifiable line (such as the D line of sodium) may be incorporated into the sample or sample electrode. The projection of this line permits the alignment and calibration of the eyepiece scale.

In a **dual source** unit excitation spectra from two samples (for example, a known and an unknown) are obtained simultaneously and the spectra of the two are visually compared. If the lines of the two spectra match, positive identification is assured. In newer models of visual spectroscopes the comparison excitation source is eliminated. Instead, images from standard films of different known spectra are projected side by side with that of the unknown. Fig. 3.19.

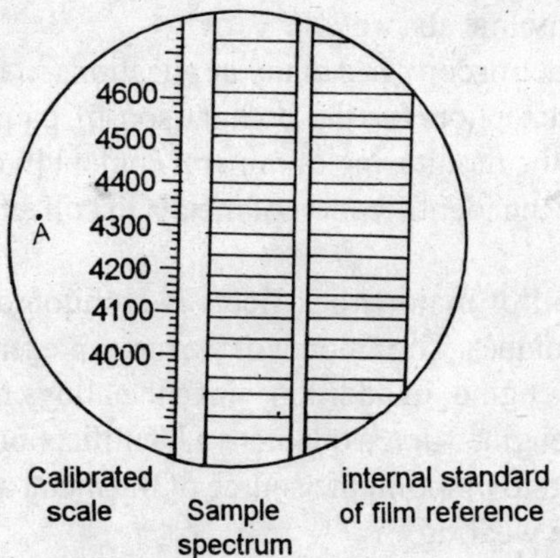

Fig. 3.19: View through the eyepiece of a visual spectroscope.

Visual spectroscopes are readily adaptable to accepting either a 35 mm or Polaroid camera at the viewing orifice. With the new instant films permanent records of the sample and a comparison spectrum are available. When a logarithmic cam is used to drive a reflective mirror, the lines which appear on the film are tapered and expanded in length. The length of each line is proportional to its intensities and to the concentration of the element from which it arises.

(vi) Spectrographic Standards

In addition to the choice of an internal standard, a most critical phase is the preparation of standard samples from which calibration curves are obtained. For accuracy the standard sample having same chemical composition and form is to be prepared which is uneconomical.

In some cases standards can be obtained from pure chemicals; solutions samples are most readily prepared by this method. Sometimes chemical analysis of a series of typical samples encompassing the expected concentration range of the elements of interest.

Applications

Emission spectroscopy is a valuable tool for rapid qualitative and semi-quantitative determination of most metal. It suffers from neither the limitations of flame emission (which was good for only a few elements) not the limitation of atomic absorption (in which a separate source lamp was required for each element). It is frequently used as a screening tool for atomic absorption.

Emission spectroscopy has many applications in mineralogy. It is used in crime detection for the comparison of paint samples, in oceanography for the monitoring of mineral content in the water, and in air pollution for the identification of metals in collected particulate matter.

The motor oil of major truck fleets is monitored on a regular basis by these techniques. The amount of the metal present is a measure of the wear on the engine. In addition, since the alloys for the various components in an engine vary, a qualitative identification of the metals present can be used to pinpoint the source of the metal and, therefore, the parts which are wearing.

It is often used for identification of samples which are not readily soluble or which contain the less common elements. Precious metals in manufacturing processes may be monitored and recovered from electronics and alloy scrap material. The techniques have been applied to the determination of zinc, copper, calcium, sodium, potassium, nickel, iron, tin, and lead in plant, animal tissue, and soil samples.

While the sensitivity of the techniques is generally not as great as those of flame emission or atomic absorption, the capability of rapid simultaneous multielement identification and analysis frequently dictates an emission spectrographic procedure. On the other hand, the destructive nature of these procedures may dictate the use of another method such as X-ray fluorescence.

Specific Analysis

Emission spectroscopy is widely used in a number of fields. For examples:

(i) *Metals and alloys.* This technique is used in iron and aluminium industries almost in all steps in their processes. This information is used to control the composition of molten metals before furthur processing : *e.g.* The presence of Ni, Cr, Si, Mn, Mo, Cu, Al, As, Sn, Co, V and lead etc., in iron and steel have been determined as a routine making use of automatic emission spectrograph. All the above listed elements have been determined in concentration ranges from 0.01% in iron 30% in steels. This method can be used to determine a large number of elements simultaneously.

(ii) *Oils.* The determination of trace element present in oils is of basic interest because it is possible that some of these if present may poison the catalyst used in cracking process. Some such elements are V, Mn, Cu, Ni, Fe etc.

Lubricating oils are analysed during use to determine the degree of wear of metal parts in motors and engines. If it is found that the concentration of a metal has increased during use it indicates excessive wear and in such a case the engine needs a overhaul. This information is used while designing engines.

(iii) *Animals and Men.* The presence of various elements like Na, K, Zn, Cu, Ca, Mg, Ni and Fe present in tissues of animals and men have been carried out by use of emission spectroscopy. The changes in the concentration of metals have been studied in relation to ageing process.

(iv) *Plants and soils.* The detection of about 40 elements in plants and soils have been carried out by using emission spectroscopy. This information has been used to diagnose deficiency problems in plants.

Other important applications of emission spectroscopy include analysis of :

(a) Ceramics (for trace elements)
(b) Metals (for oxygen and nitrogen)
(c) Aluminium (for trace of Cobalt)

(d) Graphite (for traces of Co, Ni, Mo, V)
(e) Spent nuclear fuels (for rare earths)
(f) Blood (for traces of Ca, Cu, Zn)
(g) Pancreas tissues (for Zinc)
(h) High purity acids and other analytical reagents (for a number of trace elements) .

Advantages of Emission Spectroscopy

Major advantages are :

(i) This methods can detect elements in as low concentration as 0.00017.
(ii) Almost all elements can be detected by using emission spectroscopy.
(iii) Samples are analysed in the same form as received and no concentration or separation is required.
(iv) Only a very small quantity (1–10 mg) of sample is needed.
(v) It can be accomplished in a very small time.
(vi) It can be used for routine quality control work.
(vii) It is quite suited for geological samples, metals and alloys, plants, soils, environmental and biological samples, forensic exhibits, etc.

Disadvantages of Emission Spectroscopy

Some of the disadvantages are :

(i) The emission spectrometer is a costly apparatus specially if such an apparatus is designed for analysis of heavy elements or very complex samples.
(ii) The samples gets destroyed during analysis.
(iii) It is not suitable for concentrated solutions.
(iv) It can be used only for the analysis of elements.

4

Microwave Spectroscopy

Introduction

Microwave spectroscopy is being increasingly used these days in various branches of science *e.g.*, in physics, chemistry, electronics, astronomy etc. Microwave spectroscopy has been found to be more useful in the determination of structure of those molecules which do not give good results by IR or Raman spectroscopy. The better results in microwave spectroscopy may be attributed to the more precise measurement of frequencies in the microwave region. The source of microwave radiation is a klystron tube, which is operated to produce monochromatic radiations having a frequency of about 95000 MH_2. The klystron tube is an electronic oscillator in which a beam of electrons is pulsed between the cathode and a reflector.

Molecular Spectroscopy

The spectroscopy in the microwave region (100 μm to 1 cm or 3×10^{12} H_2 to 3×10^{10} H_2) is concerned with the study of rotating molecules. Since it is *concerned with the pure rotational motion of molecules* it is also known as **rotational spectroscopy.**

Only such molecules which have resultant dipole moments on rotation can generate an electric field which is capable of interacting with the electric component of the microwave radiation. In this interaction the energy may be absorbed or radiated which then produces a spectrum. In case of those molecules which do not possess any resultant dipole moment no interaction is possible and such molecules are referred to as '*microwave inactive*' molecules. Some such microwave inactive molecules are H_2, Cl_2 etc. The examples of microwave active molecules are HCl, CH_3Cl, etc.

Rotation of Molecules

The rotation of a three dimensional body is quite complex and for simplicity let us resolve it into rotational components about three mutually perpendicular directions through the centre of gravity—the principal axis of rotation. These three components are designated as I_A, I_B, I_O.

On the basis of the relative values of their three principal moments of inertia molecules may be classified as :

(i) Linear molecules (ii) Symmetric tops.

(iii) Spherical tops (iv) Asymmetric tops.

1. *Linear molecules* refers to such molecules in which the atoms are arranged in a straight line. For example HCl, OCS (carbon oxysulphide) etc.

In case of linear molecules $I_A = 0$, and $I_B = I_C$.

2. *Symmetric tops* refers to such molecules as CH_3F in which the three hydrogen atoms are bonded tetrahedrally to the carbon.

```
H
  \
H – C – F
  /
H
```

Just as in linear molecules, the end over end rotation in and out of the plane of the paper are identical i.e., $I_B = I_C$. The moment of inertia about C – F bond axis (which is considered as the main rotational axis) is not negligible as it involves, the rotation of three hydrogen atoms off this axis. Such a molecule spinning about this axis can be considered like a top and so it is called a *symmetric top.* Thus in a symmetric top molecule

$$I_B = I_C \neq I_A,\ I_A \neq 0$$

Such molecules may be further subclassified as *prolate* symmetric top. In these $I_B = I_C > I_A$ *e.g.*, CH_3F. The other type in which $I_B = I_C < I_A$ are classified as *oblate e.g.*, BCl_3.

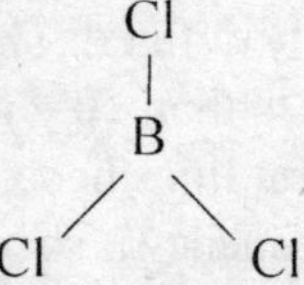

3. *Spherical tops* refers to such molecules as CH_4 which have three moments of inertia identical i.e. $I_A = I_B = I_C$.

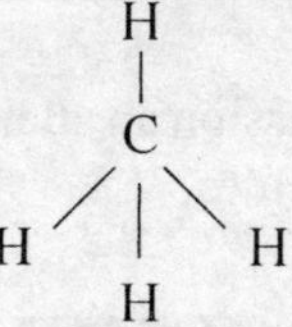

Such molecules being tetrahedral have zero resultant dipole moment and so are only of academic interest because no rotational spectrum is observed in them.

4. *Assymetric tops* refers to those molecules in which $I_A \neq I_B \neq I_C$. Most of the substances fall in this category, e.g. H_2O $CH_2 = CHCl$ etc.

Rotational spectra

In a gas phase sample, the molecules are in continual motion, appropriate to its thermal energy. Part of this is stored in the form of rotational energy, that is the kinetic energy associated with the tumbling motion of the molecules relative to an observer in the laboratory, for a single, isolated molecule, this rotational motion is quantised in accordance with quantum theory and like other forms of energy rotational energy is also quantized thus a molecule can possess certain definite values of rotational energy. The permitted energy values known as rotational energy levels can be found out by using Schrodingers equation for the system represented by that molecule. Such calculations are simple for simple molecules but complicated for complex systems.

The transitions between these individual rotational levels give rise to *rotational spectrum* of the molecule in question. The quantum states of a single molecule remain well defined as long as it is in isolation. If the molecule collides with another species, it is likely to change its rotational state and the definition of the spectrum is in danger of being destroyed. Provided the time between collisions is long compared with the inverse of the transition frequency (about 10^{-10} s), the rotational spectrum will remain sharply defined. This condition can be achieved in gases at low pressure (less than 10^{-4} atmosphere). As the gaseous pressure is raised the spectral lines become broader. In high pressure gas or liquid, the collision frequency is so high, that individual transitions can no longer be seen. In solids, the orientations of individual molecules is fixed and energy can not be stored in the form of rotational motion so rotational spectroscopy is carried out on gaseous samples at low pressure.

In the following discussion an attempt has been made to discuss each class of rotating molecules.

DIATOMIC MOLECULES

The Rigid Diatomic Molecules

These are the simplest type of linear molecules in which two masses m_1 and m_2 may be considered as being joined by a rigid bar. The length of such a bar may be taken as equal to $r_1 + r_2$ or r_0.

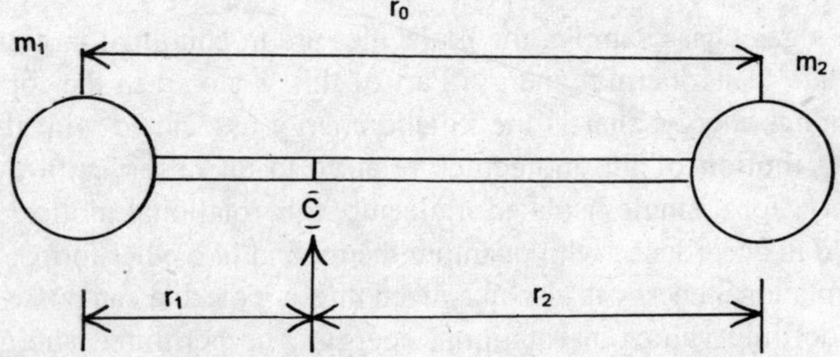

Fig. 4.1: A Rotating diatomic molecule

If such a molecule rotates end-over-end about point C, the centre of gravity, then

$$m_1 r_1 = m_2 r_2$$

The moment of inertia (I) about point C is given by

$$\begin{aligned} I &= m_1 r_1^2 + m_2 r_2^2 \\ &= m_2.r_2.r_1 + m_1.r_1.r_2 \\ &= r_1.r_2(m_1 + m_2) \qquad \text{...(i)} \end{aligned}$$

Again from the relation

$$m_1 r_1 = m_2 r_2$$

we have

$$m_1 r_1 = m_2 (r_0 - r_1) \qquad (\because r_2 = r_0 - r_1)$$

Thus

$$r_1 = \frac{m_2 \, r_0}{m_1 + m_2} \quad \text{and} \quad r_2 = \frac{m_1 \, r_2}{m_1 + m_2}$$

Substituting these values of r_1 and r_2 in (i) we get

$$I = \frac{m_1 . m_2}{m_1 + m_2} . r_0^2$$

$$= \mu . r_0^2 \qquad ...(ii)$$

where $\mu = \dfrac{m_1 . r_2}{m_1 + m_2}$ and is called *reduced mass.*

Equation (ii) gives the moment of inertia in terms of atomic masses and band length.

Assuming that a diatomic molecule behaves like a rigid rotor, it is possible to calculate rotational energy levels by making use of Schrodinger's equation.

The rotatoional energy levels allowed (E_J) to the rigid diatomic molecule are given by

$$E_J = \frac{h^2}{8\pi^2 I} J(J+1) \text{ Joules}$$

where J = 0, 1, 2...

where h_0 is Plank's constant and I is the moment of inertia which is either I_B of I_C ($I_B = I_C$). J is known as *rotational quantum number* and can have integral values including zero. The restriction to possess only integral values arises from the solution of Schrodinger's equation.

The above equation gives us energy values in Joules which can be changed into corresponding values for frequency or for wave number. For frequency we have the relation $v = \dfrac{\Delta E}{h} H_2$ and for wave number, $\bar{\nu} = \dfrac{\Delta E}{hc}$ cm^{-1}.

The above equation may be written as

$$E_J = BJ(J+1) \text{ cm}^{-1} \quad (J = 0, 1, 2...) \qquad ...(iii)$$

B is known as *rotational constant* and is equal to

$$\frac{h}{8\pi^2 I_B . C} \text{ cm}^{-1}.$$

From equation (iii) we can show allow energy level diagram as in Fig. 4.2 when J = 0 in equation (iii),

$$E_J = BJ(J+1)$$
$$= 0 \text{ cm}^{-1} \qquad ...(iv)$$

From equation (iv) it is evident that molecule is not rotating at all.

When J = 1, we get

$$E_J = B.1.(1+1)$$
$$= 2B \text{ cm}^{-1}.$$

So we see that rotating molecule has the lowest angular momentum.

In this way values for other values of J i.e., J = 2, 3, 4... etc. can also be calculated. The followed rotational energy levels of a rigid diatomic molecule are shown in Fig. 4.2.

J	E_J
4	20B
3	12B
2	6B
1	2B
0	0

Fig 4.2: The allowed rotational energy values of a rigid diatomic molecule.

Frequency of Rotational Spectral Lines

The frequency of a spectral line depends upon the difference in energy between the rotational levels between which the transition occurs. If we consider a molecule to be in the J = 0 state (i.e., the ground state, then the difference in energy between these two states can be calculated as under. The energy absorbed ΔE_J,

$$\Delta E_J = E_{J=1} - E_{J=0}$$
$$= 2B - 0$$
$$= 2B \text{ cm}^{-1}$$

Thus if the molecule is raised from J = 0 to J = 1 level an absorption line will occur at 2B cm^{-1}. In case the molecule returns to original state J = 0 from J = 1, the energy will be given out in the form of spectral lines and frequency will correspond to the energy 2B cm^{-1}.

In the same way it is possible to calculate the ΔE_J values for other transitions.

e.g. if molecule is raised from J = 1 to J = 2, we have

$$\Delta E_J = 4B \text{ cm}^{-1}.$$

These energy changes can be shown diagrammatically as under:

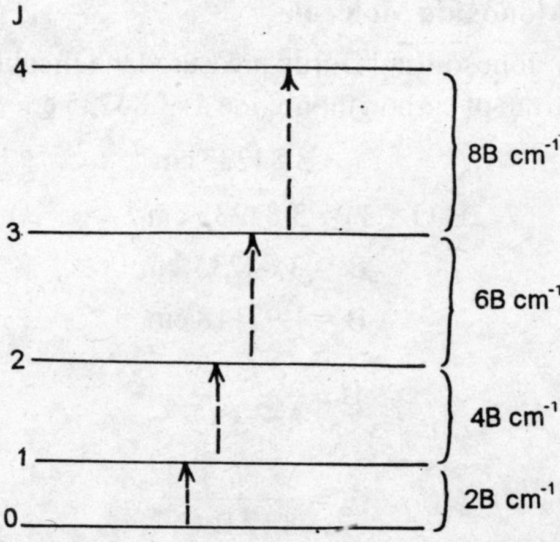

From the above pattern it is quite evident that a step-wise raising of rotational energy level will give an absorption spectrum having absorption lines at 2B, 4B, 6B...cm^{-1} and a step-wise lowering of rotational energy will give an identical emission spectrum.

To obtain the above results we have assumed that transitions can take place from a particular level only to its immediate neighbour either above or below. The same results can be obtained by using Schrodinger's equation which *forbids* any other transitions except the one in which J changes by one unit. This is known as selection *rule* and for a *diatomic rigid molecule.*

We have

$$\Delta J = \pm 1 \qquad \text{(selection rule)}$$

Validity of the Theory

We have already found ΔE_J when the molecule is raised from J to (J + 1) state. It is given by the equation:

$$J \rightarrow J + 1 = \Delta E_J = 2B\,(J + 1)\ \text{cm}^{-1}$$

This equation can be used for determination of the moment of inertia and hence the bond length. These values can then be compared with experimental results to test the validity of the theory.

(1) Carbon Monoxide Molecule

Gilliam, Johnson and Gordy measured the first line (J = 0) in the rotation spectrum of carbon monoxide as 3.84235 cm^{-1}. Thus

$$\bar{\nu}_J = 3.84235\ \text{cm}^{-1}$$

or
$$2B\,(J + 1) = 3.84235\ \text{cm}^{-1}$$

or
$$2B = 3.84235\ \text{cm}^{-1} \qquad [\because J = 0]$$

Thus
$$B = 1.92118\ \text{cm}^{-1}$$

Since
$$B = \frac{h}{8\pi^2 . I . C}$$

or
$$= \frac{h}{8\pi^2 . B . C .}$$

or
$$I_{CO} = \frac{6.626 - 10^{-34}}{8\pi^2 \times 2.99793 \times 10^{10} \times B}\ \text{kg}\,m^2$$

$$= 14.5695 \times 10^{-47}\ \text{kg}\,m^2$$

The reduced mass of CO is given by

$$\mu = \frac{m_1 . m_2}{m_1 + m_2} \times \frac{1}{N}$$

We know $m_1 = 12.0000,$ (carbon) $m_2 = 15.9994$ (oxygen)

$$\therefore \quad \mu = \frac{12 \times 15.9994}{12 \times 15.9994} \times \frac{1}{6.023 \times 10^{23}} g$$

$$= 1.1391 \times 10^{-23} \text{ g}$$

$$= 11.391 = 10^{-27} \text{ kg}$$

Hence $r^2 = \frac{I}{\mu}$

$$r^2 = 1.28 = 10^{-20} \, m^2$$

or $r = 0.1131 \, nm$

$$= 1.131 \text{ A°}.$$

These workers (Gilliam et. al.) also found the rotational absorption of ^{15}C, ^{16}O and calculated the mases of these atoms. The mass of ^{15}C was found to be 13.00007 which is in agreement with the mass of C–13 found by other methods .

Thus we conclude that it is possible to successfully use microwave spectroscopy to estimate the abundance of isotopes by comparison of absorption intensities.

Similar calculations have been made for many other molecules such as HCl, HI etc. and from the results obtained it has been established that microwave spectroscopy is an accurate method to calculate internuclear distances for simple diatomic molecules.

The results for HCl and HI are summarised

HCl

Absorption lines found at 20.7 cm^{-1} and Identified with 2B

so $B = 10.35 \text{ cm}^{-1}$, $I_{HCl} = 2.70 \times 10^{-40} \text{ g cm}^2$

μ_{HCl} (reduced mass) $= 1.627 \times 10^{24}$ g

$r_{HCl} = 1.29$ A°, $\Delta E_J = 0.405 \times 20^{-24}$ erg.

HI

Equidisatant lines with a spacing of 12.8 cm^{-1}.

Thus $2B = 12.8\ cm^{-1}$ or $B = 6.4\ cm^{-1}$

$I_{HI} = 4.372 \times 10^{-40}\ g\ cm^2$

μ_{HI} (reduced mass) $= 1.65 \times 10^{-25}$ g

$r_{HI} = 1.63$ A°.

The Non-Rigid Rotator

We have been able to calculate the internuclear distance from microwave spectra, but only with a careful selection of spectral lines. For example in case of CO molecule if we had selected any other line than the first we would not have obtained *constant* 2B separations. Generally there is a decrease in separation between successive lines with increasing J and consequently there is a decrease in apparent value of B with increasing J. It will thus result in an increase in bond length with increase in J. Hence our assumption of a *rigid bond* is only an *approximation.*

So we say that all bonds are elastic to some extent and due to this elasticity of bond following changes may occur:

(i) The elastic bond may have vibrational energy.

(ii) There is a change in values of B and r during vibrations.

These facts must be taken into consideration while interpreting the spectrum of a non-rigid rotator.

For a non-rigid molecule the rotational energy levels are found to be

$$E_J = \frac{h^2}{8\pi^2 I} J(J+1) - \frac{h^4}{32\pi^4 I^2 r^2 K} J^2(J+1) \text{ Joules}$$

Thus

$$E_J = \frac{E_J}{hc}$$

$$= BJ(J+1)\ DJ^2(J+1)\ cm^{-1}$$

where $$B = \frac{h}{8\pi^2 I.c} \quad \text{and} \quad D = \frac{h^2}{32\pi^4 I^2 r^2 k.c}$$

B is known as *rotational constant* and D is called *centrifugal distortion constant*. K is *force constant* and is given by

$$K = 4\pi^2 \overline{\omega}^2 c^2 \mu$$

Where $\overline{\omega}$ is the vibrational energy expressed in cm^{-1} (vibrational frequency of band).

The above equation for E_J is applicable to *harmonic force field* and for an *harmonic force field* the expression becomes.

$$E_J = BJ(J+1) - DJ^2(J+1)^2 + HJ^3(J+1)^3 + KJ^4(J+1)^4$$

Where H and K are small constants which depend on the geometry of the molecule.

Generally vibrational frequencies are of the order of 10^3 cm^{-1}, B is of the order of 10 cm^{-1} and D is of the order of 10^{-3} cm^{-1}.

Selection Rule

The selection rule for non-rigid body is the same as that for a rigid body *i.e.*, ($\Delta J = \pm 1$). The expression can be written as

$$\begin{aligned} \overline{\nu}_J &= B[(J+1)(J+2) - J(J+1)] \\ &\quad - D[(J+1)^2(J+2)^2 - J^2(J+1)^2] \\ &= 2B(J+1) - 4D(J+1)^3 \text{ cm}^{-1}. \end{aligned}$$

Thus we find that the only difference between the spectrum of a rigid rotator and non-rigid rotator is similar except that each spectral line undergoes displacement towards the low frequency value and the displacement increases with $(J+1)^3$.

From the values of D, it is possible to calculate the values of vibrational frequency $\overline{\omega}$ if the value of B is known. Moreover the J values of lines in an observed spectrum can also be determined if we know the value of D.

Linear Molecules (Polyatomic Molecules)

For the polyatomic linear molecules like HC ≡ CCl, OCS (carbon oxy sulphide) etc.

Since $I_B = I_C$; $I_A = 0$ so the rotational energy levels are given by a formula identical with the following equation already obtained for a diatomic molecule *i.e.*,

$$E_J = BJ(J+1) - DJ^2(J+1)^2 + \ldots \ldots \text{cm}^{-1}.$$

The simple spectrum of such a molecule that is obtained in a microwave region will show the same 2B separation modified by distortion constant.

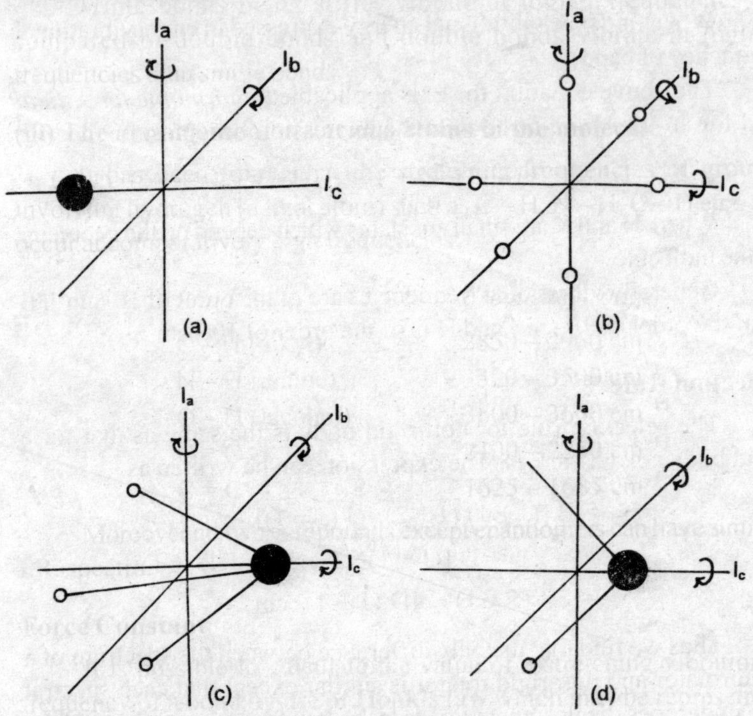

Fig. 4.3: Rotation of different types of molecules above the principal axes (a) Linear (b) Spherical (c) Symmetric top (d) Asymmetric top

It is quite clear from Fig. 4.3 that there is possibility of only one type of vibration.

The rotation about the molecular axis involves a much lower moment of inertia because only electrons will be contributing to such a motion and the very high frequency of the motion would be corresponding to electronic transition.

In case of diatomic molecules the separation between spectral lines give the moment of inertia and also the interatomic distance between the atoms. For polyatomic molecules the moment of inertia is given in terms of more than one interatomic distance and isotopic substitution is generally used.

Symmetric Top Molecule

The rotational energy levels of symmetric top molecules (*e.g.* CH_3F) are more complicated in comparison to those of linear molecules but due to their symmetry pure rotational spectra are relatively simple. For such molecules we have $I_B = I_C \neq I_A$, $I_A \neq 0$.

For such molecules there are two directions of rotation in which the molecule can emit or absorb energy. These are (i) about the main symmetry axis (*i.e.*, about C – F band in CH_3F) and (ii) that perpendicular to the axis.

In such cases two quantum numbers are needed to describes the degree of rotation, one for I_A and other for I_B or I_C. But generally a quantum number, which represent the total angular momentum and is the sum of separate angular momentum about the two different axes, is chosen and referred as J. Where J refers to the end-over-end rotation of a molecule since it is the only type of rotation allowed so generally J is used to represent *total angular* momentum. K is used for representing the angular momentum about top axis i.e., C – F band axis in case of CH_3F.

The rotational energy of a symmetric top molecule is given by

$$E = \tfrac{1}{2} I_a \omega_a^2 + \tfrac{1}{2} I_b \omega_b^2 + \tfrac{1}{2} I_c \omega_c^2$$

$$= \frac{P_a^2}{2I_a} + \frac{P_h^2}{2I_b} + \frac{P_c^2}{2I_c}$$

where,

ω = angular velocity, P = angular momentum

We know that,

$$P = \frac{h}{2\pi}\sqrt{J(J+1)}$$

and the corresponding energy will be

$$EJ = B.J.(J+1) + K^2(A - B)$$

where, $$A = \frac{h}{8\pi^2 I_a.c}\, cm^{-1}$$

and $$B = \frac{h}{8\pi^2 I_b.c}\, cm^{-1}$$

Where K is a component of J about the unique axis. (Fig 4.4)

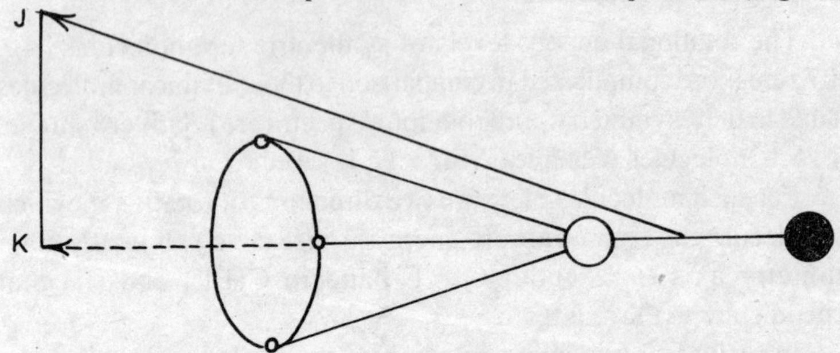

Fig. 4.4: Relation between J and K for a symmetric top molecule.

Relations between J and K for a symmetric top molecule

$$K = J, J - 1, J - 2, ...0, -(J - 1), ..., -J.$$

Thus for every value of J we have (2J + 1) values of K.

The selection rule are

$$\Delta J = 0, \pm 1, \Delta K = 0$$

When K = 0, there is no axail rotation

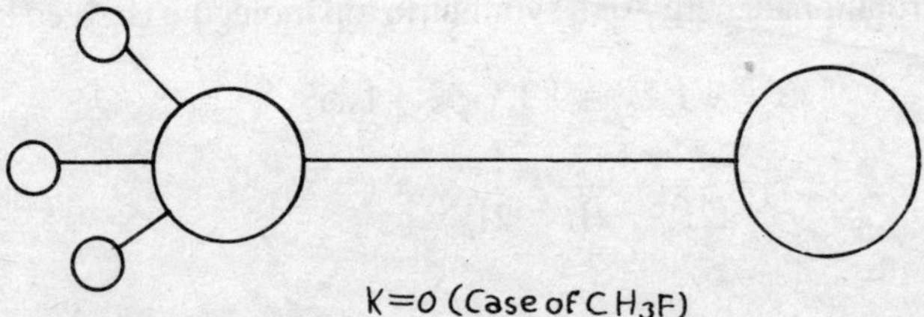

When K > 0 there is some anial rotation

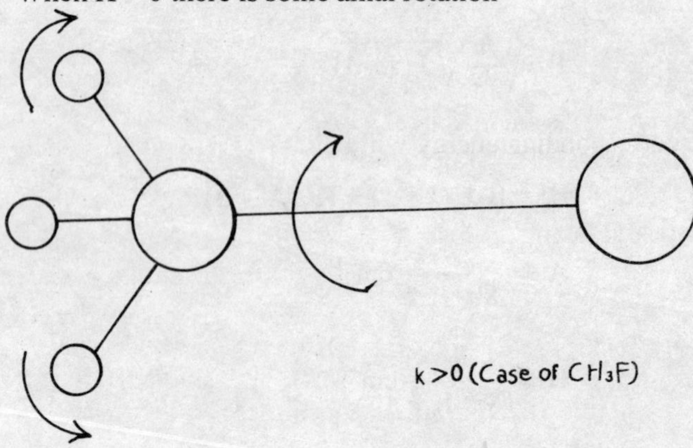

Assymetric Top Molecules

This type of molecules have all the three moments of inertia different so their rotational energy levels and spectra are very complex and no simple expression could be given for such molecules. Such molecules may be considered to lie between *oblate* and *prolate* symmetric top molecules. Each molecule and spectrum must be treated separately and certain arbitrary methods are used to get structural data for such molecules.

Spherical Top Molecules

Spherical top molecules like SF_6 or CH_4 are found to be *microwave inactive* because such molecules do not possess any resultant dipole moment.

Techniques and Instrumentation

As in other types of spectroscopy so also in microwave spectroscopy the usual pattern is followed i.e., source, monochromator, beam direction, sample and detector.

A brief discussion follows these various components of a microwave spectrometer:

1. The Source and Monochromator

Generally **klystron valve** is used as the source in microwave spectroscopy. Since it emits radiations of only a very narrow frequency range it is called 'monochromatic' so it acts as a monochromator. The frequency of emitted radiations can be varied by changing the applied voltage to klystron valve so it is possible to scan a spectrum over a limited range of frequencies making use of only a single klystron valve.

Such valves are readily available from 3000 to 5000 Mc/Sec and from such valves signals upto 2500000 Mc/Sec may be obtained.

The disadvantage of klystron valve is that total energy radiation is very small (some milliwatts only) but because of its concentration into a narrow frequency band it is possible to get strong signals using a sharply tuned detector.

2. Beam Direction

To handle the radiations emitted by klystron valve 'wave

guides' are used. *Wave guides* are hollow tubes made of copper or silver and has such a geometry that the maximum utilisation of electric and magnetic field becomes possible. These generally have a rectangular cross section-inside which the radiations is confined. It is possible to taper gently the wave guides or these may be bent to allow focusing and directing of the radiations. Since the atmospheric absorption of the beam is considerable, so the system must be efficiently evacuated. Such wave guides are commonly used in chemical research.

3. Sample and Sample Space

Thus for mostly gaseous samples are used in microwave studies. However a pressure of 0.01 mm of Hg are capable of giving a tolerable absorption spectrum so much solid or liquid substances which have their vapour pressure above 0.01 mm of Hg can be easily examined by microwave spectroscopy.

For examination of a sample by microwave spectroscopy it is placed in an evacuated 'wave guide' having their mica windows at both ends. To get a sharp absorption line it is desirable to adjust the pressure of the gas.

4. Detector

An ordinary superheterodyne radio receiver can be used as detector, for microwave spectroscopy, if it can be tuned to the appropriate higher frequency. Generally a quartz crystal is used as such a crystal is found to be more sensitive and easier to use. The crystal is mounted on a cartridge made of tungsten whisker held in point contact with it.

5. Spectrum Analyser

The spectrum analyser consists of an amplifier to amplify the detected energy and an indicator. The indicator used is either a cathode ray oscillograph or a pen and ink-recorder. The electrical signals produced by emitted vibrations are amplified and displayed or recorded by the indicator.

A simple microwave spectrometer is shown in Fig. 4.5.

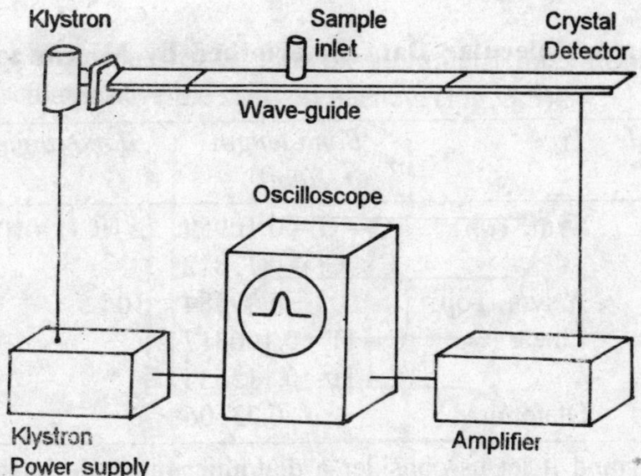

Fig. 4.5: Simplified diagram of microwave spectrometer

Working

Monochromatic radiations in microwave region obtained from klystron valve are allowed to pass through the sample under analysis. These radiation are then conducted through the 'wave guide' and then received by the detector (quartz crystal) placed at the far end of the 'wave guide'. These radiations received by the detector produce electrical signals, these signals are then amplified and displayed or recorded by the detector. From this pattern of signals it is possible to determine the frequency range of frequencies of the microwave radiations emitted.

Even this simple type of microwave spectrometer gives very accurate results because the absorption lines are narrow. However it is not possible to detect new lines for unknown substances unless their frequencies are known within a certain narrow limit.

Making use of such a spectrometer it is possible to determine moment of inertia and intermolecular distances upto ± 2A°.

Some molecular data obtained from microwave spectroscopy are summarised in Table 4.1.

Stark Effect

If the rotational spectrum is recorded in the presence of a strong electric field E, it will be observed that the lines have been split and shifted. This splitting and shifting of lives was first of all observed by Stark in atomic spectra and is known as Stark effect.

Table 4.1: Some Molecular Data Determined by Microwave spectroscopy

S.No.	*Molecule*	*Type*	*Band length (nm)*	*Band angle*
1.	CH_3Cl	Sym. Top	C – H = 0.10959 C – Cl = 0.17812	∠HCH = 108
2.	H_2O	A sym. Top	0.09584	104.5
3.	HCN	Linear	C – H = 0.106317 C – H = 0.115535	
4.	NaCl	Diatomic	0.23606	

To understand it let us consider a diatomic molecule in the presence of an external field (Fig. 4.6).

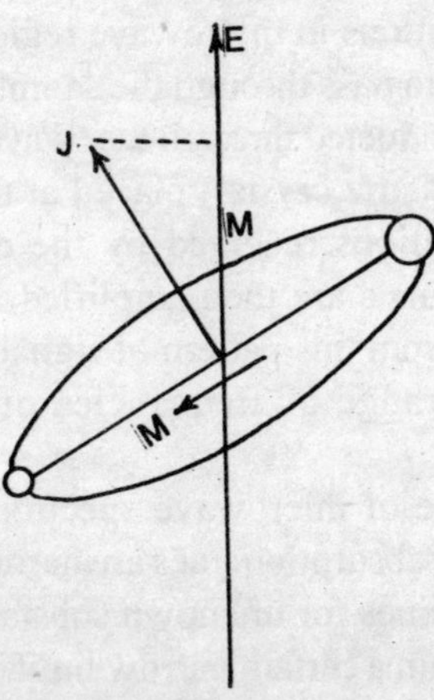

Fig. 4.6 : A diatomic molecule in the presence of an external electric field.

In this figure M is the projection of J along external electric field. The energy E_J, M in the presence of the field is given by

$$E_J, M = E_J + a_J, M\,E + \frac{1}{2}\,b^2{}_J, M\,E^2$$

Where E_J is the energy of the rotation when no external electric field is present. a_J, M and b_J, M are the first and second order Stark effects. The first order Stark effect is obtained if the dipole μ has some component along the direction of J, the second order Stark effect is obtained when μ is perpendicular to J.

The cause of difference between the first and second order Stark effect is the dependence of energy on the electric field. For a linear molecule having no component of μ in the direction of J, we have

$$E_{J,M} = \frac{h}{8\pi^2 cI} J(J+1) + \tfrac{1}{2} b_{J,M} E^2$$

The values of $b_{J,M}$ are given by

$$b_{0,0} = -\frac{16\pi^2 I c\mu}{3h} \text{ cm}^{-1} \quad (\text{for } J = 0)$$

$$b_{J,M} = -\frac{16\pi^2 I c\mu}{h} \times \frac{3M^2 - (J+1)}{J(J+1)(2J-1)(2J+3)} \text{ cm}^{-1}$$

(for $J \neq 0$)

The splitting of rotational levels when external field is present is shown in Fig. 4.7.

	J		
		------0	
		-----+1	
		-----+2	
(Without external	2 --------	------0	
field)	1 -----	-----±1	(Splitting)
	0 ---	------0	

Fig. 4.7: Splitting of rotational levels in presence of an external field

Selection Rule

$$\Delta j = \pm 1, \qquad \Delta M = 0$$

The shift in the J = 0 → 1 line is given by

$$\frac{32\pi^2 I c}{15h}$$

and the J = 1 → 2 line split is given by

$$\frac{64\pi^2 I c\mu}{105h}$$

APPLICATION OF MICROWAVE SPECTROSCOPY

Some important applications are :

1. Structural Determination

We have already seen that microwave spectroscopy can provide us useful information about the molecular symmetry and also about the molecular parameters as band length, band angle etc. This knowledge can then be utilised for determination of structure of the molecule.

The structures of $XeOF_4$, OCS, NH_3, O_3 etc., have been confirmed by microwave spectroscopy.

2. Abundance of Isotopes

This technique has also been used successfully to determine the abundance of various isotopes in molecules. This is possible because the moment of inertia possessed by each molecule is dependent upon each nuclei present in the molecule.

It is possible to locate the position of a particular isotope in the molecule by using molecular spectroscopy. For example $J = 0 \rightarrow 1$ transition for $^{14}N\ ^{15}N\ ^{18}O$ and $^{15}N\ ^{14}N\ ^{16}O$ molecule are 850 MH_2 apart.

In the end it may be concluded that microwave spectroscopy is not suitable in characterisation of new compounds because it is a property of molecule as a whole and so it is not possible to correlate certain parts of the spectrum to certain parts of the molecule.

5

Infrared Spectroscopy

Introduction

Infrared absorption of molecules is due to changes of vibrational and rotational states when subjected to infrared irradiation. Infrared absorption spectra are often called rotational vibrational spectra.

This technique is more useful as compared to other methods such as X-ray diffraction, electron spin resonance etc. because it gives more useful information regarding the structure of molecule quickly. This technique is based on the fact that after absorption of IR radiations, the molecule vibrates at many rates of vibrations and forms close-packed absorption bands known as I.R. absorption spectrum. The I.R. spectrum extends over a wide wavelength range. Different bands observed in I.R. spectrum correspond to various functional groups and bonds found in chemical molecules. The I.R. spectrum of a molecule may be considered as its finger print and can be used for the identification of the compound.

Spectral Range

The ordinary infrared region spreads over from 2.5μ to 15μ ($1\mu = 10^{-4}$ cm). The region that lies between 0.8μ to 2.5μ is known as **near infrared region** and the region that lies between 15μ to 200μ is called **far infrared region.** These two can be furthur subdivided as under Near infrared region (0.8μ to 2.5μ).

(a) *Photographic region.* This ranges from visible to 1.2μ.

(b) *Very near infrared region.* This ranges from 1.2μ to 2.5μ and is also called overtone region.

The Near infrared region between 2.5μ and 25μ is called *vibrational rotational region.*

The far infrared region from 25μ to $300-400\mu$ is called *rotation region.*

The absorption of infrared radiations can be expressed either in terms of wavelengths λ or wave number $\overline{\nu}$. In most of the cases the infrared spectra of organic compounds are plotted as percentage of transmittance against wave number. Infrared absorption takes place in a region of longer and therefore lower energy wavelengths than do ultraviolet and visible absorption. The longer wavelength requirement arises from vibrational stretching and bending of atomic bonds within the molecule as well as from molecular rotation, whereas the ultraviolet and visible absorption bonds are primarily because of electronic transitions. Whereas ultraviolet and visible spectra are generally characterized by broad absorption bands, infrared spectra are characterized by sharp bands.

While the nomenclature and graphing procedures in the ultraviolet and visible regions are almost universally standardized, considerable variation exists in the recording of infrared spectra. The horizontal coordinate is expressed in wavelength, wave number, or with a dual labelling system of both wavelength and wave number (Fig. 5.1).

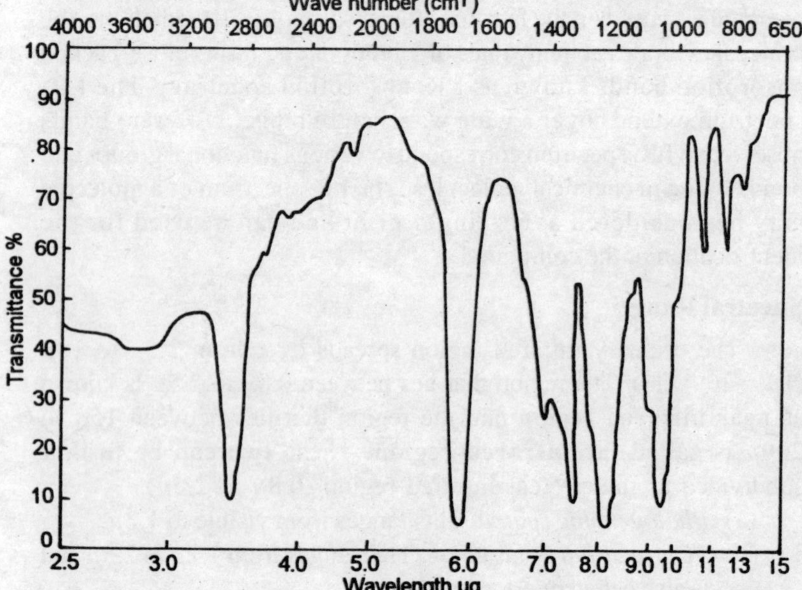

Fig. 5.1: Infrared spectrum of ethyl acetate

Although wavelengths are usually expressed in microns, μ, (10^{-6} meter), recent recommendations are that this unit be called the **micrometer,** μm. Both appear in the current literature and are identical. Because of the dual labelling it is appropriate to review the relationship

$$\bar{\nu} = \frac{1}{\lambda}$$

where $\bar{\nu}$ is the wave number in reciprocal centimeters and λ is the wavelength in centimeters. With dimensions this becomes

$$\text{cm}^{-1} = \frac{1}{\text{cm}}.$$

For example, substitution of a wave number of 1400 cm^{-1} and the conversion factor of 10^4 μm per centimeter in the equation yields a wavelength of 7.1 μm. The reciprocal relationship of wavelength and wave number prevents plotting both parameters as linear functions on the same graph.

Because of the wide range of wavelengths in the infrared and the larger relative importance of some regions, the horizontal scale is frequently nonlinear. The vertical axis is calibrated in per cent transmittance, whereas absorbance is almost universally plotted in the ultraviolet and visible regions.

In terms of wave number the traditions of this region are given as under:

$$\text{wave number}\left(\text{cm}^{-1}\right) = \frac{1}{\text{wavelength (cm)}}$$

$$\text{Thus for } 0.8\ \mu = 0.8 \times 10^{-4}\ \text{cm}$$

$$\text{wave number} = \frac{1}{0.8 \times 10^{-4}}$$

$$= 12500\ \text{cm}^{-1}$$

Similarly for 2.5μ, we have

$$\text{wave number}\left(\bar{\nu}\right) = \frac{1}{2.5 \times 10^{-4}}$$

$$= 4000\ \text{cm}^{-1}$$

for 15 μ

$$\bar{\nu} = \frac{1}{15 \times 10^{-4}}$$

$$= 667\ \text{cm}^{-1}$$

for 200 μ,

$$\bar{\nu} = \frac{1}{200 \times 10^{-4}}$$
$$= 50 \text{ cm}^{-1}.$$

Requirements for Absorption of Infrared Radiations

Those molecules which fulfil the following conditions absorb infrared radiations.

(i) Correct Wavelength of Radiation

Only those molecules can absorb radiations in which the natural frequency of vibration of some part is the same as the frequency of incident radiations.

Consider the case of a molecule of HCl. This molecule has a natural frequency of vibration as 8.7×10^{13} sec^{-1}. When I.R. radiations are allowed to pass through such a system then the analysis of transmitted radiations shows that the part of I.R. radiations with frequency 8.7×10^{13} sec^{-1} has been absorbed and so it has not been transmitted. This frequency (8.7×10^{13} sec^{-1}) is thus the characteristic frequency for HCl molecule.

The I.R. radiations that is absorbed causes the molecule to vibrate at an increased amplitude.

(ii) Electric Dipole

Only those molecules can absorb I.R. raidations, in which, such an absorption produces some changes in the electric dipole of the molecule.

A **dipole moment** is a quantitative measure of this charge distribution; the less symmetrical the distribution, the greater the numerical magnitude of the moment.

In the vibration of a molecule, the continual fluctuation of the charge distribution sets up an alternating electrical field which interacts with the electrical vector of the impinging radiation, provided the molecule has a natural vibration which involves a periodic variation of the dipole moment, and also provided the frequency of the impinging light is the same as this natural molecular frequency. The light will strongly excite those specific vibrations, lose energy in the process, and an absorption band will be observed. The molecule absorbs the incoming energy by changing its amplitude and its electrical dipole moment. The intensity of absorption is proportional to the square of the rate of change of the dipole.

It follows that *only those vibrations which involve an oscillation of the dipole moment* will record themselves in the infrared spectrum.

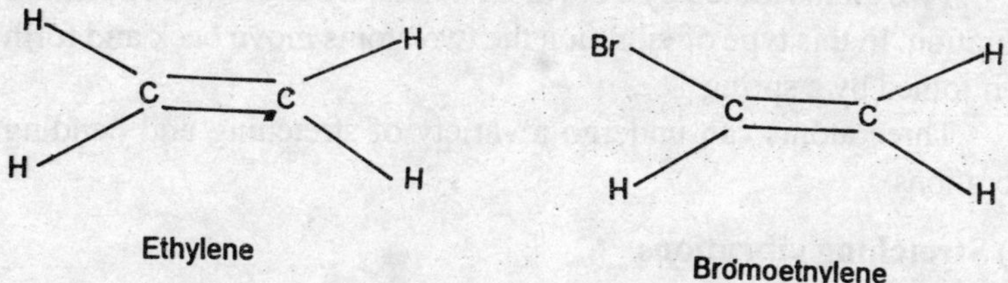

Fig. 5.2: The carbon-carbon double bond stretch in ethylene (left) and bromoethylene (right). No change in the dipole moment is produced by stretching ethylene but a change is produced by stretching Bromoethylene.

For example no change in the dipole moment is produced by the carbon—carbon double bond stretching of the symmetrical molecule, ethylene (Fig. 5.2). Since there is no change in the dipole moment, the bond does not absorb infrared radiation. On the other hand, substitution of a bromine for a hydrogen atom to form bromoethylene destroys the symmetry around the double bond.

When such a molecule is placed in an electric field, some force is exerted on the molecule and this increases or decreases separation between the opposite charges present in the molecule. Due to this change in spacing between the charged atoms there occurs a change in the dipole of the molecule. Due to this periodic change in dipole, the charged atoms of the molecule, when they vibrate absorb some I.R. radiations from the radiation source. If the rate of change in vibration is fast then the absorption of radiation is intense which results in an intense I.R. band.

Since the resultant dipole moment for symmetrical diatomic molecules such as O_2, N_2 etc., is zero, these molecule will not give I.R. absorption spectra.

Air mainly consists of O_2 and N_2 which do not have any I.R. absorption so it is not essential to evacuate the infrared spectrometers. The small quantities of CO_2 and H_2O present in air produce only weak I.R. spectra and such a spectra cancels out between the sample beam and reference beam when a double beam instrument is used.

Theory of I.R. Absorption Spectroscopy

We have already discussed that the atoms constituting a molecule are not stationary. They are continuously rotating and vibrating in a number of ways.

Two atoms joined by a covalent bond may undergo a stretching vibration. In this type of vibration the two atoms move back and forth as if joined by a spring.

Three atoms can undergo a variety of stretching and bending vibrations.

(a) Stretching vibrations

This type of vibrations arise because of stretching and contracting bonds without producing any change in bond angles.

Two types of stretching vibrations are (i) symmetric stretching and (ii) Asymmetric stretching.

(i) *Symmetric stretching*

In this type of strecthing, a movement of atoms with respect to a particular atom in a molecule is in the same direction. (Fig. 5.3).

(ii) *Asymmetric strecthing*

In this type of stretching one atom approaches the central atom whereas the other atoms moves away from it. (Fig. 5.3).

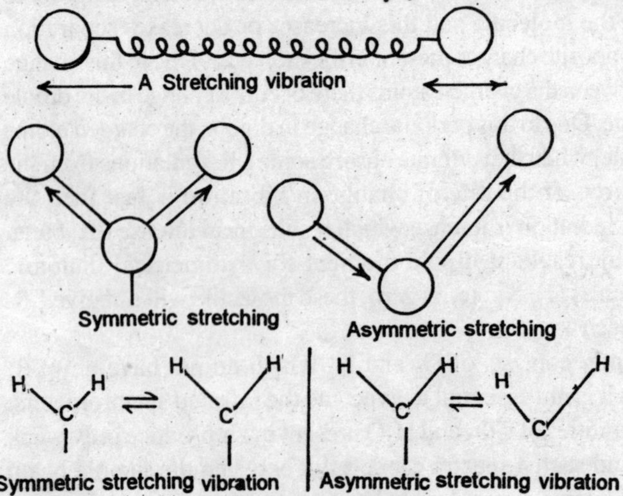

Fig. 5.3: Stretching and stretching vibrations

(b) Bending Vibrations or Deformations

In case of bending vibrations there is a deformation of bond angles and there is no change in bond lengths.

Bending vibrations are further classified as (i) Scissoring (ii) Rocking (iii) Wagging (iv) Twisting.

(i) Scissoring

In this type of vibration the two atoms approach each other in the same plane. (Fig. 5.4).

(ii) Rocking

In this type of viration the movement of atoms occur in the same direction as also in the same plane. (Fig. 5.4).

(iii) Wagging

In this type, the two atoms move up and below the plane with respect to the central atom. (Fig. 5.4).

(iv) Twisting

In this type, one of the atoms move up the plane and the other moves down the plane with respect to the central atom. (Fig. 5.4).

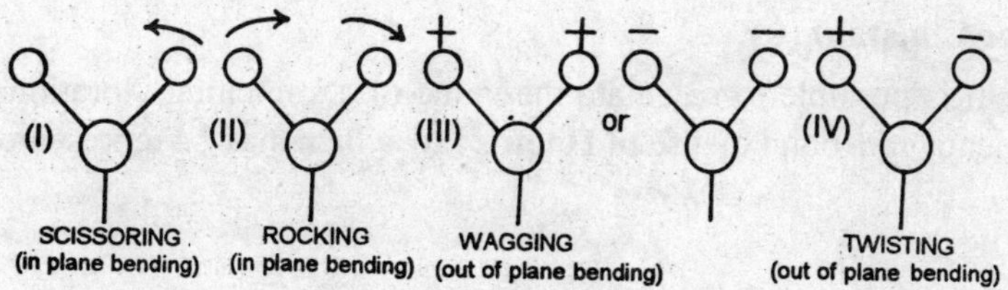

Fig. 5.4: Bending vibrations

(+ and – signs represent motion above and below the plane of the paper respectively).

Since the energy needed to stretch a spring is more than that needed to bend it so the stretching absorption of a bond will appear at a higher frequency than the bending absorption of the same bond.

Vibrational energy depends upon :

(i) The mass of atom present in a molecule

Light elements vibrate at higher frequencies as compared to

heavier elements.

(ii) Strength of the bonds

Triple bonds being stiffer vibrate at higher frequencies as compared to double bonds and double bonds vibrate at higher frequencies than single bonds.

(iii) The arrangement of various atoms in the molecule

It has been found that the stretching frequencies of groups involving hydrogen (a light atom) such as C – H, N – H, O – H etc., all occur at comparatively high frequencies.

Bond	*Frequency range*
C – H (alkyl)	2850 – 2960 cm^{-1}
N – H (amine)	3300 – 3500 cm^{-1}
O – H (alcohol)	3600 – 3650 cm^{-1}
C ≡ C	2100 – 2250 cm^{-1}
C = C	1625 – 1685 cm^{-1}

Moreover no two compounds except enantiomers can have similar I.R. spectra.

Force Constant

It is possible to calculate the value of a stretching vibrational frequency of a bond by use of **Hook's law** which may be represented as

$$\bar{v} = \frac{v}{c} = \frac{1}{2\pi c}\left[\frac{k}{\dfrac{m_1 \times m_2}{m_1 + m_2}}\right]^{\frac{1}{2}}$$

$$= \frac{1}{2\pi c}\sqrt{\frac{k}{x}}$$

where

$$x = \frac{m_1 m_2}{m_1 + m_2}$$

x is the reduced mass, m_1 and m_2 are the masses of atoms in a particular bond, k is the force constant. It is regarded as a measure of stiffness of the bond. For a single bond it is approximately 5×10^5 dynes/cm. Its value becomes double and triple for a double and a triple bond respectively.

If we consider a diatomic molecule having resultant dipole moment, the vibratory motion of the nuclei of such a molecule may be similar to that of a linear harmonic oscillator. If the bond between to nuclei of diatomic molecule is distorted from equilibrium length r_0 to a new length r. Then the restoring forces on each atoms of diatomic molecule will be given by

$$m_1 . \frac{d^2 r_1}{dt^2} = -k(r - r_0) \qquad \text{...(i)}$$

$$m_2 . \frac{d^2 r_2}{dt^2} = -k(r - r_0) \qquad \text{...(ii)}$$

Where r_1 and r_2 are positions of atoms 1 and 2 relative to the centre of gravity of the molecule.

Thus,

$$r_1 = \frac{m_1}{m_1 + m_2} . r \qquad \text{...(iii)}$$

and

$$r_2 = \frac{m_1}{m_1 + m_2} . r \qquad \text{...(iv)}$$

From (i) and (iii) we get

$$\frac{m_1 . m_2}{(m_1 + m_2)} . \frac{d^2 r}{dt^2} = -k(r - r_0) \qquad \text{...(v)}$$

Since r_0 is constant so $\frac{dr}{dt}$ will be zero.

Hence

$$\frac{d^2 r}{dt^2} = \frac{d^2 (r - r_0)}{dt^2} \qquad \text{...(vi)}$$

From (v) and (vi), we get

$$\frac{m_1 . m_2}{(m_1 + m_2)} \cdot \frac{d^2 (r - r_0)}{dt^2} = - k (r - r_0) \qquad ...(vii)$$

Putting $(r - r_0) = x$ and $\frac{m_1 m_2}{m_1 + m_2} = \mu$ in (vii)

We get

$$\mu . \frac{d^2 x}{dt^2} = - kx$$

or,

$$\frac{d^2 x}{dt^2} + \frac{k}{\mu} x = 0$$

or,

$$\frac{d^2 x}{dt^2} + \omega^2 x = 0$$

where

$$\omega^2 = \frac{k}{\mu} \quad \text{or} \quad \omega = \sqrt{\frac{k}{\mu}}$$

Where k, the force constant is expected in dynes/cm and μ is the reduced mass.

So the value of vibrational frequency (or wave number) depends upon (i) bond strength and (ii) reduced mass. Its value increases with an increase in bond strength or decrease in reduced mass.

It was expected that O – H bond should absorb at a higher frequency than F – H bond but actually F – H bond absorbs at higher frequency due to more electronegativity of F than that of O.

Energy Levels

The energy values for linear harmonic oscillators can not be given by equations of ordinary mechanics. The solutions of such equations for potential energy are found when

$$E = \left(v + \tfrac{1}{2}\right) \frac{h}{2\pi} \sqrt{\frac{k}{\mu}}$$

Where v is the vibrational quantum number and can takes only positive integral values including zero. Thus only certain discrete values of energy are possible for a vibrator.

The term $\sqrt{\frac{k}{\mu}}/2\pi$ appears both in mechanical equations and quantum equations so, we can write

$$E = \left(v + \tfrac{1}{2}\right) h v_m$$

Thus the energy values for the energy of a linear harmonic oscillator are of the following type:
where

$$\overline{\omega} = \frac{\overline{v}}{c}$$

$$= \frac{\omega}{2\pi c}$$

$\overline{\omega}$ is the vibrational frequency of the oscillator in terms of wave number.

In case the oscillating molecule is in its lowest state of vibration, the $v = 0$ and so energy E_0 is given by

$$E_0 = \frac{1}{2} hc\,\overline{\omega}.$$

E_0 is called the **zero point energy** of the molecule. So even at O*k* (– 273°C) when translational and rotational motion cease, vibrational motion still exists.

In case the diatomic molecule undergoes transition from a higher vibrational state v' to some lower vibrational state v'', the change in vibrational energy ΔE_v is given by

$$\Delta E_v = E'_v - E''_v = \left(v' + \tfrac{1}{2}\right) hc\,\overline{\omega} - \left(v'' + \tfrac{1}{2}\right) hc\,\overline{\omega}$$

or,

$$\Delta E_v = \left(v' - v''\right) hc\,\overline{\omega}$$

In such a case the frequency of transition is

$$v = \frac{\Delta E_v}{hc}$$

$$= \frac{\left(v' - v''\right) hc\,\overline{\omega}}{hc}$$

$$= \left(v' - v''\right) \overline{\omega}$$

For a transition from $v = 1$ to $v = 0$, it becomes

$$\bar{v}_{1\to 0} = \bar{\omega} = \frac{v}{c} = \frac{1}{2\pi c}\sqrt{\frac{k}{\mu}}\ \text{cm}^{-1}.$$

For a diatomic molecule possessing some dipole moment, the change in vibrational energy is given by

$$v = \frac{E_v' - E_v''}{h}$$

or,

$$\bar{\omega}_B = \frac{v}{c} = \frac{E_v' - E_v''}{hc}$$

Where $\bar{\omega}_B$ is the wave number of radiation given out.

Hence,

$$\bar{\omega}_B = \frac{\left(v' + \frac{1}{2}\right)hc\,\omega - \left(v'' + \frac{1}{2}\right)hc\,\omega}{hc}$$

[Substituting E_v' & E_v'']

or

$$\bar{\omega}_B = \left(v' + \tfrac{1}{2}\right)\bar{\omega} - \left(v'' + \tfrac{1}{2}\right)\bar{\omega}$$

But

$$v' - v'' = 1$$

or

$$v' = v'' +$$

$\because$

$$\bar{\omega}_B = \left(v' + 1 + \tfrac{1}{2}\right)\bar{\omega} - \left(v'' + \tfrac{1}{2}\right)\omega$$

or

$$\bar{\omega}_B = \bar{\omega}$$

or

$$\frac{\bar{\omega}_B}{c} = \frac{\bar{\omega}}{c}$$

or

$$\bar{v}_B = \bar{v}.$$

It means that vibrational spectrum should be made up of single lines if no rotational energy changes are involved. Thus we find that the frequency of radiated light is equal to the frequency of the oscillator for any values of υ, thus all vibrational lines from any harmonic oscillator have the same frequency. (Fig. 5.5).

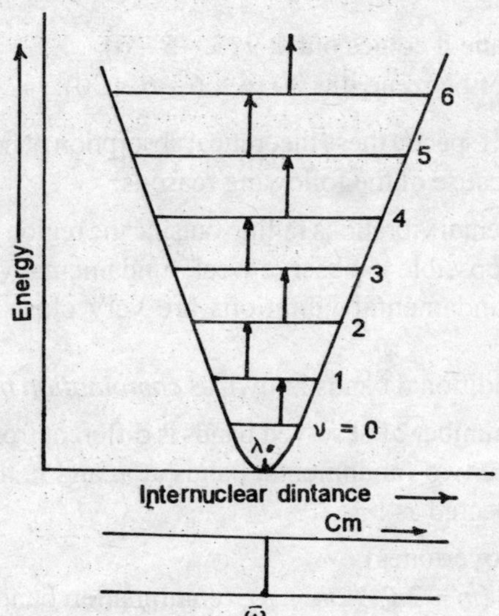

Fig. 5.5: The allowed-vibrational energy levels and transitions between them for a diatomic molecule undergoing simple harmonic motion.

Selection Rules

It has been found that the energy for a transition from energy level 1 to 2 or from energy level 2 to 3 should be same and must be identical to that of transition from energy level 0 to 1, so we can say that the only trnasitions that can take place are those in which vibrational quantum number changes by unity *i.e.*, $\Delta\upsilon = \pm 1$. The strongest band in the vibrational spectrum, called *fundamental band* in the $\upsilon = 1 \leftarrow 0$ transition.

Number of Fundamental Vibrations

In case of poly atomic molecules more than one fundamental absorption bands are observed.

We can find out that for a linear molecule of *n* atoms there could be $(3n - 5)$ theoretical possible fundamental absorption bands *e.g.* In case of CO_2 the number of theoretical absorption bands could be $(3 \times 3 - 5 = 4)$ four.

Similarly in case of non-linear molecules the number of theoretical absorption bands is given by $(3n-6)$.

For example :

(i) In methane it comes out at 9 $(3 \times 5 - 6)$.

(ii) In case of benzene it is 30 $(6 \times 6 - 6 = 30)$.

In actual I.R. spectra these theoretical absorption bands are seldom obtained. It is because of the following reasons:

(i) Fundamental vibrations falling outside the region $(2.5\mu - 15\mu)$.

(ii) It is not possible to observe weak fundamental vibrations.

(iii) Some fundamental vibrations are very close and so they overlap.

(iv) Some additional bands known as *combination bands* appear.

Hence the number of observed bands is different from theoretical bands. If there are two fundamental bands at m and n, the additional bands can be expected as :

(i) $2m$, $2n$ (overtones)

(ii) $(m + n)$, $(m + 2n)$, $(2m + n)$, (combination bands)

(iii) $(m - n)$, $(m - 2n)$, $(2m - n)$, (Difference bands).

It should be kept in mind that additional bands are 10 to 100 time less intense than fundamental bands.

Following important points should be kept in mind:

(i) For a diatomic molecule as harmonic oscillator, it is possible to describe the lowest vibrational state but not possible to describe higher vibrational states.

(ii) When observed experimentally it was found that the vibrational energy levels are less and less separated as v increases and these are not evenly spaced as expected.

(iii) Overtones and combination terms are also observed in vibrational spectra. These can not be explained by harmonic oscillator.

Because of these discrepancies a diatomic molecule is considered as anharmonic oscillator.

Vibrating Diatomic Molecule as Anharmonic Oscillator

Fig. 5.6 shows the energy curve for a typical diatomic molecule which behaves as anharmonic oscillator.

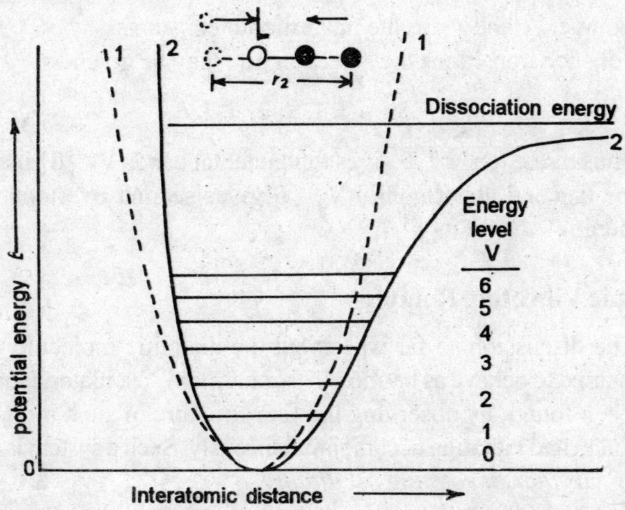

Fig. 5.6 :

These curves depart from harmonic behaviour by varying degrees which depend upon the nature of bond and the atoms involved. It is to be noted that both the harmonic and anharmonic curves are nearly alike at lower energies.

The deviations in case of anharnonic oscillators can be explained on the basis of an empirical expression given by P.M. Morse and known as Morse function. It is as under

$$E = D_{eq}\,[1 - e^{a/r}\, r_{eq}]^2$$

where $D_e r$ = Dissociation energy for the molecule

a = Constant for the diatomic molecule

r_{eq} = internuclear distance that corresponds to a minimum of Morse curve.

Making use of this equation in Schrodinger's equation, the various allowed values of vibrational energy levels are

$$E_v = \left(u+\tfrac{1}{2}\right) hc\,\overline{\omega}\,e - \left(u+\tfrac{1}{2}\right)^2 hcx\,\overline{\omega}\,e + \left(u+\tfrac{1}{2}\right)^3 hcy\,\overline{\omega} + \ldots\ldots$$

Where $\overline{\omega}e$= equilibrium frequency of molecule as expressed in wave number, x and y are the anharmonic constants.

For such transitions the selection rule may be given as

$$\Delta v = \pm 1, \pm 2, \pm 3......$$

Thus transition $V_{1\rightarrow}0$ gives fundamental bands $V_{2\rightarrow}0$ gives first overtone (second anharmonic) $V_{3\rightarrow}0$ gives second overtone (third anharmonic).

Diatomic Vibrating Rotator

The discussion so far was about the diatomic molecule which was assumed to behave as harmonic or anharmonic oscillator. However it has been found, by observing the fine structure of such molecules, that rotation and vibration occurs simultaneously. Such a system is called *rotating vibrator* or *rotating oscillator*.

In case no interaction exists between rotation and vibration, then the total energy of the system (vibrating rotator) is given by the sum of vibrational and rotational energies.

$$E_{total} = E_{rot} + E_{irb}$$

The two cases assuming the diatomic molecule as (i) harmonic oscillator and rigid rotator and (ii) as anharmonic oscillator and non-rigid rotator, are discussed.

(i) Diatomic molecule as a harmonic oscillator and a rigid rotator

The total energy of such a system is given by

$$E_{total} = \left(v + \tfrac{1}{2}\right)hc\overline{\omega} + \frac{h}{8\pi^2 I} J(J+1)$$

The selection rule for combined vibrational and rotational motions are the same as those for such motions separately. Thus

$$\Delta v = \pm 1, \pm 2, \pm 3,$$

and $$\Delta J = \pm 1$$

It is therefore possible to make calculations for the frequency of transitional lines between two states. If we assume that the simultaneous transition occurs from vibrational level v' to v'' and from rotational level J' to J", then we have

$$E'_{vib\cdot rot} - E''_{vib\cdot rot} = (v' - v'')hc\overline{\omega} + \frac{h}{8\pi^2 I}\left[J'(J'+1) - J''(J''+1)\right]$$

The frequency corresponding to this energy will be given by

$$\overline{v} = \frac{E'_{vib\cdot rot} - E''_{vib\cdot rot}}{hc}$$

$$= \frac{(v' - v'')hc\,\overline{\omega} + \frac{h}{8\pi^2 I}\left[J'(J'+1) - J''(J''+1)\right]}{hc}$$

$$= (v' - v'')\,\overline{\omega} + B\left[J'(J'+1) - J''(J''+1)\right]$$

where

$$B = \frac{h}{8\pi^2 Ic}$$

again
$$v = (v' - v'')\,\overline{\omega} + B\left[J^2 + J' + 1 - J''^2 - J''\right]$$

$$= (v' - v'')\,\overline{\omega} + B\,(J' - J'')(J' + J'' + 1)$$

Since we have assumed vibrations to be harmonic so the possible vibrational transitions that can occur are those in which $v' - v'' = 1$, and in such a case the above equation becomes

$$\overline{v} = \overline{\omega} + B\,(J' - J'')(J' + J'' + 1)$$

This equation makes it quite clear that the frequencies of the lines will vary with the values of J' and J". Thus

(i) If $\Delta J = 1$, then $J' = J'' + 1$ or $J' - J'' = 1$

Then the above equation becomes

$$\overline{v}(R) = \overline{\omega} + 2B(J'' + 1)\ \text{cm}^{-1}$$

where $J'' = 0, 1, 2$ etc.

(ii) $\Delta J = -1$, then $J'' = J' + 1$ or $J' - J'' = -1$

and in such a case it becomes,

$$\overline{v}(R) = \overline{\omega} - 2B(J'' + 1)\ \text{cm}^{-1}$$

where $J'' = 0, 1, 2$, etc.

(ii) If $\Delta J = -1$, then $J'' = J' + 1$ or $J' - J'' = -1$

and in such a case it becomes,

$$\bar{\nu}(P) = \bar{\omega} - 2B(J'+1)\text{cm}^{-1} \text{ (by combining them)}$$

where $J'' = 0, 1, 2$, etc.

From the equation obtained in the above two cases, we get,

$$\bar{\nu} = \bar{\omega} + 2B\,m\,\text{cm}^{-1}$$

where $m = \pm 1, \pm 2, \pm 3.......$

m can not have a zero value because that would imply negative value for J' or J" which is not possible.

Thus we find that the spectrum will consist of lines which are equally spaced. The spacing being 2B on each side of band origin.

In case $v' - v'' = 1$ and $J' - J'' = 0$

Then we have $\bar{\nu} = \bar{\omega}$

It means that in such a case the frequency (in wave number) of the corresponding spectral line in the same as the vibrational frequency of the molecule expressed in cm^{-1}. The spectral line with $J = 0$ is generally missing.

Limitation

This concept of diatomic molecule as a harmonic oscillator and rigid rotator had to be dropped because the predictions made by use of above equations were not found to be in agreement with the experimental results. The diatomic molecule is now considered as anharmonic oscillator and a non-rigid rotator.

(ii) Diatomic molecule as anharmonic oscillator and a non-rigid rotator

In such a case it is assumed that no interaction exists between vibration and rotation and the energy can be written as

$$E_{vr} = E_{rot} + E_{vib.}$$

$$= hc\left[\left\{B(J+1) - DJ^2(J+1)^2 + ... + \left(v+\tfrac{1}{2}\right)\bar{\omega}_e - x\left(v+\tfrac{1}{2}\right)\bar{\omega}_e\right\}\right]$$

Since the selection rules for combined motion remains the same as that for each motion separately i.e.

$$\Delta v = \pm 1, \pm 2........$$

and $\Delta J = \pm 1$

It is possible to find out the frequency of spectral line because of transition between two states. In case a good spectrometer is used (resolving power = 0.5 cm) we can neglect the value of D and then E_{vr} is given by

$$E_{vr} = hc\, BJ(J+1) + \left(v + \tfrac{1}{2}\right)\overline{\omega}_e hc - x\left(v + \tfrac{1}{2}\right)^2 \overline{\omega}_e hc$$

In case only the transition from V = 0 to V = 1 is considered, then the frequency of transition will be given by

$$\overline{v} = \frac{E' - E''}{hc}$$

or if the rotational quantum numbers are J' and J" corresponding to v = 0 and v = 1 respectively, then we have

$$\overline{v} = \frac{1}{hc}\Big[\Big\{hc\, BJ'(J'+1) + \left(v + \tfrac{1}{2}\right)\overline{\omega}_e hc - x\left(v + \tfrac{1}{2}\right)^2 \overline{\omega}_e hc\Big\}$$
$$-\Big\{hc\, BJ''(J''+1) + 0 + \tfrac{1}{2}\overline{\omega}_e hc$$
$$-x\left(0 + \tfrac{1}{2}\right)^2 \overline{\omega}_e hc\Big\}\Big]$$

or

$$v = BJ'(J'+1) + \frac{3}{2}\overline{\omega}_e - \frac{9}{4}x\,\overline{\omega}_e - BJ''(J''+1)$$
$$-\tfrac{1}{2}\overline{\omega}_e + \tfrac{1}{4}\overline{\omega}_e x$$
$$= BJ'(J''+1) + (1-2x)\overline{\omega}_e - BJ''(J''+1)$$
$$= \overline{v}_0 + B\left[J'(J'+1) - J''(J''+1)\right]$$
$$= \overline{v}_0 + B\left[(J'-J'')(J'+J''+1)\right] \text{cm}^{-1}$$

where $\overline{v}_0 = \overline{\omega}_e(1-2x)$

There is no change in value of B in the two states i.e., it remains same in upper and lower vibrational states. It is so because rotation has been considered as being independent of any vibrational change. Thus we can write

(i) When ΔJ = 1i.e., J' = J" + 1 or J' – J" = + 1

This equation represents the vibrations-rotation spectrum which would contain equally spaced lines having a spacing of 2B on either side of band origin. Generally no linear at v_0.

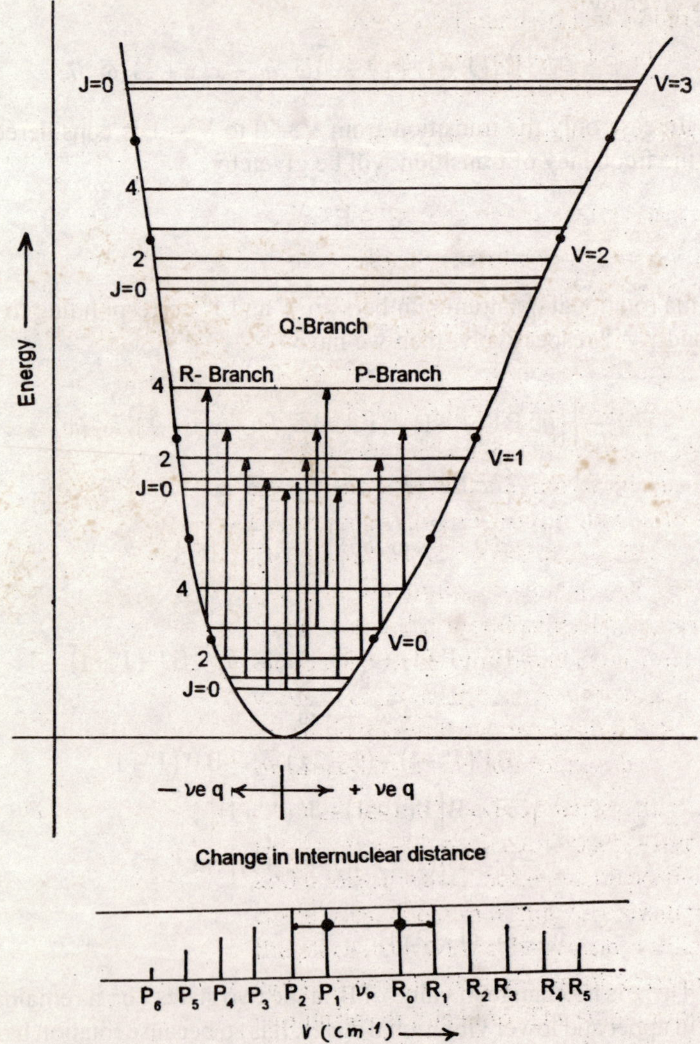

Fig. 5.7: Transitions between the rotational vibrational energy levels of a diatomic molecule (for the states v = 0 and v = 1). Note that first line in P branch is P_1 where in R branch it is R_0.

then we have

$$\bar{\nu}(R) = \bar{\nu}_0 + 2B(J''+1)\ cm^{-1},$$

where $J'' = 0, 1, 2.......$

(ii) When DJ = – 1 *i.e.*, J" = J' + 1 or J' – J" = – 1

$$\bar{\nu}(P) = \bar{\nu}_0 - 2B(J'+1)\ cm^{-1},$$

where $J' = 0, 1, 2,........$

Combining the above two equations, we get

$$\bar{\nu} = \bar{\nu}_0 + 2Bm\ cm^{-1}, \quad \text{where} \quad m = \pm 1, \pm 2,........$$

m can not have a zero value otherwise values of J' or J" will become negative which is not possible. The frequency $\bar{\nu}_0$ is generally referred to as band origin.

Linear Molecules

All the linear molecules (those with or without any permanent electric dipole) are found to be I.R. active if some of their vibrations can produce an oscillating dipole. This can be easily observed in case of CO_2.CO_2 gives two types of I.R. active vibrations.

(i) The oscillating dipole moment is parallel to the molecular axis.

←→ ← → ← → ←

O..............C..............O O..............C..............O

Due to this the parallel band arises. The selection rule is $\Delta\nu = +1$, $\Delta J = +1$

(ii) The oscillating dipole moment is perpendicular to the molecular axis.

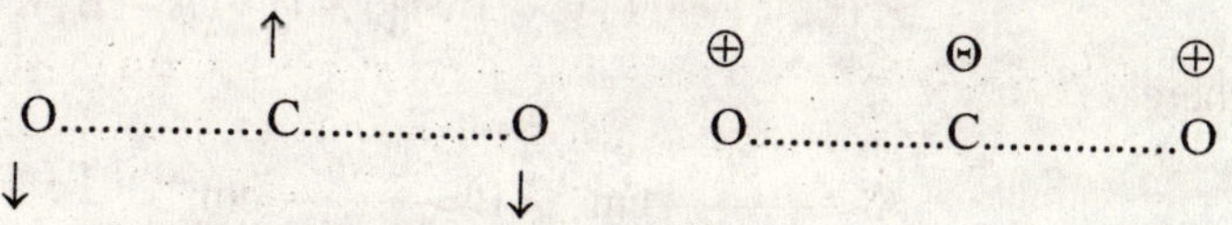

Due to this the perpendicular band arises. The selection rules are $\angle\nu = \pm 1$, $\Delta J = O, \pm 1$

It is quite evident that $\Delta J = 0$ transition arises because a component of electric field ε is perpendicular to the molecular axis is capable of exciting the vibration without producing any rotational changes. For a diatomic molecule the component of ε can not produce vibration unless the molecule has resultant electronic orbital momentum. Nitric Oxide (NO) is such a compound.

The $\Delta J = 0$ transitions produce a new group of lines which are reffered to as the Q. **branch.**

$$\text{Q branch } (J' = J''); \qquad \Delta E\, vr = \bar{v}_e\,(v' - v'') = \bar{v}_e.$$

It should be remembered that polyatomic molecules with zero dipole moment do not give rise to pure rotation spectra in the microwave region (*e.g.*, CO_2, C_2H_2, CH_4 etc.) but such molecules do produce some vibrational spectra in the I.R. region, if these spectra exhibit resolved fine structure, the moment of inertia of molecule can be found out.

Symmetric Top Molecules

For molecules possessing a centre of symmetry (*e.g.*, CO_2, C_2H_2 etc.), the centre of symmetry affects the alternate lines in P and R branches. This is due to the existence of nuclear spin. Thus in case of CO_2 it leads to the line spacing of 4B instead of 2B. It can be experimentally demonstrated from the spectrum of $^{18}O — C — O^{16}$ and $O^{16} — C — O^{16}$. In case of $^{18}O — C — O^{16}$ the symmetry does not exist and so nuclear spin does not affect the spectrum and the line spacing is found to be just half of that in $^{16}O — C — O^{16}$.

Neglecting the anharmonicity and centrifugal distortion, the energy due to vibration-rotation interaction is given by

$$\angle E_{vr} = \left(v + \tfrac{1}{2}\right) v_e + Bv\,J(J+1) + (A - B)\,K^2 \text{ cm}^{-1}$$

where

$$A = \frac{h}{8\pi^2 I_a c}\text{ cm}^{-1}, \quad B = \frac{h}{8\pi^2 I_b c}\text{ cm}^{-1}$$

and

$$K = 0, \pm 1, \pm 2 \ldots\ldots \pm J$$

The selection rules are $\Delta K = 0$, $\Delta J = 0, \pm 1$ for parallel band and $\Delta K = \pm 1$, $\angle J = 0, \pm 1$ for perpendicular band.

Fig. 5.8 shows the parallel band of CH_2Br

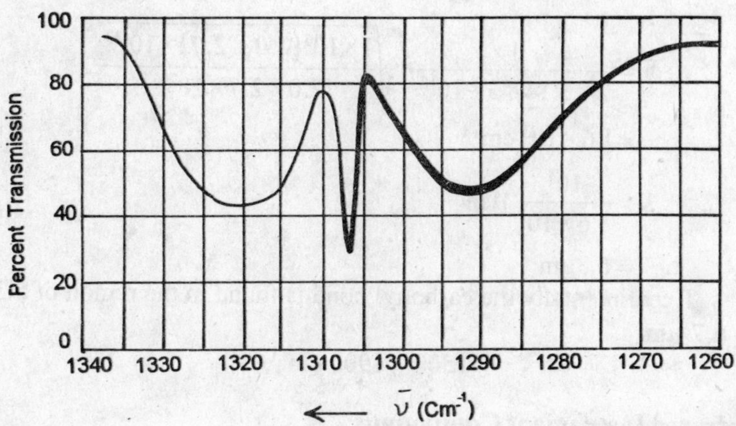

Fig. 5.8:

Asymmetric Molecules

The study of such molecules is very complex and is not appropriate to discuss it here.

Example

Calculate the approximate wave number and wavelength of fundamental vibration peak due to vibration of C = 0 (carbonyl) group.

Given that the force constant value for a double bond is 1×10^6 dynes/cm.

$$\text{mass of C – atom} = 2 \times 10^{-23}\ \text{g}$$

and

$$\text{mass of C – atom} = 2.7 \times 10^{-23}\ \text{g}\ ,$$

Solution:

Using the equation

$$\bar{\omega} = \frac{1}{2\pi C}\sqrt{\frac{K}{\mu}}$$

$$= \frac{1}{2\pi C}\sqrt{\frac{K(m_1 + m_2)}{m_1 m_2}}$$

we get

$$\overline{\omega} = \frac{1}{2\times 3.14\times 3\times 10^{10}} \times \sqrt{\frac{1\times 10^{6}(2.0+2.7)\times 10^{-23}}{2.0\times 2.7\times 10^{-46}}}$$

$$= 1.6\times 10^{3}\ \text{cm}^{-1}$$

$$\lambda = \frac{10^{4}}{1.6\times 10^{3}}\ \mu\text{m}$$

$$= 6.3\mu\text{m}$$

Experimentally the carbonyl bond is found in the region of 5.3 to 6.7 μm

or 1500 – 1900 cm^{-1}.

Infrared Instrument Compounds

The double beam infrared spectrophotometer is shown in Fig. 5.9.

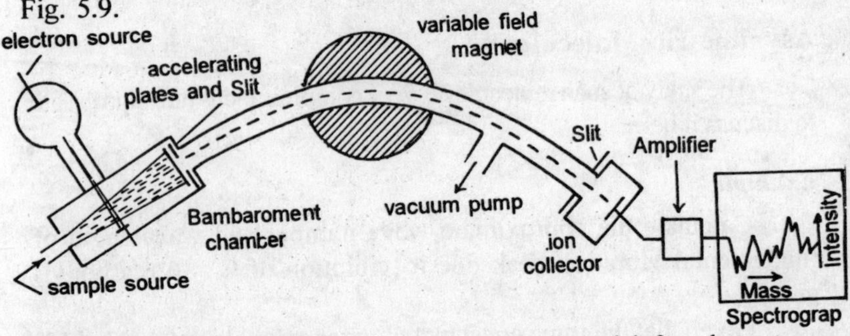

Fig. 5.9: Schematic representation of a 'double beam' infrared spectrophotometer.

The main parts of an I.R. spectrometer are :

(i) I.R. radiation sources
(ii) Monochromators
(iii) Sample cells
(iv) Detectors

(i) I.R. Radiation Sources

A common light source of infrared radiation is a *ceramic glower* (a mixture of zirconium oxide, yttrium oxide and erbium oxide, heated to 1500°C by electrical means). The other popular source of I.R. radiations are.

(a) Incandescent lamps

Such lamps are used as source in the near infrared instruments. These fall in far infrared.

(b) Wernst Glower

It is made up of a hollow rod having a diameter of about 2 mm and a length of about 30 mm. It is known as *ceramic glower.* Since it is non-conducting at ordinary temperature so it is heated electrically to about 1500°C to bring it to the conducting state. It does not undergo oxidation in air.

It emits I.R. radiations over a wide wavelength range and its energy to mainly concentrated in the visible and near infrared region of the spectrum. It emits maximum radiations at 7100 cm^{-1} (1.4 μ).

(c) Globar source

It consists of a red of sintered silicon carbide. It is about 50 mm long and 4 mm in diameter. On being heated to 1300–1700°C it emits radiations in the I.R. region. It emits maximum radiations at 5200 cm^{-1}. It has a high positive coeficient of resistance.

It is self starting and can be controlled with a variable transformer. However, it is not as intense as the Wernst glower.

(d) Mercury Arc

High pressure mercury arc is also used as a source of I.R. radiations. Backman formulated the quartz mercury lamps for I.R. region. The advantage of such a formulation is that at shorter wavelengths, the heated quartz envelope emits the radiations and at longer wavelengths the mercury plasma produces radiations through the quartz.

(e) Sometimes a *carbon arc* is also being used as a source of I.R. radiations

In case the wavelength range and intensity required are not too high, an electrically heated *nichrone wire* can also be used as a source of I.R. radiations. A *rhodium wire* heater sealed in a cylinder is also used.

For work in near I.R. region a *tungsten filament lamp* is found to be satisfactory source.

Monochromators

Either optical prism or gratings are used to obtain approximately monochromatic light. Two main types of monochromators in use are

(i) single pass monochromators and

(ii) double pass monochromators.

Both of these are prism monochromators—

(i) Single pass monochromators

Such a monochromator is shown in Fig. 5.10.

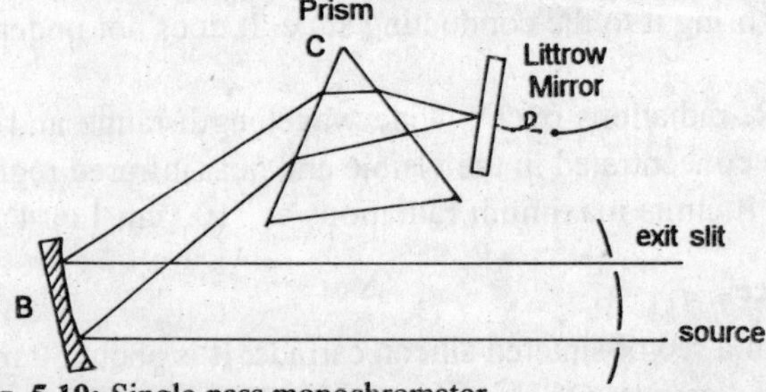

Fig. 5.10: Single-pass monochromator

In this the sample is placed near the focus of the beam just before the entrance slit 'A' to the monochromator. The radiations from I.R. source after passing through the sample and entrance slit falls on the off axis parabolic Littrow mirror B. The mirror renders them parallel sends them to prism C. The dispersed radiations are then reflected by a plane mirror D in such a way that they pass through the prism a second time and then finally focussed into the exist slit of the monochromator, and then pass to the dector.

(ii) Double pass monochromators

Such a monochromator is shown in Fig. 5.11.

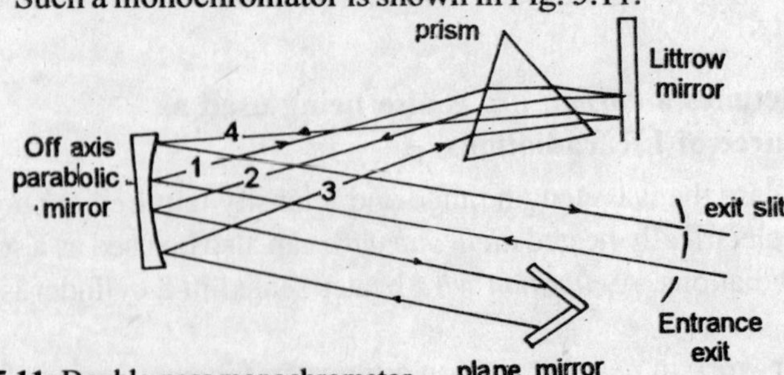

Fig. 5.11: Double-pass monochromator.

In this type of monochromator the I.R. radiations from the source make four passes shown as 1, 2, 3, 4 through the prism. This type of monochromators are capable of producing more resolution as compared to single pass monochromators.

For both these types of monochromators sodium chloride (rock salt) prism is used for the entire region 4000 – 650°cm^{-1}. Prisms of LiF and CaF_2 are found to produce more resolution in the region in which significant stretching vibrations are located.

Beyond 20μ sodium chloride absorbs strongly and can not be used. For *far* I.R. (15 – 40 μ) region *crystalline kBr* and *cerium bromide* are quite satisfactory. For *near* I.R. (1 – 5 μ) *Lithium fluoride* is preferred.

Since sodium chloride is hygroscopic, so when used, it is protected from the condensation of moisture by working at a suitable temperature.

Grating monochromators

Such monochromators produce more resolution. Grating is a series of parallel straight lines cut into a plane surface.

Dispersion through such a grating are governed by mathematical relationship

$$n\lambda = d\,(\text{Sin } i \pm \text{Sin } \theta)$$

Where n is the order, λ is the wavelength, d the distance between grooves, i the angle of incidence and θ the angle of dispersion of light of a particular wavelength.

For radiations of different wavelengths (λ) the angle of dispersion (θ) is different.

Advantages of grating monochromator

It has the following advantages as compared to prism monochromators:

(i) Grating can be made from aluminium which remains unattacked by moisture where as prisms made of metal salts get etched by atmospheric moisture.

(ii) Grating monochromators can be used over a wider wavelength range.

(iii) They provide greater resolution than a prism.

(iv) They provide nearly linear dispersion.
(v) They are not liable to attack by water.

In most of the instruments gratings are used in combination with a small prism which acts as *order sorter.*

Sample cells and Sampling of Substances

The sample used for recording I.R. spectrum may be in gaseous, liquid, solid or solution form. However, these samples have to be treated differently. The only common point being that the material containing the sample must be transparent to I.R. Radiations.

(a) Sampling of solids

Any one of the following methods may be used :

(i) Solids run in Solution

In the solid can be easily dissolved in a suitable solvent then the solution is run in one of the cells for liquids. The solids are dissolved in solvents because solids result in scattering of I.R. radiations too much. Generally they are dissolved in organic solvents such as CCl_4, $CHCl_3$, CS_2 etc. However this method can not be used for all solids because of availability of a limited number of solvents. Moreover no single solvent is available which is transparent through out the I.R. region.

While making investigations by this method, the solvent is kept in a cell of same thickness as that containing the sample to compensate for the absorption due to solvent. The solvent cell is placed in the path of the reference beam of double beam spectrometer.

(ii) Solid Film

This technique is found to be suitable for qualitative analysis of amorphous solids. In this technique the solid is deposited over the surface of a K Br or NaCl cell by evaporation of a solution of solid.

(iii) Mull technique

In this technique the sample is ground with **nujol** (liquid paraffin or some other mineral oil) to make a thick past **(Mull).** This mull is then spread between I.R. transmitting windows and mounted in the path of infrared beam and the spectrum is run. Though *nujol* is transparent through out I.R. region get its absorption maximas occur it 2915, 1462, 1376 and 719 cm^{-1} (Fig. 5.12).

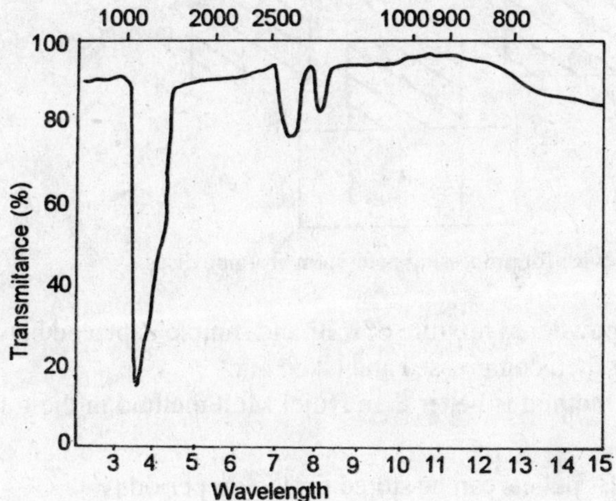

Fig. 5.12 : I.R. spectrum of Nujol Mull

The absorption bands of the sample that coincide with the absorption bands of Nujol Mull will therefore remain hidden. Generally the solid sample in Nujol Mull is used in combination with hexachlorobutadine which absorbs in the region 1630 – 1510, 1200 – 1140 and 1010 – 760 cm^{-1}. Since the absorption bands of Nujol Mull and hexachlorobutadine appear in different regions so use of such a combination allows the recording of I.R. spectrum.

(iv) Pressed Pellet technique

In this technique sample is used as a *pellet* formed by pressing of finely powdered intimate mixture of solid sample with 100 times of its weight of KBr under high pressure.

A device used to press the mixture of KBr and solid sample is shown in Fig. 5.13.

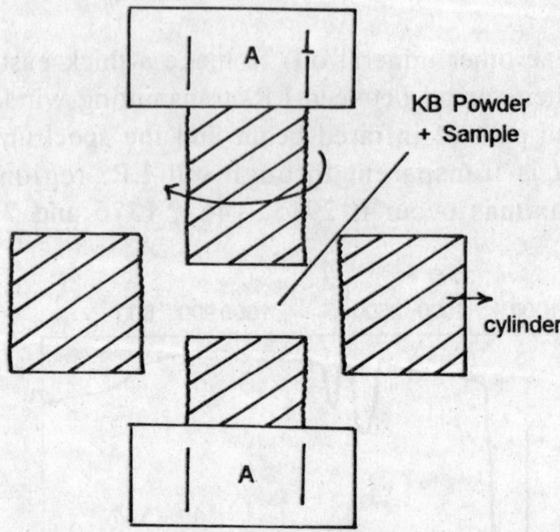

Fig. 5.13: Device for processing potassium bromide discs.

The powdered mixture of KBr and sample is placed between A and A', it is then compressed and taken out.

This method is better than Nujol Mull method in the following ways:

(i) KBr pellets can be stored for longer periods
(ii) It is possible to suitably adjust the concentration of the sample.
(iii) The resolution of KBr spectrum is better.

The disadvantages associated with the technique are:

(i) A band always appears at 3450 cm^{-1}. This is due to O – H group of moisture which is always present in the sample.
(ii) Some times the high pressure used during pellet formation may bring about polymorphic changes in the crystallinity in the sample which may produce many complexities in the I.R. spectrum of the sample.
(iii) In those cases in which polymess can not be easily grinded with KBr, this method can not be used.

(b) Sampling of Liquids

Liquids can be used as such. They are taken into rectangular cells made up of NaCl, KBr on Th Br and their I.R. spectra are recorded. The thickness of the sample is so chosen that trnasmittance is about

15 – 20%. Generally a thin layer (0001 – 0005 mm thick) is used.

For a good cell, the thickness is calculated using the relation

$$2t = \frac{N}{\omega_1 - \omega_2}$$

where N is the number of fringes between wave numbers ω_1 and ω_2

'Matched Cells' are commonly used for double beam work. In 'matched cell' one cell contains the sample and the other contains the solvent. Both these must have the same thickness. It is desirable to protect these cells from moisture in which they are soluble.

(c) Sampling of Gases

For gases a special cell about 10 cm in length is used. The gas is closed in this cell which is then placed across the path of infrared radiations. The end walls of cells are generally made of NaCl. To measure very dilute gases long path cells are commonly used. The effective path length of about 40 cm is attained making use of multiple reflections through mirrors.

The sample used in all these forms is very small (a few milli-grams). The sample used should be dry as water gives strong absorption bands near 2.7 μm (or 3710 cm^{-1}) and near 6.15 μm (or 1630 cm^{-1}). This absorption can overlap with absorptions due to the sample.

4. Detectors

A photo conductivity cell is generally used for detection of I.R. spectra in the near infrared region. Thermal detectors are used for other parts of the region.

A few type of detectors used in the I.R. spectroscopy are described below :

(i) Bolometers

It is a their metal conductor whose temperature changes when I.R. radiations fall on it. This change in temperature of the conductor brings about a change in the resistance of this conductor and the degree of change in resistance corresponds to the amount of I.R. radiations that fall on it.

A bolometer is made as one arm of a Wheatstone Bridge (Fig. 5.14). The balancing arm of the bridge is made of a similar metal strip which is not exposed to I.R. radiations. The bridge is in balanced position when no I.R. radiations are falling on the bolometer. When I.R. radiations fall on bolometer it becomes unbalanced which is indicated by a flow of current through the glavanometer (G). The intensity of I.R. radiations falling on bolometer can be measured by measuring the deflection in the galvanometer.

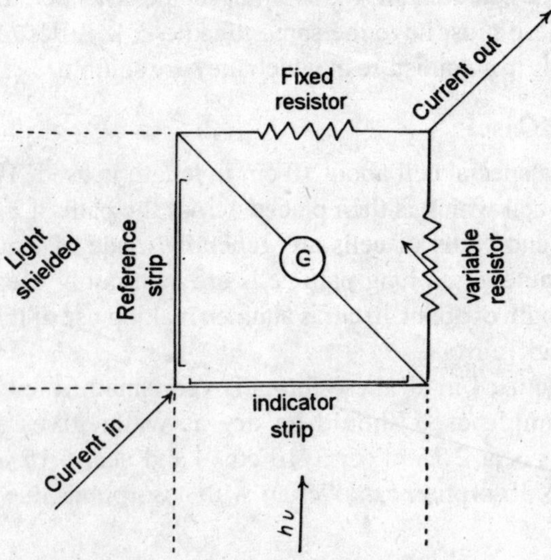

Fig.5.14: Schematic diagram of a bolometer. Wheatstone Bridge for infrared detection.

(ii) Thermocouples

To get a thermocouple, both the ends of a two wires of different semiconductor materials of high thermoelectric efficiency are welded so as to get two welded joints. Now if these welded joints are maintained at different temperatures it will produce a small electric current. To avoid any loss of energy the thermocouple is enclosed in an evacuated steel tube.

One joint known as cold junction is kept at constant temperature and is not exposed to I.R. radiations where as the other welded joint known as hot junction is exposed to I.R. radiations. Due to I.R. radiations the temperature of this junction increases and due to temperature difference between the junctions a small potential difference is developed.

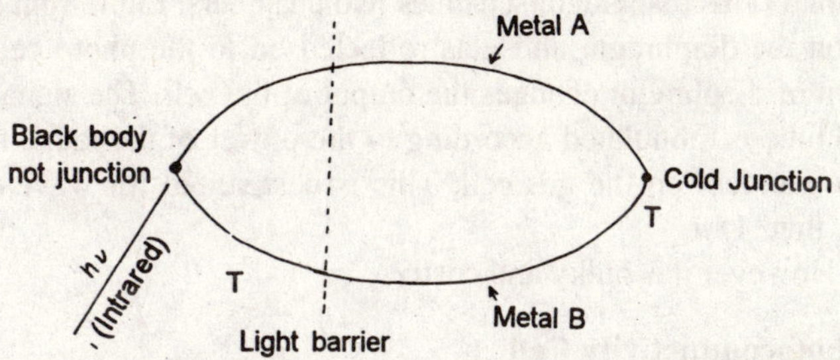

Fig.5.15: Schematic diagram of a thermocouple. The "hot junction" is exposed to the infrared radiation. The "cold junction" is insulated from it. Current is proportional to $T_1 - T_2$.

(iii) Thermistors

These are made up of metal oxides and have a negative thermal coefficient of electrical resistance. The working of thermistors is similar to that of bolometers.

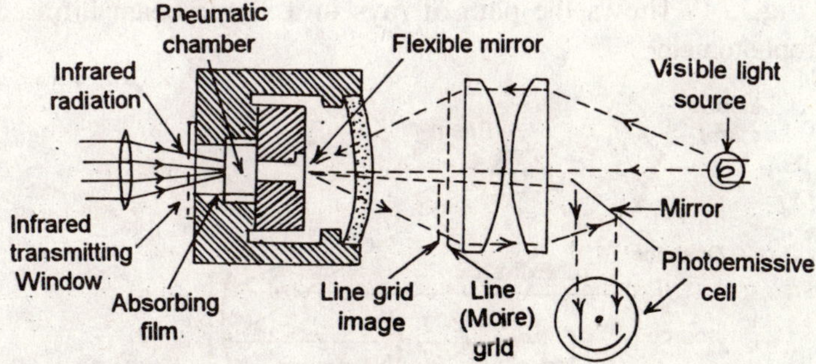

Fig.5.16: Golay pneumatic infrared detector.

(iv) Golay Cell

This is shown in Fig. 5.16: and is commonly used in many commercial spectrophotometers. A Golay Cell is made up of a small metal cylinder that is closed by blackened metal plate at one end and a flexible metallized diaphragm at the other. It is filled with xenon and then sealed. When I.R. radiations fall on the blackened metal plate, the gas get heated up and so it expands which increases the pressure and due to this increased pressure the metal diaphragm gets deformed. This diaphragm separates two chambers. Light from source falls on the diaphragm and gets reflected on to the photo cell. The motion of diaphragm changes the output of the cell. The signal from photo tube is modulated according to the power of the radiant beam that is incident on the gas cell. This is best suited for wavelengths larger than 15μ.

However it is bulky and costly.

(v) Photoconductivity Cell

It is a non-thermal conductor and is more sensitive. It is made up of a thin layer of PbS or lead telluride which is supported on glass and kept enclosed in a glass envelope. On receiving the I.R. radiations it shows an increase in conductance. It is quite sensitive but has some practical disadvantages.

At room temperature it has only a limited range.

Single Beam and Double Beam Spectrophotometers

Fig. 5.17 shows the path of rays in a single beam infrared spectrophotometer.

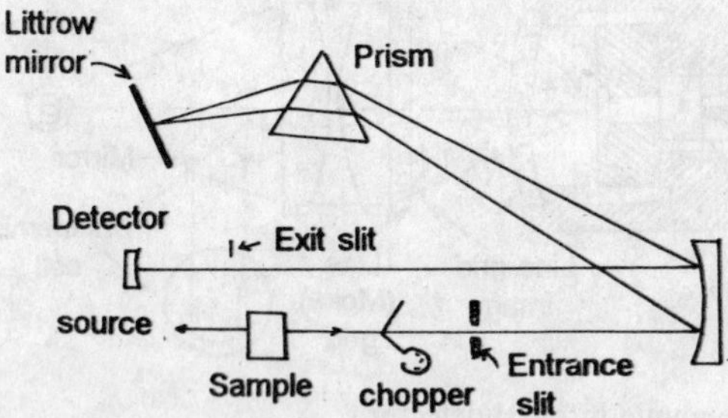

Fig.5.17: Schematic diagram showing the optical path in a single beam infrared spectrophotometer.

As is evident the emitted radiation pass through the sample and then through the fixed prism and a rotating Littrow mirror. Both these (prism and mirror) select the desired wavelength and then allow it to pass through the detector, which measures intensity of transmitted radiation. If the intensity of original radiation is known then the intensity absorbed can be calculated and thus the absorption spectrum of the sample can be detected.

This has the following disadvantages :

(i) The intensity of emitted radiation changes with source so the spectrum obtained is quite deformed.

(ii) Bands of solvent appear in the spectrum if a solution of sample is analysed. A correction has to be applied for the same.

To over come these difficulties a double beam spectrohotometer as shown in Fig. 5.18 is used.

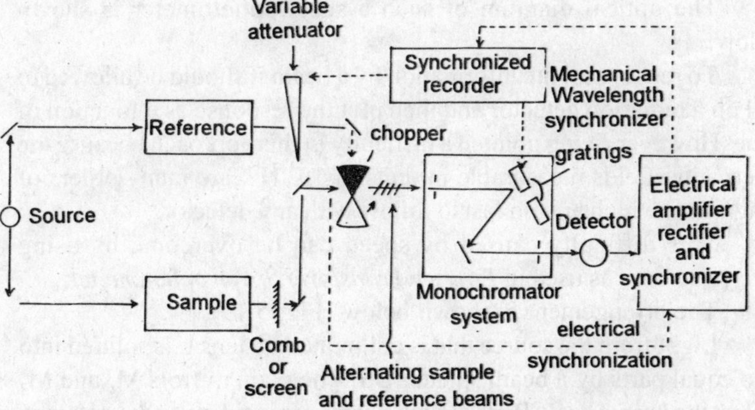

Fig.5.18: Schematic diagram of a double beam double-grating recording infrared spectrophotometer

In this the emitted radiations are split up into two beams. These two are optically and energetically identical. One of these passes through the sample and is called sample beam and another through reference sample and is called reference beam. Then the two half beams recombine and then pass along the optical path to the detector.

In the absence of the sample the reference beam reaches the detector unabsorbed and the sample beam is also not absorbed and is equal to the reference beam. A steady signal reaches the detector when two such half beams combine.

When the sample is placed in the path of sample beam. Its intensity decreases after passing through the sample and when such a half beam combines with a reference half beam and oscillating signal is produced on the detector. This signal is passed by the detector to the recording unit through a servomator.

Advances in Spectrophotometers

Now a days a variety of *automatic scanning spectrophotometers* for *mid I.R. region* are available for research laboratories.

Rapid scan spectrophotometers have also been introduced which can be directly connected to a gas chromatograph.

Fourier Transform Methods (F.T. Methods),

Interferometric Spectrophotometers

The optical diagram of such a spectrophotometer is shown below:

To retrieve all the information by a beam it should be allowed to fall on a radiation detector and then plot the response as a function of time. However *Eving* pointed a difficulty in this approach because the alternating fields measurable in units of 10^4 Hz, are many orders of magnitude which are too fast to follow with any detector.

This difficulty caused by speed can be overcome by using *Interferometer* as used in *Fourier Transform Spectrophotometer.*

The arrangement is shown below (Fig. 5.19).

Light from the source that is collimated by lens L is splitted into two equal parts by a beam spletter BS. The plain mirrors M_1 and M_2 reflect the beams back. Parts of both beams are incident on the detector. When the distance of the mirrors M_1 and M_2 is equal from the spletter BS, the detector will see the same time changing electromagnetic fields that it would have been without the interferometer, but they will only be half intense. If mirror M_2 is then permitted to move to the right along the optical axis, the phases of two beams reaching the detector are expected to differ. Thus, interference will occur. The retardation of the phase for a given increase in path length is dependent on the wavelength of radiation, and detector observes them as a series of successive maxima and minima of intensity.

For use as a spectrophotometer, the mirror M_2 is moved at a constant rate for a distance long compared to wavelength. Under such conditions, the detector response fluctuates at a rate that depends upon the rate of movement of mirror and the wavelength of radiation.

Any combination of frequencies with corresponding amplitudes can produce an *interferogram* that contains all spectral information about the original radiations. From arthematical point of view the interferogram is the *Fourier Transform* of the spectrum and the task of computer is to apply *inverse Fourier Transform.* For optical diagram see fig. 5.19.

Interferometric or *Fourier-Transform Spectroscopy (F.T.S.)* uses all frequencies from the source simultaneously. It has greatly increased sensitivity.

The other improvements include

(i) Hadamard spectroscopy

(ii) Abridged spectrophotometers

(iii) Non-disperse photometers etc.

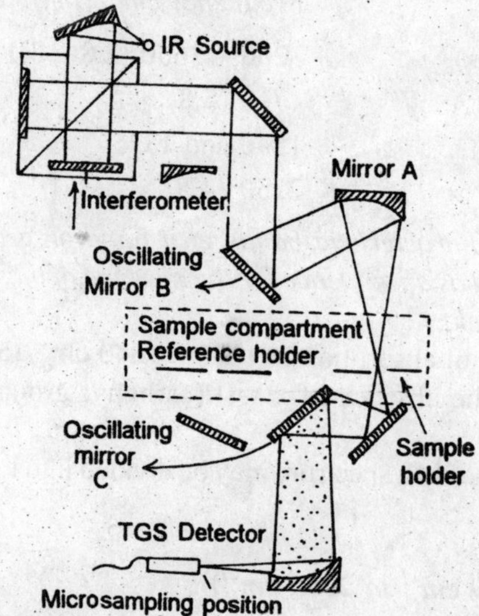

Fig. 5.19: Optical diagram of a F.T. Spectrometer.

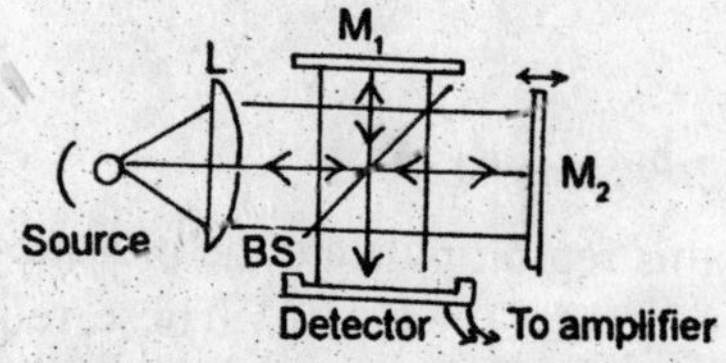

Fig. 5.20: Michelson interferometer as used in a F.T. Spectrometer.

Interpretation of I.R. Spectrum

Different types of bands present in a molecule absorb different wavelengths of frequencies in the infrared region of electromagnetic radiations giving rise to different specific absorption bands. *Thus an absorption band in the infrared spectrum refers to an absorption of an infrared radiation of a frequency characteristic of a band or a group and indicates the presence of that bond or group in the molecule.* For example, the absorption band 3 μm (3333 cm^{-1}) is indicative of the presence of –OH group. Each band or group of atoms (functional groups) absorb different frequencies in the I.R. region. Following table lists some groups and the frequencies absorbed by them:

Group	*Frequency* cm^{-1} *(μm)*
O – H (alcohols)	3200 – 3600 (2.8 – 3.1)
C = O (Ketones)	1710 (5.8)
– CH_3 (methyl)	1540 and 1375
– CN	2250 (4.5)

The absence of characteristic bands of a band or group can generally be considered as an evidence for the absence of these bonds or groups in the molecule.

Thus the absence of absorption in 1850 – 1640 cm^{-1} (5.4 – 6.5 μm) region indicates the absence of $x = 0$ (carbonyl group) in the molecule.

For interpretation of I.R. spectrum may be divided into following four regions:

(i) 2.5 – 4.5 μm (4000 cm^{-1} to 2200 cm^{-1})

Absorption bounds appearing in this region are due to bonds to hydrogen *i.e.* X – H region (C – H, N – H, O – H, S – H, O – D, H – X etc.).

(ii) 4.5 – 5.0 μm (2200 – 2000 cm^{-1})

Absorption bands in this region indicate the presence of triple bands such as C ≡ C, C ≡ N, or C = C = C. These require high frequency (low wavelength) and high energy.

(iii) 5.5 – 6.5 *μm (1800 – 1550 cm^{-1})*

This is double bond region and absorption bands in this region indicate the presence of double bonds such as C = C, C = O, C = N, N = N etc.

These three regions of I.R. spectrum are very important because these show up more clearly in the spectrum and are uncomplicated by other absorptions.

This short wavelength region less than 6.5 μm of the spectrum is called **Functional group region** because it reveals the principal functional groups of the molecule.

(iv) 6.5—15 μm *(1500–650 cm^{-1})*

A large number of sharp peaks can be observed in this region which are due to stretching vibration of single bands such as C – C, C – O, C – N, C – S, C – X etc. These are further complicated due to bending vibrations and it becomes difficult to assign each peak to a known vibration. The region is so characteristic of each molecule that even minor vibrations in the structure cause characteristic I.R. spectrum. This region is mostly used for comparisons and identification of organic molecules.

If a peak by peak comparison of an unknown spectrum is carried out with the spectrum of a known suspected compound, this will either confirm reject as such a possibility. It is just like a finger print to confirm the identity.

Some Additional Points about Infrared Spectrum

(i) Because of absorption of I.R. radiations the covalent bands present in the molecule get promoted from one vibrational energy level to a higher vibrational energy level.

(ii) More energy is needed by stronger bands to vibrate (stretch or band). Thus they absorb radiations of higher frequencies or shorter wavelengths.

(iii) Since different functional groups absorb infrared radiations at different wavelengths so their presence or absence in a molecule can be determined by looking at the I.R. spectrum.

(iv) No two compounds can have an exactly identical I.R. spectra.

(v) It is advisable to start from high frequency end of the spectrum and to make the use of finger print-region only for confirmation.

(vi) Generally most of the valuable infromation for the molecule can be provided by the absence or presence of a few bands and so the total interpretation of the spectrum is seldom required.

(vii) It is always desirable to rely more negative evidence than on positive evidence.

A study of I.R. spectrum gives us most of the required information about presence or absence of certain bands or groups in a molecule and thus it helps us to establish the presence or absence of certain groups and also to establish the nature of band or group present in the molecule. For example a spectrum with a strong band at 5.82 μm (1717 cm^{-1}) indicates the presence of a C = 0 (carbonyl groups in the molecule and the absence of this band indicates the absence of C = 0 (carbonyl) group in the molecule.

QUALITATIVE METHODS

Functional Group Analysis

While the force constants and frequency calculations are generally used and refined by the theoretician, they are approximations only. They, combined with a lot of experimental information, which has been obtained over the years, have led to the development of charts which correlate the frequency, wavelengths, and functional groups.

With experience, some of the specific functional group frequencies may be recognized. Any strong band between 5.5 μm and 5.9 μm is almost centainly due to the carbonyl. While the different classes of organic compounds containing a carbonyl group generally show a maximum at characteristic frequencies, that is not always the case. If an acid suspected, the high frequency region is examined for hydroxyl absorption. Esters in general show a strong band from 7.7 to 8.3 μm. Phthalates in particular show characteristic triples at 7.85, 8.8, and 10.7 μm.

Carbon—carbon double bonds exhibit little change in dipole moment if the bonds occur in a perfectly symmetrical molecule such as ethylene, or in the middle of a long hydrocarbon chain where the atomic structure at each end of the bond is well balanced. Therefore, the C = C group must terminate the chain or be situated so that an N or O atom enters into some mode of vibration to produce a dipole moment which yields an intense band. All amides show a very strong band at 6.0 –

6.2 μm. There are mostly two bands at 3 μm due to the strecthing of the NH_2 group. Because aniline has two N – H bonds, it shows strong absorption at both 2.9 and 2.7 μm. Di-*n*-propylamine, on the other hand, shows only one absorption band. This, at 7.05 – 7.15 μm is attributed to a C – N stretching vibration.

If the phenyl group represents more than a tenth of the size of a molecule it is usually comparatively easy to establish from the three absroptions which it produces. These are the CH stretching bands at 3.25 – 3.30 μm; and the ring vibration bands at 6.18 – 6.23 μm and 6.67 – 6.72 μm. When the phenyl ring is conjugated, a new ring vibration of about the same intensity is made active, and its absorption will be found at 6.30 – 6.33 μm in spectra such as banzamide, benzoic acid, diphenyl, benzal derivatives, etc. The presence of all four of these band may be taken as a sure indication of the presence of a phenyl ring conjugated to some other double bond in the molecule.

The following list of compounds may often be recognized from a single absorption frequency; esters, salts, acid halides, thioesters, disubstituted amides, lactones, ketones, and unconjugated C = N and C = C compounds.

If a given group is repeated in a molecule and no interactions and symmetries are involved, the characteristic absorptions may be indistinguishable from those of a single group. If the two groups are symmetrically placed, it will be necessary to recognize the cases individually without much assistance from the infrared methods.

The presence of conjugation is accompanied by a shift from normal positions for the bands of both bonds involved. This shift is from 0.1 to 0.5 μm and is always toward longer wavelengths. This effect is applicable to all the cases of conjugation examined except that between two carbonyl groups. In compounds like ketones a new band is introduced at very short wavelengths. Its indicated bond strength corresponds to that of a triple bond, about 4.75 μm.

It is possible for a double bond to be placed on a molecule in such a fashion that it could pertain of two structural interpretations. In the urethane series, the carbonyl can be considered as an ester and an amide simultaneously. The carbonyl band appears at approximately 5.85 μm midway between the 5.75 μm (ester group) and 6.0 – 6.2 μm (amide) bands.

The correlation charts might be adequate for the identification of one compound from a group of five or six possibilities, but the complexity of organic molecules limits their usefulness. Absoluted identification of a complex molecule by means of keying out all the absorption bands is frequently more an idealistic than a realistic objective.

QUANTITATIVE METHODS

Theoretically, infrared absorption should follow Beer's Law; on a practical basis it is seldom this straightforward. A number of factors may contribute to both real and apparent chemical and instrumental deviations. For example :

1. The required short path lengths of the sample cells are difficult to reproduce.
2. The low energy radiation requires wide slit widths. These in turn lead to lower molar absorptivities and sensitivities. Molar absorptivities must be empirically determined.
3. Scattered radiation at high concentrations may cause considerable error.
4. Most infrared instruments are recording double beam instruments. While this is useful for rapid qualitative applications, the extra-electronic noise which is introduced (as compared to a single beam unit) may cause loss of sensitivity, precision, and accuracy.
5. It is difficult to obtain and maintain an instrumental reference point equivalent to the "zero adjust" of visible and ultraviolet spectrophotometry.

In spite of these limitations, quantitative analysis in the infrared is generally possible provided the problems are recognized and conditions are stadardized. One such technique, the **baseline method,** is illustrated in Figure 5.21. In this procedure the spectrum of the sample is analysed as follows:

1. An absorption band of the standard which is isolated from those of the other matrix components is selected.
2. A stright line (baseline) is drawn tangent to the spectral absorption curve on both sides of the chosen absorption, peak (transmittance minimum)
3. The transmitted energy. P., and the incident energy. P_0, are

determined from the graph. P_0 is the interception of the frequency line with the tangent line and P is the point of maximum absorbance (minimum transmittance).

4. Absorbance ($\log P_0/P$) is calculated and a Beer's Law plot for a nuymber of standards is obtained. The concentration of the sample is then determined from the graph.

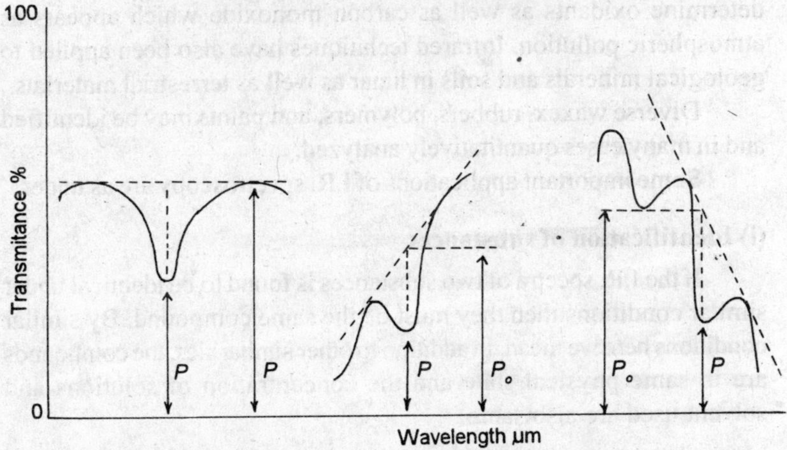

Fig.5.21: The baseline method for calculation of the transmittance ratio in quantitative analysis.

Because of close or overlapping bands and the continually changing baseline, the choice of a baseline is not always as easy as it first seems. Considerable ingenuity must be employed. Best results are obtained in laboratories where repeated quantitative analysis of the same components is required. With considerable experience on a particular system, reliable results may be obtained provided all variables are standardized.

APPLICATIONS

Infrared analysis finds the majority of its applications in the identification of the functional groups in organic compounds. This identification has been extended to such diverse applications as the determination of steroids, hormones, and pharmaceutical chemicals. Vitamin B_{12} (cyanobalamin) has been determined by infrared by the measurement of the absorption of the cyano group. The lipids, carbohydrates, amino acids, proteins, nucleic acids, enzymes, and many other biochemical compounds have been extensively studies.

In pollution studies sub-micro amounts of asbestos were collected on filters and then identified. The procedure has also been used to determine the styrene-butadiene rubber extract from the dust that is collected on filters along the highways. Long path infrared is used to determine oxidants as well as carbon monoxide which appears as atmospheric pollution. Infrared techniques have also been applied to geological minerals and soils in lunar as well as terrestrial materials.

Diverse waxex, rubbers, polymers, and paints may be identified and in many cases quantitatively analyzed.

Some important applications of I.R. spectroscopy are as under.

(i) Identification of substances

If the I.R. spectra of two substances is found to be identical under similar conditions then they must be the same compound. By similar conditions here we mean, in addition to other similarities, the compounds are in same physical state and the concentration of solutions and solvent used are also same.

(ii) Confirmation of molecular structure

A careful examination of I.R. spectrum, appearance or absorption of bands in particular region or their absence is almost a sure evidence about the absence or presence of some particular band or group in the molecule.

(iii) Checking purity of compounds

A pure compound will always give a clear I.R. spectrum where as an impure sample gives a blurred spectrum along with many bands which cannot be interpreted. For example the presence of a ketone as an impurity in an alcohol gives a poor I.R. spectrum which contains additional absorption bands due to C = O (carbonyl) group.

Similarly a small quantity of ketone in a hydrocarbon can be detected because the hydrocarbons do not have any absorption band near 1720 cm^{-1} which is the characteristic frequency of a carbonyl group.

Such methods have been used an industrial scale for control of a large number of chemicals.

(iv) Shape of a molecule

I.R. method can be used to find the shape and symmetry of a molecule. For example in I.R. spectra of NO_2, if considered linear should give only two bands which are active in I.R. region (like those of CO_2) and if it is considered as a bent molecule it should give three bands which are active in I.R. region (like those of H_2O). Actual I.R. spectrum of NO_2 when studied it was found that it contains three peaks at 750, 1325 and 1615 cm^{-1} respectively. Thus NO_2 is a bent molecule. Many compounds have been assigned their structures on such considerations.

(v) Presence of water in a compound

If water is present in lattice of a compound, three bands at 360 – 3200 cm^{-1}, 1650 cm^{-1}, and 600 – 300^{-1} will appear in its I.R. spectrum. In case water is coordinated to the metal another band appears at 880 – 65 cm^{-1}. This method has been used to detect water molecules in clatharate compounds.

(iv) Studying the progress of the reactions

A careful examination of I.R. spectrum can be used to follow the progress of a reaction (most organic reactions) readily. This is achieved by examining spectra of portions with drawn from reaction mixture. For example to study the oxidation of a secondary alcohol to a ketone, it is expected that O – H stretching near 2.8 μm will diasappear and a stretching vibration near 5.8 μm will appear due to C = O stretching.

Thus, it is possible to predict whether the expected compound has been formed or not. It is desirable to examine the whole of I.R. spectrum and to find if it is completely consistent with the anticipated structural formula.

Most of the absorption bands that organic compounds show cannot be interpreted with assurance. This is because of the following reasons:

(i) Shifting of a group by structural features—conjugation, electron withdrawl by neighbouring substitutent, angle strain, Vander Walls strain, hydrogen banding etc.

(ii) Overtones appear at just twice the frequency of fundamental bands.

The infrared position of various bond vibrations are given in Table 5.1.

Table 5.1: Infrared Positions of Various Bond Vibrations

Bond	Wavelength (μm)	Frequency cm^{-1}	Intensity
—C—H (in alkanes)	3.38–3.51	2850–2960	*m-s*
=C—H (in alkenes)	3.22–3.32	3010–3100	*m*
≡C—H (in alkynes)	3.03	3300	*s*
Ar—C—H (aromatic ring)	3.22–3.33	3010–3100	*m*
—C—H (C—H bond in aldehyeds)	3.47–3.77	2650–2880	(generally two bands)
>C=C< (C—C bond in alkenes)	6.66–16.66	600–1500	seldom used
—C = C—(C = C bond in alkenes)	5.95 – 6.2	1620 – 1680	ω
—C ≡ C—(C ≡ C in alkyenes)	4.42 – 4.76	2100 – 2660	ω
—C = C—(C = C in aromatic rings)	6.25 – 6.66	1500 – 1600	ω
—C—O (C—O bond in) alcohol, ether etc.)	7.7 – 10	1000 – 1300	s
>C = O C = O in aldehydes, ketones etc.)	5.68–5.95	1680–1760	s
—O—H (O—H in alcohols phenols etc.)	2.74–2.78	3600–3650	ω
Hydrogen bonded alcohols, phenols	2.94 – 3.11	3200 – 3400	s (broad)
O—H in hydrogen bonded carboxylic acids.	3.33 – 4.0	2500 – 3000	ω
—N—H (N—H in amines)	2.86–3.03	3300–3500	ω
—C—N (N—H in amine)	7.36–8.49	1130–1360	ω
—C ≡ N (C ≡ N in nitrites)	4.42–4.52	2230–2260	*m*
—NO_2 (nitro compounds)	6.5–7.5	1345–1385	s

Here s = strong, m = medium, w = weak

It is possible to make certain important distinctions in the O – H, N – H stretching region (3650 – 3000 cm^{-1}).

Free (non hydrogen bonded). O – H stretching vibrations appear in the 3640 – 3650 cm^{-1}. Free phenol O – H stretching show absorption at 3610 cm^{-1}. The intensity of free O – H bonds is low and will be evident in dilute solutions (or in gas phase).

A hydrogen bonded O – H absorption appear in the range 2500 – 3000 cm^{-1} as a broad and intense band.

Non-bonded amines show bands in the region 3500 – 3200 cm^{-1}. In case of primary amines two bands are seen, in case of secondary amines only are band appears and in case of teritary amines no band could be seen.

Carboxylic acids in the solid state and even in dilute solutions exist as dimers and show non O – H bands, instead they show a strong band in 3300 – 2500 cm^{-1} region.

I.R. Spectrum of Some Simple Molecules

An attempt will now be made to study the identifications of significant absorption bands in the spectra of some simple molecules.

1. I.R. Spectrum of C_3H_6O

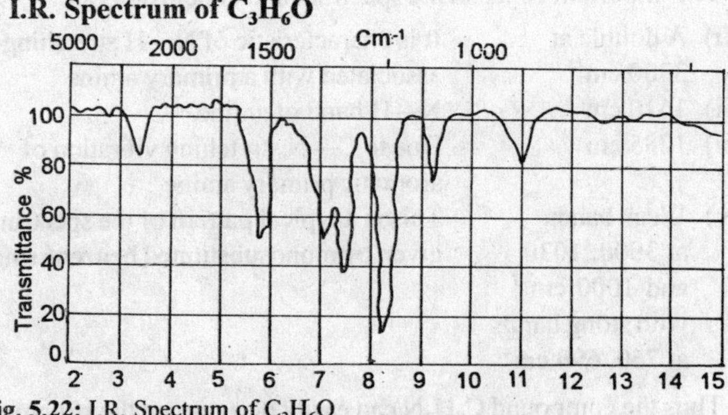

Fig. 5.22: I.R. Spectrum of C_3H_6O

The Important bands in the spectrum and conclusions are:

(i) 3000 cm^{-1} – CH_3 (methyl) stretching vibration

(ii) 1370 cm^{-1} Symmetric methyl banding vibration

(iii) 1430 cm^{-1} Asymmetric methyl stretching vibration

(iv) 1724 cm^{-1} Presence of C = O (cabonyl group)

(v) 1218 cm^{-1} It is a combination band for

$$\begin{array}{c} O \\ \| \\ C—C—C. \end{array}$$

With the above information it is quite simple to assign this spectrum

$$\text{to acetone } H_3C—\overset{\overset{\displaystyle O}{\|}}{C}—CH_3$$

2. I.R. Spectrum of C_8H_7N

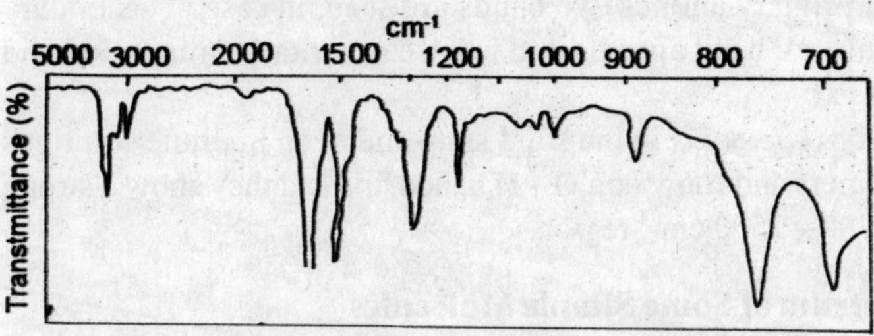

Fig. 5.23: I.R. Spectrum of C_8H_7N

The important bands in the spectrum and conclusions are:

(i) A double at 3300 cm^{-1}	It is characteristic of N—H stretching associated with a primary amine
(ii) 1610 cm^{-1}	N—H band of amine
(iii) 1285 cm^{-1}	Due to C—N stretching vibration of aromatic primary amine
(iv) Weak bands at 3000, 1030 and 1000 cm^{-1} with stong bands at 760, 690 cm^{-1}	This is a typical pattern of the spectrum given by monosubstituted benzene ring.

Thus the compound C_8H_7N can easily be assigned the structural formula as

NH$_2$

$C_6H_5NH_2$ or

so the compound is anline.

3. I.R. Spectrum of $C_9H_5O_4N$

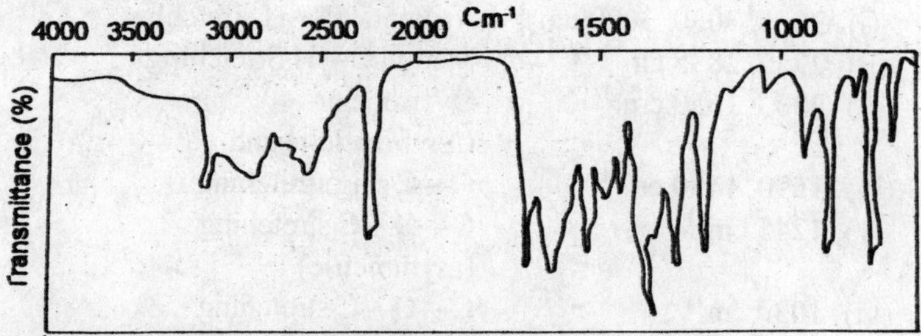

Fig. 5.24: I.R. Spectrum of $C_9H_5O_4N$

Important bands and the conlusions are as under:

(i) 3100 cm^{-1}	Aryl C—H stretching band
(ii) 3200 –3400 cm^{-1}	Hydrogen banded O—H of a carboxylic acid
(iii) 2225 cm^{-1}	Conjugated C ≡ C
(iv) 1690 cm^{-1}	Conjugated —COOH
(v) 1605 cm^{-1}	Benzene ring
(vi) 1520 – 1350 cm^{-1}	Conjugated —NO_2 group
(vii) 950 – 650 cm^{-1}	Substituted ring
(viii) 3550 cm^{-1}	O—H of water because of moisture in KBr.

Using the above information the compound $C_9H_5O_4N$ may be assigned the structural formula as

O_2N —⟨benzene ring⟩— C≡C — COOH

4. I.R. Spectrum of C_7H_8O

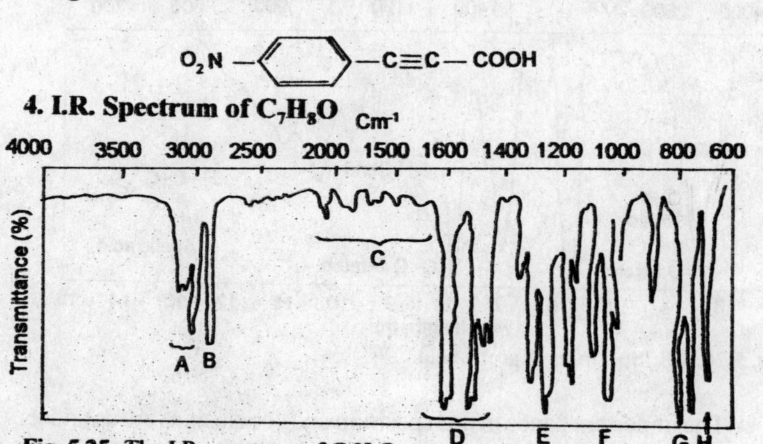

Fig. 5.25: The I.R. spectrum of C_7H_8O

Important bands and the conclusions are as under:

(i)	3660, 3030, 3000 cm^{-1}	Aromatic C—H stretching
(ii)	2950, 2835 cm^{-1}	Methyl C—H Stretching
(iii)	200 – 1650 cm^{-1}	Overtone or Combination band
(iv)	1590, 1480 cm^{-1}	C—C ring stretching
(v)	1245 cm^{-1}	C—O—C stretching (asymmetric)
(vi)	1030 cm^{-1}	C—O—C stretching (symmetric)
(vii)	800 – 740 cm^{-1}	C—H (out of plane) bending
(viii)	680 cm^{-1}	C.... ....C (out of plane) bending

Making use of this information the compound C_7H_8O can be assigned the structure

OCH_3

So the compound is anisole.

5. I.R. Spectrum of $C_2H_4O_2$

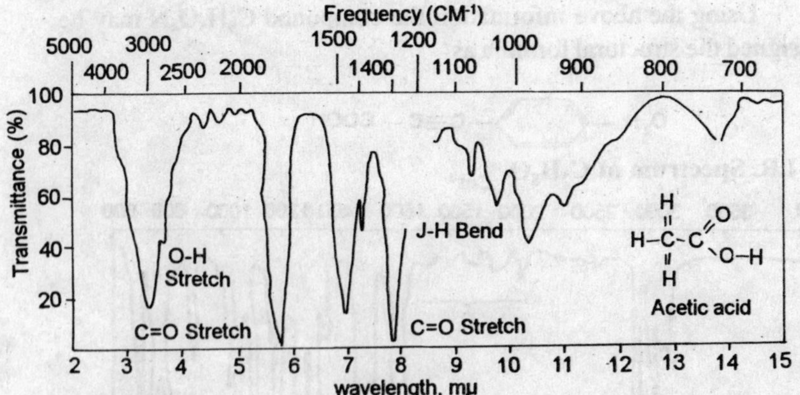

Fig. 5.26: I.R. Spectrum of acetic acid

The important bands in this spectrum and conclusions are as under Hydrogen banded.

(i) 3030 cm^{-1} — O—H stretching of carboxylic acid

(ii) 1710 cm^{-1} — C = O stretching

(iii) 1280 cm^{-1} — O—H stretching of carboxylic acid

(iv) 945 cm^{-1} — O—H bending of carboxylic acid.

Thus the compound with the empirical formula $C_2H_4O_2$ is CH_3COOH or acetic acid.

6. I.R. Spectrum of C_7H_6O

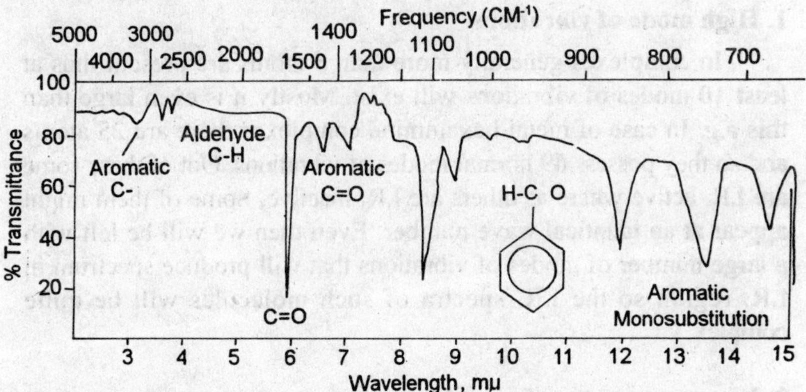

Fig. 5.27: I.R. Spectrum of C_7H_6O

The important bands in the above spectrum are :

(i) 3080 cm^{-1} — C—H stretching due to aromatic ring

(ii) 2860 cm^{-1}, 2775 cm^{-1} — C—H stretching due to an aldehyde

(iii) 1700 cm^{-1} — C = O stretching due to an aromatic aldehyde

(iv) 1595 cm^{-1} — C = C stretching due to an aromatic ring

(v) 475 cm^{-1}, 685 cm^{-1} — Bands due to characteristic aromatic monosubstitution (C—H deformation)

With this information the compound C_7H_6O can be assigned the structure as

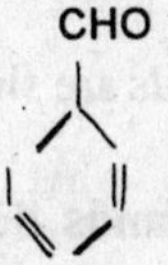

CHO

Thus the compound is Benzaldehyde

Applications of Infrared Spectroscopy to Inorganic Complexes

To carry out the spectral analysis of complexes even if we know their structure is not so easy. This is due to the following reasons:

1. High mode of vibrations

In complexes generally more than 5 atoms are present thus at least 10 modes of vibrations will exist. Mostly it is even large than this *e.g.* In case of metal hexammine complexes there are 25 atoms and so they possess 69 normal modes of vibrations. Out of these some are I.R. active where as others are I.R. inactive. Some of them might appear at an identical wave number. Even then we will be left with a large number of modes of vibrations that will produce spectrum in I.R. region so the I.R. spectra of such molecules will be quite complex.

2. Lower symmetry of complexes

It has been observed that when complexes with mixed ligands are formed there is always a decrease in symmetry. Such decrease in symmetry also accompanies the formation of polynuclear complexes. Due to this decrease in symmetry there occurs an increase in number of I.R. bands of such complexes.

3. Formation of Chelates

In those complexes in which chelation occurs and large number of bands appear in their I.R. spectrum and so the I.R. spectra of such complexes become still more complex.

Thus in most cases the I.R. spectra of complexes are difficult to interpret. However, to interpret spectra of such complexes following assumptions are kept in mind:

(i) The vibrations of ligands are slightly affected by the formation of the complex.

(ii) The vibrations of ligands do not undergo coupling with vibrations of other ligand or with skeletal vibrations.

(iii) There should be no change in symmetry if a ligand is coordinated to the central metal atom.

When the I.R. spectra of coordination complexes have been interpreted by keeping these assumptions in mind the results have been quite encouraging.

It is possible to explain geometrical isomerism by making use of I.R. spectroscopy in case of inorganic complexes.

Geometrical Isomerism

In some cases I.R. spectra of the compounds had been used to make a distinction between the trans- and is forms of a compound. This was used by space *et al.* to distinguish cis- and trans-isomers of bipyridyl cobalt (III) chloride. It was on the basis of I.R. spectra that these two compounds with empirical formula [CO (bipy) Cl_2Cl] having violet and pink colour were identified as trans- and cis-isomers respectively.

Considering the structure of the two isomers with distorted tetrabedral symmetry (trans-forms) and octabedral symmetry (cis-form) it can be easily concluded that lowering of symmetry is more is cis-form as compared to that in trans-form. Thus there will be more I.R. bands in cis- form than in trans-form. The bands of low intensity are obtained in trans form and these are not seen in spectra but in cis-form the case reverse.

Clark *et al.* also applied I.R. spectra to differentiate cis- and trans-isomers of $[M(en)_2X_2]^+$ type of complexes e.g. $[Rh\,(en)_2\,Cl_2]$ Cl.

In some cases it has not been possible to make a distinction in the cis- and trans- isomers by used I.R. spectra, however, in such cases it has been successfully achieved by using X-ray diffraction method.

Applications of I.R. Spectroscopy to Quantitative Analysis

It is possible to make quantitative determination of a compound making use of its I.R. spectra. For such determination the concentration of one of the functional groups of the compound is determined considering the case of *n*-hexane and hexanol, it is possible to find concentration of hexanol if we measure the absorption of the O—H band. From this measurement the concentration of hexanol is obtained by use of Bear-Lamberts' Law, *i.e.*

$$A = -\log \frac{I_1}{I_0} = abc$$

where A = absorbance
I_0 = Intensity of incident radiation
I_1 = Intensity of transmitted radiation
a = absorptivty of cell
b = initial path length of the cell
c = concentration of solution.

For any cell b is constant, a can also be considered as constant because it is a property of the sample under investigation, hence A α C. Any of the following two methods may be used to determine A:

1. Cell in-cell out method

To determine A by this method the instrument is set at desired wave number and the cell containing the sample is introduced in the sample beam. The reading thus obtained on a linear transmittance paper is then directly proportional to I_1. Now the cell is filled with pure solvent and the linear transmittance again found. This time the reading is proportional to 10. Making use of Beer's Law the value of A is calculated corresponding to this concentration (C). Various values of A for different values of C are found and then a plot is drawn between the values of A against C. In this way a calibration curve is obtained. This curve is then used to determine the concentration of the unknown sample.

2. Base line method

In this method a tangent is drawn to the background level of the spectrum on either side of the band [Fig. 5.28 (*a*) and Fig. 5.28 (*b*)]. Some times, the band under consideration may not be present, then the background line is drawn that may approximate the appearance of the spectrum [Fig. 5.28 (*c*)]. For multi component mixtures a line to nearby minima is chosen as background line [Fig. 5.28 (*d*)]. If linear transmittance paper is used then I_0 is measured at the bottom of the band and I_0 is measured at the same wave number and absorbance is calculated. This method automatically makes corrections for cell absorbance and reflection losses.

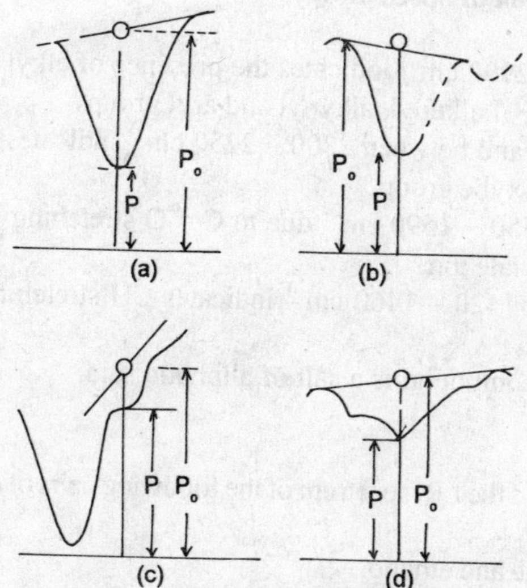

Fig. 5.28: Quantitative analysis involving baseline construction.

Limitations of I.R. Spectroscopy

Some of the limitations of I.R. spectroscopy are as under :

(i) This can not be used to find the molecular weight of the compound.

(ii) No information can be obtained regarding the relative position of various groups present in the compound.

(iii) It is not possible to find whether the given compound is pure or is a mixture of compounds by taking single I.R. spectrum.

SOLVED EXAMPLES

Example 1 :

Classify the compound from I.R. spectrum given below:

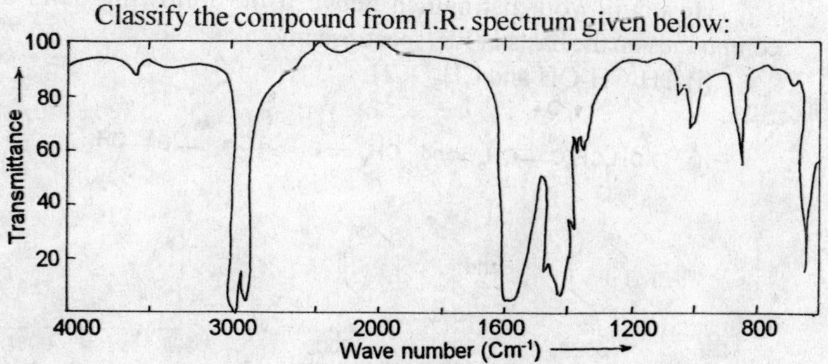

(i) Peak ~ 2900 cm^{-1} indicates the presence of alkyl group and absence of alkenyl, alkynyl and aryl groups.

(ii) Broad band between 2800 ~ 2250 cm^{-1} indicates OH group of carboxylic group.

(iii) Peak 1550 ~ 1590 cm^{-1} due to C = O stretching indicates a carboxylate ion.

(iv) Peak at 1420 – 1460 cm^{-1} indicates CH stretching for CH_3 and CH_2.

Thus the compound is a salt of aliphatic acid.

Example 2 :

How would the I.R. spectrum of the following pairs of compounds differ:

(i) Acetone and ethanol.

(ii) Acetic acid and methanol.

Solution :

(i) Acetone will show absorption band at 1710 cm^{-1} because of C = O stretching.

Ethanol will show absorption band at 3200 – 3600 cm^{-1} because of hydrogen bonded —OH.

(ii) Acetic acid shows absorption band at 1700 – 1725 cm^{-1} because of C = O group while at 2500 – 3000 cm^{-1} hydrogen bonded —OH.

Methanol will show absorption just like ethanol as given above in (i).

Example 3 :

How will you distinguish between the following pairs of compounds on the basis of I.R. spectroscopy?

(i) CH_3CH_2OH and CH_3OCH_3

(ii) $CH_3CH_2\overset{O}{\overset{\|}{C}}—CH_3$ and $CH_2 = CH—CH_2—O—CH_3$

(iii)  cyclohexanone and cyclohexenone (ring structures with C=O)

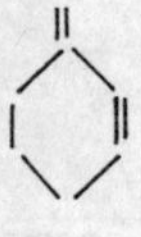

Solution:

(i) CH_3CH_2OH will show absorption band at 1050 cm^{-1} because of primary alcohol group and at 3200 – 3650 cm^{-1} because of hydrogen bond.

Alkyl ether will show absorption band at 1100 cm^{-1}.

(ii) $CH_3CH_2COCH_3$ will show absorption band at 1725 – 1705 cm^{-1} because of a saturated butane and $CH_2 = CH—CH_2OCH_3$ will show absorption bands at 3040 = 3010 cm^{-1} because of alkene part and 1100 cm^{-1} because of alkyl ether.

(iii)

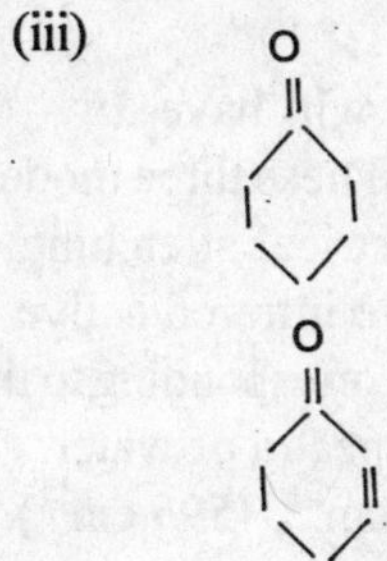

will show absorption band at 1725 – 1705 cm^{-1} being saturated cyclic keto compound.

will show absorption band at about 40 cm^{-1} less being α, β-unsaturated keto compound.

(iv) $ClCH_2CH_2CH_2COOH$ will show absorption band at 800– cm^{-1} because of a saturated acid.

$CH_3OCH_2CH_2COCl$ will show bands at 1100 cm^{-1} because of alkyl ether and at 1795 cm^{-1} because of acid chloride.

Example 4:

A compound has the molecular formular $C_7H_6O_2$. It gives an I.R. absorption band at 1771 cm^{-1}. On treatment with $LiAlH_4$ it gets converted into compound B which shows characteristic I.R. absorption bands at 3330 cm^{-1} and 1050 cm^{-1}. Assign structures to compounds A and B.

Solution :

Absorption band 1771 cm^{-1} is because of —COOH group.

∴ Structure of A is C_6H_5COOH.

On reduction with LiAlH4 the I.R. band at 3330 cm^{-1} is because of hydrogen bonding and 1050 cm^{-1} is because of primary alcoholic group,

$$\text{Hence} \quad \underset{\text{Benzoic acid}}{C_6H_5COOH} \xrightarrow{LiAlH_4} \underset{\text{Benzyl alcohol}}{C_6H_5CH_2OH}$$

Example 5 :

How many fundamental vibrational frequencies would you expect to observe in the infrared absorption spectrum of H_2O?

Solution :

Water molecule is a bent molecule so it will have $3n - 6$ or $3 \times 3 - 6 = 3$ fundamentals modes of vibrations. These three modes of vibrations are symmetrical stretching, Asymmetrical stretching and Bending. All these three modes of vibtations are infrared active and are non degenerate. Thus three absorption bands corresponding to these three modes of vibrations are expected in I.R. spectra of water. (The observed absorptions occur at 3652 cm^{-1}, 3756 cm^{-1}, 1595 cm^{-1}).

Example 6 :

Indicate the absorptions expected in the I.R. spectrum of CH_4—CH_2—OH.

Solution :

If a spectrum of heat sample of CH_3—CH_2—OH is taken then the following absorptions may be expected.

Band frequency	*Stretching*
3400 – 3250 cm^{-1}	Banded O—H stretching
3000 – 2700 cm^{-1}	To bands due to —CH_3— and —CH_2 stertching

Example 7 :

How many fundamental vibrational frequencies would you expect to observe in the infrared spectrum of CO_2?

Solution :

CO_2 is a linear molecule so it will have $3n - 5$ or $3 \times 3 - 5 = 4$, fundamental vibrational modes. These are symmetrical stretching,

Asymmertrical stretching, bending (on the plane of the paper), bending (perpendicular to the plane of the paper).

Out of the four modes the symmetrical stretching mode will be infrared inactive because in this no change in dipole moment takes place. The two bending modes (*i.e.*, on the plane of paper bending and perpendicular to the plane of paper) are degenerate and only one absorption band will appear which will be a combination band. Thus in I.R. spectra of CO_2 there will be only two fundamental absorption bands. Actually these two bands are observed at 2349 cm^{-1} and 667 cm^{-1} respectively.

Example 8.

Give approximate portions of the characteristic infrared bands in the following compounds:

(i) $CH_3—\overset{\overset{\displaystyle O}{||}}{C}—CH_4$ (ii) $CH_2=CH—CH_2—\overset{\overset{\displaystyle O}{||}}{C}—CH_3$

Solution :

(i) C—H stretching peak at 2925 – 2855 cm^{-1}
C = O stretching peak at 1725 – 1705 cm^{-1}

(ii) C—H stretching peak at 3040 – 3010 cm^{-1}
C—H and C = O stretching peak at 2925 – 2855 cm^{-1} and 1725 – 1705 cm^{-1} respecitvely.
C = C stretching peak at 1650 cm^{-1}

Example 9 :

Which of the following vibrations will be active in the infrared region ?

	Molecule	*Mode of vibration*
(i)	$CH_3—CH_3$	C—C stretching
(ii)	$CH_3—CH_3$	C—C stretching
(iii)	SO_2	Symmetrical stretching
(iv)	$CH_2=CH_2$	C—H stretching.

Solution :

(i) Inactive (ii) Active
(iii) Active (iv) Active

6

Ultraviolet and Visible Spectroscopy

Introduction

Ultraviolet spectroscopy deals with the recording of absorption of light in the visible and ultraviolet regions of the spectrum. Light of wavelength between 4000 A° and 7500 A° (400 – 750 nm) is visible. Just beyond the red end of visible spectrum (wavelength more than 750 nm) starts the infrared region and just beyond the violet end of the visible spectrum (wavelength less than 300 nm) lies in the ultraviolet region. Thus, ultraviolet region starts at the blue end of the visible light (about 4000 A°) and ends at 2000 A° (200 nm). This region is further sub-divided into *near ultraviolet* region (200 – 400 nm) and *far* or *vacuum ultraviolet region* (region below 2000 A°).

When radiant energy impinges upon a solution it may be absorbed, transmitted, reflected, or refracted. In spectrophotometry the absorbed light is determined. However, because of the diffuculty of directly measuring the absorbed energy, the transmitted energy is measured and the amount absorbed is indirectly determined by subtracting the transmitted from the initial energy.

The relationship

$$E = h\nu$$

where E is the transition energy from a stable to excited state, h is Planck's constant, and ν is the frequency. The energy of ultraviolet and visible radiation corresponds to that required for electronic transitions of the outer electrons. Since these are the valence electrons of the molecules or ions, the absorption spectra reflect the bonding characteristics of these particles. It is frequently impossible to obtain resolution of the numerous close absorption lines, and the plot of absorbance vs. wavelength yields a broad, smooth band spectrum.

The absorbance at various wavelengths is determined with a **spectrophotometer.** This instrument in its simplest form consists of

1. a radiant energy source.
2. a means of obtaining monochromatic or nearly monochromatic light.
3. a cell for the sample or blank.
4. a detection and readout assembly.

Actually, more sophisticated instruments are generally used. In order to obtain a broad overview of the instrument, to establish nomenclature, and to define system parameters, the schematic diagram (shown below) will be considered in more detail.

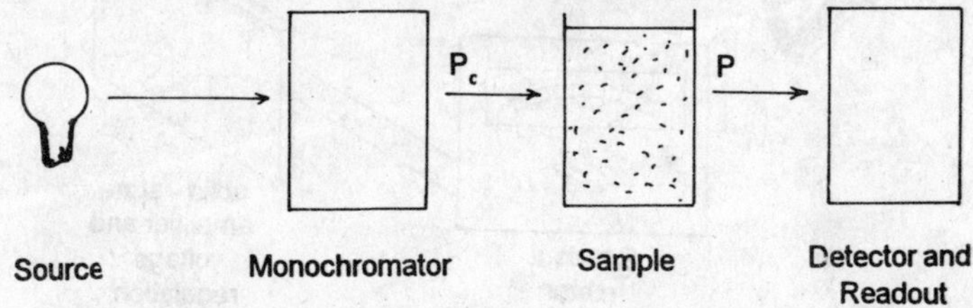

Fig. 6.1: Schematic diagram of a simple spectrophotometer. P_0 is the power of the radiation entering the sample cell and P is the transmitted power.

After originating at the source and passing through he monochromator system the radiant energy. P_0, enters the sample where a portion of it is absorbed. The remainder of he energy, P, *i.e.*, that which is transmitted, strikes the detector and a quantitative readout is obtained.

The **radiant power,** P_0, which is striking the sample, is reduced in intensity while passing through the sample and the emerging power, P, is measured. **Transmittance, T,** is the ratio of the radiant power, P, transmitted by a sample to the radiant power, P_0, entering the sample and **percent transmittance** is this ratio multiplied by 100, *i.e.*

$$T = P/P_0 \quad \text{and} \quad \%\,T = 100T = 100\,(P/P_0)$$

Absorbance, A, is the logarithm to the base 10 of the reciprocal of the transmittance :

$$A = \text{lon } l/T = \text{long } P_0/P.$$

Fig. 6.2: is a schematic diagram for the spectrophotometer. It show the design of a basic spectrophotometer. The source and sample cell are apparent from the drawing. The interference filter is the monochromator and the silicon photocell is the detector.

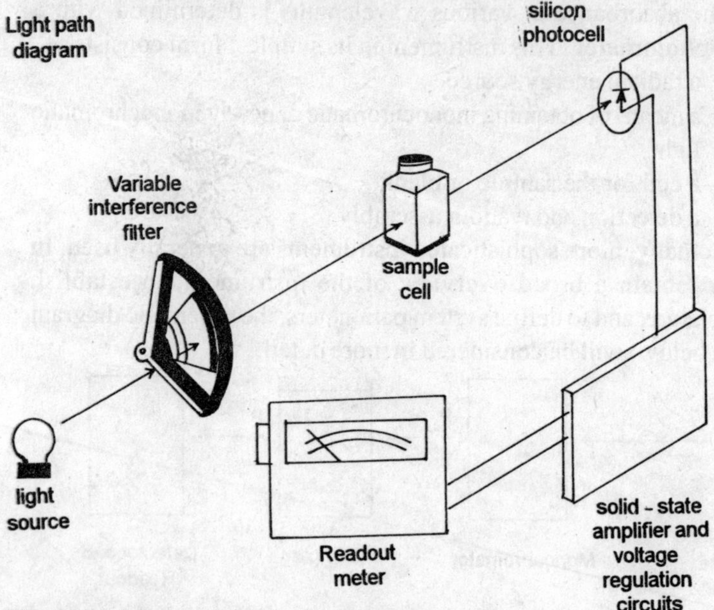

Fig. 6.2: Schematic diagram of Spectrophotometer

ABSORPTION LAWS

Path Length

The relationship between the relative intensity of the radiant energy emerging from a solution and the depth of the solution was independently discovered by **Lambert and Bougér.** They found that the intensity of the transmitted energy decreases exponentially as the depth (path length of the beam through the sample) increases. This can be expressed as

$$A = \log P_0/P = -\log P/P_0 = \log 1/T = kb$$

where A is the absorbance, P and P_0 are the respective powers of the transmitted and incident beams. T is the transmittance, k is constant for the chemical system, and b is the path length of the light.

The **Lambert-Bouger's Law** states, "When monochromatic light passes through a transparent medium, layers of equal thickness of that homogeneous absorbing medium absorbs equal portions of incident radiations."

Thus, a beam of monochromatic radiations of radiant power P_0 upon passing through an absorber of thickness *b* decreases in radiant power according to the following equation:

$$dp = k_1. P_0 .db$$

For *homogeneous systems, there are no exceptions to this rule.*

This reciprocal logarithmic function and the continual decrease in power as the radiant energy passes through the solution is shown in Fig. 6.3 (a). The transmittance is not only decreasing but is decreasing in an exponential function. At a depth of 1 cm the exiting radiant energy has decreased to one-half of the initial energy, and at 2 cm it has decreased to one-fourth of the initial or one-half of the energy which was present at the 1 cm thickness. Fig. 6.3 (b) shows the equivalent linear function which can be obtained when absorbance is plotted against the path length.

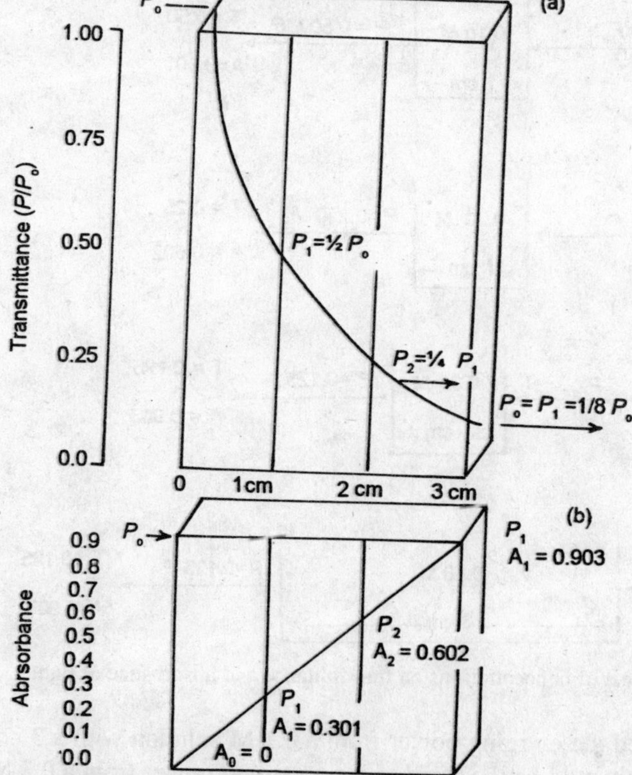

Fig.6.3: Effect of path length on transmittance and absorbance or light

Concentration

The relationship between the concentration of an absorbing species in a solution and the transmission of radiant energy through the solution was discovered by Beer in 1852. Concentratin, like depth, is a reciprocal exponential function of transmittance and a linear function of absorbance. This can be expressed as

$$A = -\log P/P_0 = -\log T = k'c$$

where P and P_0 are again the respective energy powers of the emerging and incident beams, k' is a constant for the chemical system, and c is the concentrating of the absorbing species.

Considering Fig. 6.4, if the emerging power, P, from a 0.1 solution is 0.50 P_0 the emerging power from a 0.2 M solution of the same species would be 0.25 P_0 and that from a 0.3 M solution would be 0.125 P_0.

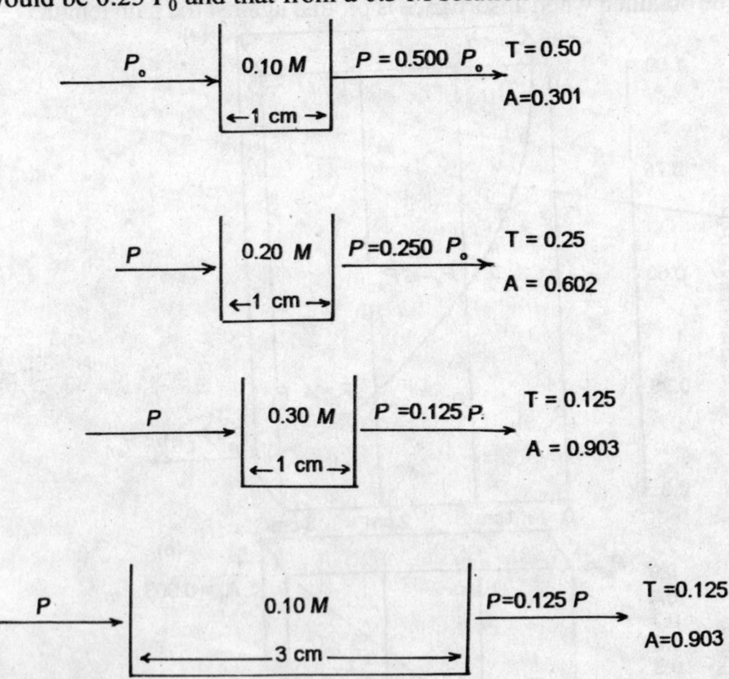

Fig. 6.4: Effect of concentrations on transmittance and absorbance of light

As expected the emrging power from a 0.1 M solution with a 3 cm path length would be the same as the emerging power from a 0.3 M solution with a 1 cm path length provided the P_0 is the same in both cases and that identical species are present in both systems. In both cases the same number of absorbing molecules would be struck by the radiation of the same energy which is passing through the identical solution.

Absorptivity

Absorptivity, *a* is defined as

$$a = A/bc.$$

It is a *fundamental physical constant for a specific chemical species at a specific wavelength.* Unlike depth and concentration, it cannot be varied by the operator. Its dimensions are of liters/(gram centimeter) when the depth of the solution (path length), *b*, is designated in its usual units of centimetre and the concentration, *c*, has units of grams/liter. If the concentration units are moles/liter, absorptivity is known as **molar absorptivity** and is designated as e, and possesses units of liters/(mole centimeter). The relationship between the two different absorptivities is

$$\varepsilon = aW$$

which with dimensions becomes

$$\frac{\text{liters}}{\text{mole centimetre}} = \frac{\text{liters}}{\text{gram centimetre}} \times \frac{\text{gram}}{\text{mole}}$$

where W is the respective molecular or atomic weight of the compound or element being determined. For example, a copper chelate[1] with an absorptivity of 245 and a molecular weight of 120 has a molar absorptivity of 29,400.

Absorptivity reflects the **sensitivity** of the procedure; the greater the absorptivity the greater the sensitivity. While it is a constant and cannot be changed, a different absorptivity and thus, sensitivity may be obtained by the selection of either.

1. a different wavelength of measurement or
2. a different absorbing species.

For example, the Pd-PAN chelate (Fig. 6.5) in chloroform shows strong absorption bands at both 620 and 675 nm. The respective molar absorbancies are 1.31×10^4 liter/mole cm and 1.44×10^4 litre/mole cm; thus, at any given concentration the absorbance at 675 nm is approximately 9% greater than that at 620 nm. Other things being equal, the 675 nm wavelength would be chosen for use in quantitative procedures.

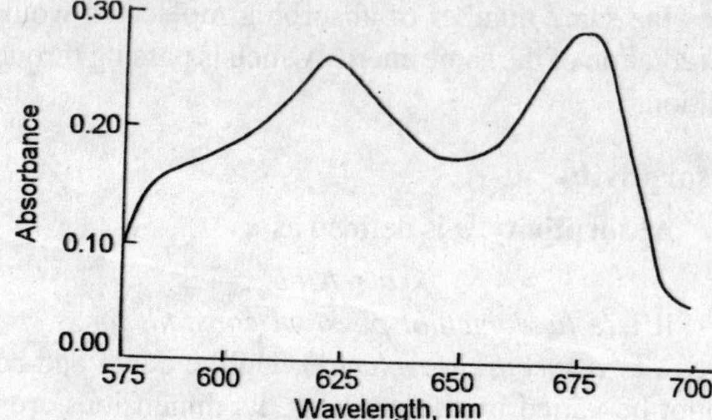

Fig. 6.5: Absorption spectrum of Pd-PAN chelate in $CHCl_3$, PAN is 1-(2-pyridylazo)-2-naphthol

With respect to the change in sensitivity with a change in absorbing species, consider two copper systems, one containing Cu^{2+} in water and the other $Cu^{2\pm}$ in concentrated ammonium hydroxide.It is difficult to visually detect the blue colour of a 0.001 M aqueous cupric nitrate solution. Upon the addition of concentrated ammonia to the previous solution an intense blue colour develops, the change being brought about as the copper-aquo complex ion is changed to the copper-ammonia complex ion. the reaction may represented by the equation.

$$\underset{\text{light blue}}{Cu\,(H_2O)_4^{\,2+}} + 4NH_3 \rightarrow \underset{\text{intense blue}}{Cu\,(NH_3)_4^{\,2+}} + 4H_2O$$

The absorptivity of the copper ammonia complex ion is much greater as compared to that of the copper-aquo complex when measurements are made in the blue transmission (approximately 600 nm) region. Incidentally, this is the reason that the addition of concentrated ammonia to a solution is such a sensitive test for the copper ion.

Combined Law

Since the traction of radiation absorbed depends upon the three factors: absorptivity, path length, and concentration. These parameters are mostly combined to yield the Beer—Bouger—Lambert law, more commonly known as Beer's Law. It states that

Beer—Bouger—Lambert Law commonly called **Beers Law** states, "The fraction of the monochromatic radiant energy absorbed as passing through a solution is directly proportional to the concentration of the absorber."

It is thus a relationship between the light absorptive capacity and the concentration of the absorber in solution and can be expressed as

$$dP = k_2 . P . dc$$

Combining the two laws gives

$$P = P_0 . e^{-k_2 bc}$$

or $$P = P_0 . 10^{-abc} \qquad \left(\text{where a} = \frac{k_2}{2.303}\right)$$

or $$\log \frac{P_0}{P} = abc = A \qquad \textbf{(Absorbance)}$$

$$A = abc$$

where A is absorbance, a is absorptivity in liters (gram centimetre), *b* is the path length or depth of the solution in centimeters, and *c* is the concentration in grams/ liter.

Alternatively, the combined law is expressed as

$$A = \varepsilon bc$$

Again A is absorbance but ε (molar absorptivity) now has the units of litres /(mole centimetre) and *b* still possesses units of centimetre. The concentration, *c*, is now expressed in molarity.

Beer's law is applicable not only to solutions but to gases and solids as well.

Beer's law is also applicable to a solution containing more than one kind of absorbing substance, provided there is no interaction among the various species for **multi-component system.**

$$A_{total} = A_1 + A_2 + ... + A_n$$
$$= a_1 \, b \, c_1 + a_2 \, b \, c_2 + ... a_n \, b \, c_n$$

where 'a' is **absorptivity,** which is characteristic of a particular combination of solute and solvent for a given wavelength. It is defined as $\frac{A}{bc}$

It must be noted the absorptivity (a) is a property of substance (*an intensive property*), while absorbance (A) is a property y particular sample (an extensive property), and will therefore vary with the concentration and dimension of the cantainer.

Deviations from Beer's Law

Beer's Law, states that the absorbance of a particular chemical species is linearly proportional to the concentration of that particular species at a specific wavelength. It assumes that each absorbing particle is absorbing independently of every other absorbing particle. These conditions are quite strict and at times the lack of ability to meet them will lead to deviations. Because of the possibility of deviations the entire concentration range of interest should be checked for adherence to the *absorbance proportional to concentration relationship* at the chosen wavelength.

The deviations from Beer's Law may be due to interaction of the solute molecules with each other or with the solvent or may be due to instrumental factors.

The real limitation of Beer's Law is that it is successful is describing the absorption behaviour of dilute solutions only.

As the degree of interaction depends upon the concentration, the occurrence of this phenomenon causes deviations from the linear relationship between absorbance and concentration.

Most of the **chemical deviations** are apparent rather than real, and are brought about by a change in the nature or concentration of the absorbing species. Any changes in the system brought about by factors such as pH, oxidation—reduction, reaction with the solvent, or even dilution which cause a shift in equilibria might lead to increased or decreased concentration and a change in the measured absorbance. However, the concentration of the measured species is changing and this change is reflected in the absorbance. It is not a true deviation from Beer's Law.

For example, considering a chelation equilibria system used in the determination of copper, the copper being determined as a chelate. Also assume the chelating agent exists in three equilibria species (H_2KE^+, HKE and KE^-) as follows :

$$H_2KE^+ \rightleftharpoons HKE + H^+$$ ...(1)

green yellow

$$HKE \rightleftharpoons H^+ + H^-$$ ...(2)

yellow red

where H_2KE^+ is the acidic from HKE is the neutral form, and KE^- is the basic form of the chelating agent, HKE. It is desired to measure copper as the red $Cu(KE)^+$ which is produced as under:

$$HKE + Cu^{2+} \rightleftharpoons Cu(KE)^+ + H^+$$ ...(3)

yellow blue red

It is apparent from Equation (3) that an excess of acid (low pH) would shift the equilibrium to the left and so the measured concentration of $Cu(KE)^+$ would not accurately reflect the total Cu(11) in the solution. On the other hand, an excess of base would shift the equilibrium in Equation (2) to the right and an excess of KE^- would be produced. Since KE^- and $Cu(KE)^+$ are both red and absorb at the same wavelength, any measurements at this wavelength would reflect the concentrations of both the species, KE^- and $Cu(KE)^+$. High absorbances would be obtained.

In an unbuffered potassium dichromate solution the following equilibrium exists :

$$Cr_2O_7^{2-} = H_2O \rightleftharpoons 2H^+ + 2CrO_4^{2-}$$

Upon dilution at low hydrogen ion concentrations, the concentration of CrO_4^{2-} increases, the concentration of $Cr_2O_7^{2-}$ decreases, and the ratio of CrO_4^{2-} to $Cr_2O_7^{2-}$ increases. An apparent deviation from Beer's Law may result from an attempt to utilize this reaction in a chromium determination at high pH. The shift may, however, be controlled by the use of large hydrogen ion concentrations.

Care should be taken to assure the presence of a true solution. A **colloidal suspension** or a **precipitate** in the cell will not only cause a lowering of the concentration of the measured species but may lead to spurious results because of scattering and reflecting of the impinging light. Impurities that absorb or fluorences at the measured wavelength should also be taken into considerations.

The *presence of impurities* that **fluoresce** or **absorb** at the required wavelength may also cause deviation from Beer's Law.

Beer's Law is not applicable in case of suspensions. The latter can be estimated colorimetrically after preparing a reference curve for known concentration.

Deviations from Beer's Law also arise, because it depends upon the *refractive index* of the solutions. Thus if concentration change causes significant alterations in the reftractive index of the solution, deviation from Beer's Law is observed. A correction for this effect may be made by substitution of the quantity.

$$\frac{an}{\left(n^2+2\right)^2} \text{ for } a = \frac{A}{bc} \text{ in equation } A = a.b.c.$$

In general this correction is not significant at concentrations of less than 0.01 F.

Instrumental deviation from Beer's Law may also arise because Beer's Law requires monochromatic radiation since absorptivity is a constant at a single wavelength and changes with a change is wavelength. The possibility of error due to the practical impossibility of obtaining monochromatic radiation may be minimized by the selection of a spectral region where the change in absorptivity with a change in wavelength is very small. This dictates a wavelength selection from a broad band rather than from a sharply rising or sharply falling section of the absorption curve. This may be clarified by reference to Fig. 6.6.

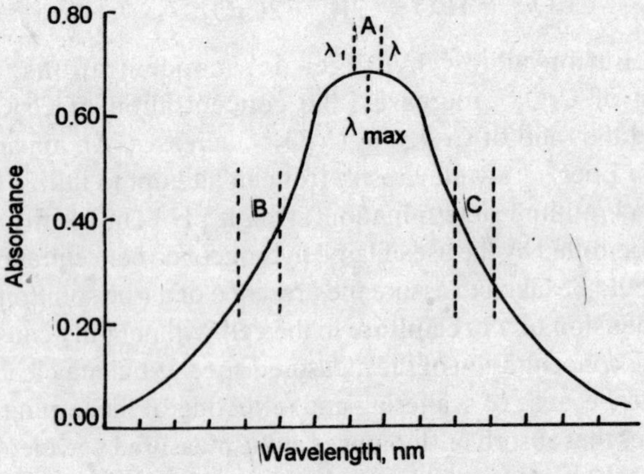

Fig.6.6: Choice of wavelength and slit widths

The energy which passes through the sample and activates the detector is a function of the weighted average of the absorptivities of all the wavelengths which are transmitted by the monochromator system. At the top (point A) of the absorption band, the absorptivities at λ_1 and λ_2 are close to that of the desired wavelength (λ_{max}). Minimum error will arise from the use of this wavelength. On the sharply rising (point B) or falling (point C) positions the weighted average yields low results, these low results becoming more and more apparent at increasing concentrations.

Because of the numerous possibilities of an error arising from deviations it is necessary in practice to prepare a calibrated curve for the absorbance—concentration relationship at the chosen wavelength. Beer's Law should be verified in each application on each instrument. It should not be taken for granted.

Instrumental Error Curve

Since absorbance, A, is defined by

$$A = abc = \log 1/T$$

the readout scales are calibrated in both absorbance and percent transmittance. This eliminates the process of first obtaining percent transmittance readings and then coverting these values to absorbances. While this is convenient, it may lead to an error which is not readily apparent. If absolute error in transmittance readings is assumed to be constant over the entire scale, the relative error would be different at different percent transmittance and absorbance values.

Fig. 6.7 is a relative error curve for a single beam instrument. The relative error reaches a minimum at 36.8% transmittance and 0.434 absorbance. On a practical basis this means that the most precise readings are obtained near these values. The curve also demonstrates that the relative error increases drastically at both ends of the scale. Procedures should be designed to give readings between 15% T (0.824A) and 80% T (0.097A) so as to avoid these problems.The uncertainties which arise from estimating the logarithmic absorbance scale may be overcome by reading the percent transmittance scale and then converting this reading to absorbance.

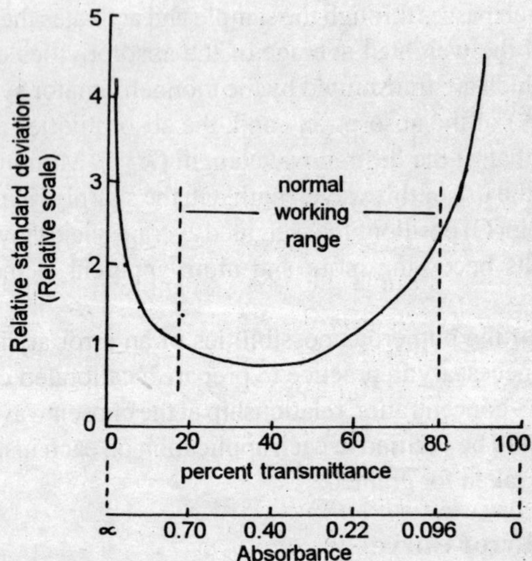

Fig. 6.7: Theoretical error curve of precision vs. percent transmittance and absorbance. The normal working range is between 15% and 80% T (0.82 to 0.10A).

Principle of u.v. Spectroscopy-Electronic Excitation

Ultraviolet absorption spectra arises from transition of electron or electrons within a molecule or an ion from a lower to a higher electronic energy level and the ultraviolet emission spectra arises from the reverse type of transition. For radiation to cause electronic excitation, it must be in u.v. region of electromagnetic spectrum.

Ultraviolet absorption spectra are attributed to a process in which the outer electrons of atoms or molecules absorb radiant energy and undergo transitions to higher energy levels.

These electronic transitions are *quantized* and depend on the electronic structure of the absorber. The wavelength at which an ultraviolet absorbance maximum is found depends on the magnitude of the energy involved for a specific electronic transition.

Electronic Excitation

The absorption of light energy by organic compounds in the visible and ultraviolet region involves promotion of electrons in σ, π and *n*-orbitals from the ground state to a higher energy state. An *n*-

orbital electron means a non-bonding electron *i.e.* one of an unshared pair. These higher energy states are described by molecular orbitals which are vacant in the unexcited or ground state. These are generally refered to as *antibonding orbitals*. The antibonding orbital associated with a σ bond is called σ* (sigma star) orbital and the one associated with a π-bond is called a π* (Pi star) orbital. Since *n* electrons do not form bonds, no antibonding orbitals are associated with them.

The electronic transitions (→) that are involved in the ultraviolet and visible region are of the following types:

$$\sigma - \sigma^*, n \to \sigma^*, n \to \pi^* \text{ and } \pi \to \pi^* \text{ (Fig. 6.8)}$$

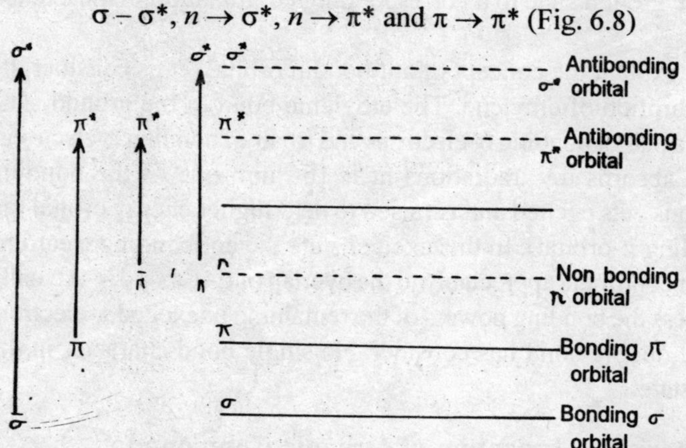

Fig. 6.8: Energy required for electronic transition of s, p and *n* electrons

The energies required for the various transitions are in the order

$$\sigma - \sigma^* > \pi \to \sigma^* > \pi \to \pi^* > n \to \pi^*$$

Thus $n \to \pi^*$ energy requires less energy

Then $\pi \to \pi^*$ or $\sigma - \sigma^*$ transitions.

Transition	*Region*	*Wavelength*
σ − σ*	Far u.v	< 200 nm
π − π*	—	—
π − σ*	u.v	= 200 nm
π − π*	Near u.v. and visible	300 − 600 nm

From energy the fig. it can be easily seen that $n \rightarrow \pi^*$ needs less energy compared to the energy needed for $\pi \rightarrow \pi^*$ or $\sigma \rightarrow \sigma^*$ transition.

Since ultraviolet energy is quantised, the absorption spectrum obtained due to a single electronic transition will show a single discrete line. Thus discrete line is not obtained because electronic absorption is superimposed upon rotational and vibrational sub levels. Due to this the spectrum of a simple molecule in gaseous state consists of narrow absorption peaks and each peak represents a transition from some particular combination of a vibrational and rotational levels in the electronic ground state to a corresponding combination in the excited state.

To explain the concept of antibondin orbital let us consider the u..v. absorption of ethylene. The ethylenic bond in the ground state contains a pair of bonding σ-electrons and a pair of bonding π-electrons. When it absorbs u.v. radiations near 165 nm, one of the bonding π-electrons gets excited and is raised to next higher energy orbital (an antibonding π-orbital). In this excited state the antibonding electrons no longer contribute appreciably to the overlap of C–C bond.—Actually it weakness the bonding powers of the remaining unexcited π-electron. Thus the double bond has considerable single bond character in the excited state.

Characteristic Absorption of Organic Compound

The absorption characteristics of organic molecules in the ultraviolet region depend upon the electronic transitions that can take place in the molecule and also the affect of atomic environment on the transitions. The summary of electronic structures and the electronic transitions involved in ultraviolet absorption is given in Table 6.1.

(i) Compounds containing only σ-electrons

($\sigma - \sigma^*$ *transitions*)

Saturated hydrocarbons contains σ-electrons only. Since a very high energy (185 Kcal/Mole) is needed for the $\sigma \rightarrow \sigma^*$ transitions so such transitions take place at shorter wavelengths i.e., they take place only in *far ultra-violet* region and no absorption spectra is obtained in the *near ultra-violet region.*

Table 6.1 : Summary of Electronic Structure and Transitions

Example	*Electronic Structure*	*Transition*	λ_{max} *nm*	ε_{max}	*Band*
1	2	3	4	5	6
Ethane	σ	$\sigma \rightarrow \sigma^*$	135	... ...	
Water	n	$n \rightarrow \sigma^*$	167	7,000	
Methanol	n	$n \rightarrow \pi^*$	183	500	
Ethylene	π	$\pi \rightarrow \pi^*$	165	10,000	
Acetylene	π	$\pi \rightarrow \pi^*$	173	6,000	
Ecetone	π and n	$n \rightarrow \pi^*$	150	... ...	
		$n \rightarrow \sigma^*$	188	1,860	
		$n \rightarrow \pi^*$	272	15	R-Band due to carbonyl group
1, 3-Butadiene	$\pi - \pi$	$\pi \rightarrow \pi^*$	217	21,000	Conjugated band

1	2	3	4	5	6
Benzene	Aromatic π	Aromatic π → π*	180	60.000	Ethylene band
		” ” ”	200	8.000	
		” ” ”	256	215	Benzenoid band
Acetophenone	Aromatic π→π	Aromatic π → π*	240	13.000	Conjugated
	and *n*	π → π*	278	1,110	Benzenoid band
		π → π*	319	50	R-band due to
				16	carbonyl group
Cetaldehyde	π and *n*	π → π*			
		n → σ*	290	60	
		n → π*			
Acetic acid	π and *n*	*n* → π*	240		
Phenol	Aromatic π	π → π*	210	6,200	Ethylenic band
	Aromatic π	π → π*	270	1,450	Benzenoid band

(ii) Saturated Compounds containing n-Electrons
($n - \sigma^*$ *transitions*)

Saturated compounds which contain betero atoms like oxygen, nitrogen, sulphur or halogens posses non-bonding electrons (*n*- or *p*-electrons) in addition to σ-electrons.

Since the energy needed for $n - \sigma^*$ transition is less than the energy needed $\sigma - *$ transition it is expected that compounds in which $n \rightarrow \sigma^*$ transitions occur should show absorption bands in *near ultra-violet* region, but the majority of compounds in this class show no absorption in the *near ultra-violet* region. Alcohols and ether absorb at wavelength shorther than 185 nm and, therefore, are commonly used as solvents for work in the *near ultra-violet* region. When these compounds are used as solvents, the intense absorption extends into the *near ultra-violet* region producing end absorption or cut off in the 200 – 220 nm region. Sulphides, disulphides, thiols, amines, bromides may exhibit weak absorption in the *near ultra-violet* region.

In case of allcyl halides, the energy required for $n \rightarrow \sigma^{**}$ transition decreases with increase in the size of the halogen atom or decrease in the electronegativity of the atom.

The **dipole moment** also plays a significant role. Greater the transition dipole the greater in the absorption intensity. For symmetrical molecules, smaller is the probability of transition dipole and the light is absorbed by molecule very slowly. When symmetry of the molecule is decreased by an introduced group, the transition dipole and hence the intensity of light absorbed will also be increased. The dipole moment changes as a result of oscillation of electron pairs. Ease of excitation of different groups follows the order

$$N = O > C = S > N = N > C = O > C = N > C = C$$

The $n \rightarrow \sigma^*$ transitions are sensitive to hydrogen bonding. The **hydrogen bonding** shifts the absorption towards shorter wavelengths.

(ii) Compounds containing π-Electrons (chromophores)
($\pi - \pi^*$ *transitions*)

A **chromophore** is a covalently unsaturated group responsible for electronic absorption *e.g.*, C = C, C = O, NO_2 etc.

All such compounds contain π-electrons and may also contain non-bonding electron pairs. Many of the single chromophoric groups absorb strongly in the *far ultra-violet region* with no absorption in *near ultra-violet region*. Those groups which contain both π and *n*-electrons can undergo the following transitions.

$$n \rightarrow \sigma^*, \pi \rightarrow \pi^*, n \rightarrow \pi^*$$

Absorption of single chromophore in the *near ultra-violet* region is due to the low energy, forbidden $n \rightarrow \pi^*$ transition.

The transitions $n \rightarrow \sigma^*$, $\pi \rightarrow \pi^*$ and $n \rightarrow \pi^*$ are of prime concern to an organic chemist. These discussed below :

1. π → π* Transitions

Most applications of absorption spectroscopy to organic compounds are based upon transitions of *n* or π-electrons to π* excited state. Such transitions need energies whose absorption peaks appear in the region 200–700 nm which is experimentally convenient spectral region. Strictly speaking the term chromophore applies to these unsaturated absorbing centres.

A $\pi \rightarrow \pi^*$ transition can take place even in a simple alkene such as ethylene (ethene) but in this case the absorption occurs in *far ultra-violet* region. In vapour phase ethylene absorbs at 165 nm (ε_{max} = 10,000). A second band has been observed near 200 nm which has been attributed to the elevation of two π-electrons to the π* orbital.

The alkyl substitution of the parent compound shifts the absorption to a longer wavelength **(bathochromic effect)**. With increase in the number of alkyl groups this effect also increases. If a hetero atom that contains a non-bonding electron pair is attached to the ethylenic linkage, it brings about a shift of absorption to the longer wavelength.

Nitrogen and sulphur atoms are more effective in bringing the absorption well into the near ultraviolet region e.g. methyl vinyl sulphide exhibits absorption at 228 nm (ε_{max} = 8,000), conjugated dienes (having alternate single and double bands) exhibit intense $\pi \rightarrow \pi^*$ transition bands in the longer wavelengths, 215 – 250 nm region. 1, 3-Butadiene exhibits absorption at 217 (ε_{max} = 21,000). Further conjugation in open chain trienes and polyenes results in, additional shift of absorption to the longer wavelength region. It also brings about an increase in the intensity of absorption. β-carotene which contains π carbon-carbon bonds in conjugation owes its yellow colour to the absorption at the violet end of the spectrum (λ_{max} = 451 nm).

$\pi \rightarrow \pi^*$ transition also occurs in carbonyl compounds and the electron goes from a stable (bonding) π-orbital to and unstable π* (antibonding) orbital.

$$\rangle C = \ddot{O}: \quad \rightarrow \quad \rangle C \overset{\cdot}{\underset{\cdot}{-}} \ddot{O}:$$

In case of carbonyl compounds absorption occurs at about 150 nm.

The wavelength of absorption peaks for conjugated systems when two carbonyl or carboxylate groups are conjugated as in α, β unsaturated aldehydes and ketones the absorption peaks are sensitive to the type of groups attached to the doubly bonded atoms.

Absorption by Aromatic system

Benzene exhibits three absorption bands: 184 nm (ε_{max} = 10,000), 204 nm (79,000) and 256 nm (ε_{max} = 200). The absorption at 256 nm is a weak absorption and is termed as **B band.** The band at 184 nm is a strong band and the band at 204 nm is a weak band called **E_2 band.**

Auxochromic groups refers to a group which when attached to a chromophore, changes both the wavelength and the intensity of absorption, *e.g.*, –OH, –NH_2 –Cl etc. Such a group when substituted on the benzene ring shifts the band to the longer wavelength region. This is because of π-π conjugation.

Compound		λ_{max}	ε_{max}	
1. Phenol	π → π*	210	6200	(E_2 band)
	π → π*	270	1450	(B-band)
2. Aniline	π → π*	230	8600	(E_2 band)
	π → π*	280	1430	(B-band)
1. Chlorobenzene				
	π → π*	210	7600	(E_2 band)
	π → π*	265	240	(B-band)

In case the unsaturated chromophore is directly attached to the benzene ring it produces a strong shift to the longer wavelength region. This may be attributed to extended conjugation.

π → π* Conjugated Structures

Compound	λ_{max}	ε_{max}
1. Styrene	244	12,000
2. Acetophenone	240	10,000
3. Benzoic acid	210	10,000
4. Benzaldehyde	236	13,000
5. Benzophenone	252	20,000

In case both auxochromic group and chromophore appear on the same ring they both influence absorption. The influence is more pronounced in case an electron releasing group and electron with drawing groups are para to one another.

Absorption Maximum in Dienes

With the increase in the conjugation of chromophore the wavelength of the absorption maximum increases, ε_{max} also increases with increase in the conjugation of chromophore. If there are 5 or more double bonds in conjugation in a molecule it appears coloured.

Double bonds can be present in conjugation as follows:

(i) Acylic dienes

1, 3-Butadiene is such a diene with the structural formula.

$$H_2C = CH - CH = CH_2$$

(ii) Homo-annular conjugated dienes

In such dienes both conjugated double bonds are in the same ring *e.g.*

(iii) Hetero annular conjugated dienes

In such dienes the conjugated double bonds are not present in the same ring *e.g.*

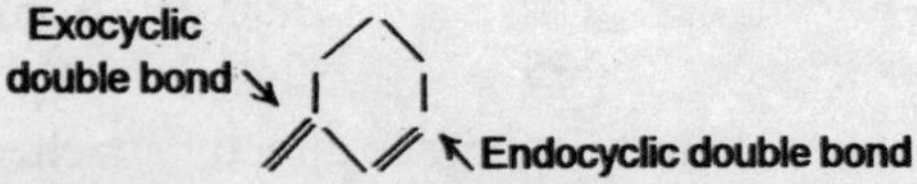

Exocyclic double bond

Endocyclic double bond

Empirical Rules for Calculating Absorption Maximum in Dienes Woodward Fieser Rules

Woodward in 1942 found that the position of the absorption maximum depends on the number and type of substituents in a diene system.

The empirical rules for calculating λ_{max} from molecular structure of the compound were developed by woodward and were later (1948) modified by Fieser.

(Solvent $\rightarrow$ Ethanol, Transition $- \pi \rightarrow \pi^*$)

Values assigned to :

(i) Parent homoannular conjugated dienes = 253 nm
(ii) Parent heteroannular conjugated dienes = 214 nm
(iii) Butadiene system = 217 nm
(iv) Each double bond extending conjugation = 30 nm
(v) Each alkyl substitutent or ring residue = 5 nm
(vi) Each exocyclic double band = 5 nm
(vii) Auxochromos

$—OCOCH_3$	= 0 nm
—OR	= 6 nm
—SR	= 30 nm
—Cl, —Br	= 5 nm
$—NR_2$	= 60 nm

These values have been illustrated by the following examples.

(a) $CH_3—CH = CH—CH = CH—CH_3$
(2, 4-hexadiene)

Basic value	= 217 nm
Two alkyl substituents	= 2 × 5 = 10 nm
Calculated value	= 227 nm
λ_{max} observed	= 227 nm

(b) CH-CH=CH_2
3 cyclohexyl dienepropene

Basic value	= 217 nm
2 ring residue	= 10 nm (2 × 5 = 10)
1 exocyclic double band	= 5 nm
λ_{max}, calculated value	= 232 nm
λ_{max}, observed	= 236.5 nm.

(c) =C—CH =C—H, CH_3, CH_3
4 cvclopentylidene-pent-2-ene

Basic value	= 217 nm
Two methyl substituents	= 10 nm
Two ring residue	= 10 nm
One exocydic double bond	= 5 nm
λ_{max} calculated	= 242 nm.

Woodward-Fieser rules for calculating Absorption Maximum in α, β-unsaturated carbonyl compounds.

Woodward and Fieser gave an empirical set of rules which were later modified by Scot to calculate λ_{max} α, β-unsaturated carbonyl compounds. E values are generally above 10,000 and increase with the length of the conjugated system.

Values assigned to:

(i) Parent α, β-unsaturated six membered ring or acyclic ketone = 215 nm

(ii) Parent α, β-unsaturated five membered ring ketone = 202 nm

(iii) Parent α, β-unsaturated aldehyde = 207 nm

(iv) Increments for

- (a) Each double band extending conjugation = 30 nm
- (b) Each alkyl group or ring residue at α-position = 10 nm
- (c) Each alkyl group or ring residue at β-position = 12 nm
- (d) Each alkyl group or ring residue at γ-position = 18 nm
- (e) Each exocyclic double band = 5 nm
- (f) Hanoannular diene component = 39 nm

(v) For auxochromes

(a) —OH	α-position	= 35 nm
	β-position	= 30 nm
	γ-position	= 50 nm
(b) —OAc	α-position	= 6 nm
	β-position	= 6 nm
	γ-position	= 6 nm
(c) —OMe	α-position	= 35 nm
	β-position	= 30 nm
	γ-position	= 17 nm
	δ-position	= 31 nm
(d) —Cl	α-position	= 15 nm
	β-position	= 12 nm
(e) —Br	α-position	= 25 nm
	β-position	= 30 nm
(f) —NR_2	β-position	= 95 nm

λ_{max} depends upon the nature of the solvent. More polar solvent from hydrogen bond with carbonyl group and thus $n \rightarrow \pi^*$ transition shifts towards shorter wavelength (blue shift). Solvent correcrtion is as under :

Hexane	= 11nm
Dioxane	= 7 nm
Methanol	= θ nn
Water	= – 8 nm
Chloroform	= – 1 nm.

Following examples illustrate these

(a) 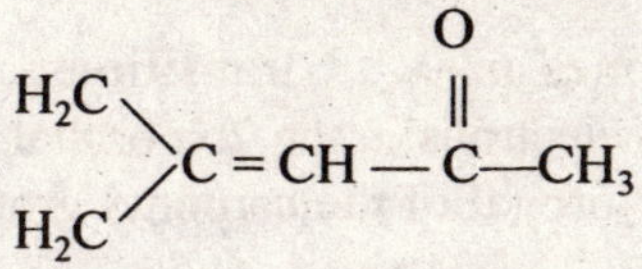 $(H_2C)_2C=CH—C(=O)—CH_3$

4-methylpent-3-en-2-one

Basic value	= 215 nm
Two alkyl substituents at β-position	= 24 nm
λ_{max}, calculated	= 239 nm

(b)

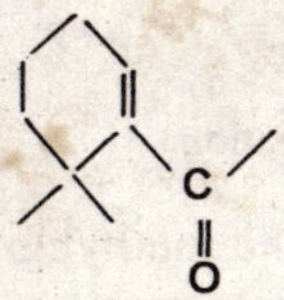

Basic value	=215 nm
α-ring residue	= 10 nm
β-ring residue	= 12 nm
λ_{max}, calculated	=237 nm
λ_{max}, observed	=232 nm

2. *n* → π* Transitions

These transitions are exhibited by unsaturated molecules containing some hetero atoms like oxygen, sulphur, nitrogen etc.

The carbonyl compounds containing C = O group also exhibit such transitions *e.g.*, *p*-nitrophenol shows absorption at 318 nm (ε_{max} = 10,000) and *p*-nitroaniline at 381 nm ($\varepsilon^{\partial}_{max}$ = 13,500). CH_3CHO exhibits absorption at 293 nm (εe_{max} – 11.8) etc.

Compounds containing a carbonyl group in conjugation with an ethylenic group show an intense absorption band in the 215 – 250 nm region ($\varepsilon^{\partial}_{max}$ = 10,000 – 20,000) which is due to π → π* transition and a weak band at 310 – 330 nm which is due to $n \rightarrow \pi^*$ transitions.

These are examplified as under :

Compound	$\pi \rightarrow \pi^*$ λ_{max}	$\pi \rightarrow \pi^*$ $\log \varepsilon_{max}$	$n \rightarrow \pi^*$ λ_{max}	$n \rightarrow \pi^*$ $\log \varepsilon_{max}$
1. Methyl ketone	212.5	3.85	320	1.43
2. Acrolein	210	4.96	315	1.41
3. Crotonaldehyde	220	4.17	322	1.39

In aldehydes and ketones $n \rightarrow \pi^*$ transitions are due to excitation of one lone pair of electrons in the 2*py* orbital of oxygen atom into the antibonding π^*-orbital of the carbonyl group. This is shown in Fig. 6.9.

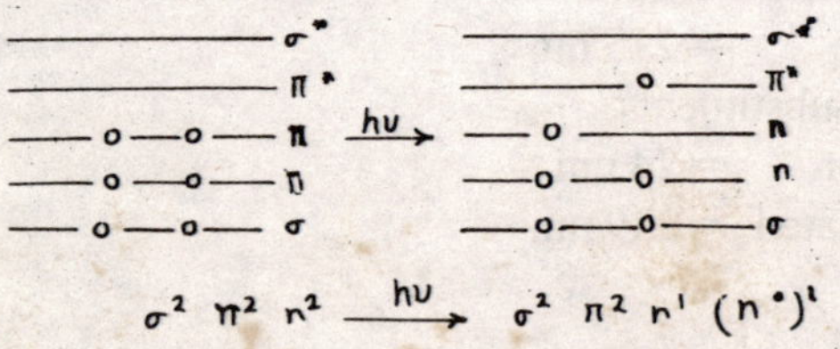

$$\sigma^2 \pi^2 n^2 \xrightarrow{h\nu} \sigma^2 \pi^2 n^1 (\pi^*)^1$$

Fig. 6.9: Representation of a $n \rightarrow \pi^*$ transition

If hydrogen atom is replaced by a methyl group (—CH_3) in an aldehyde group, a shift in band occurs, because of $n \rightarrow \pi^*$ transition, towards shorter wavelengths.

To characterise $n \rightarrow \pi^*$ transition two methods are used.

(i) One method to characterize $n \rightarrow \pi^*$ transition is to observe the spectrum in acid solution. In acidic solution bands due to $n \rightarrow \pi^*$ transitions disappear because of the formation of a bond between acidic proton and lone pair of electrons on the hetero atom. For example band due to $n \rightarrow \pi^*$ transition in pyridine disappears in acidic solution.

$$C_5H_5N: + H^+ \rightarrow C_5H_6N^+ - H$$

or (pyridine) + H^+ → (pyridinium ion, N^+–H)

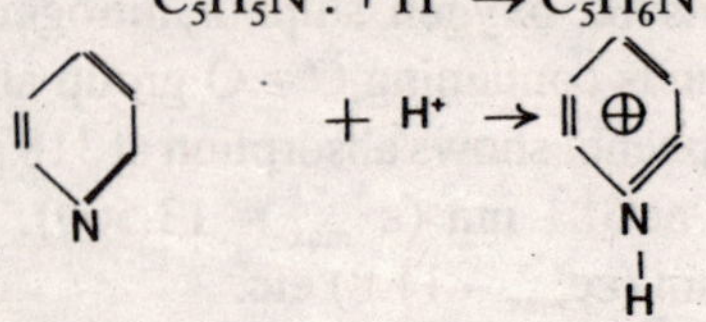

(ii) The second method involves the comparison of the spectrum of the compound containing the hetero atom with a similar compound not containing the hetero atom. For example a band appears at about 3000 A° in the spectrum of pyrazine but no such band is observed in the spectrum of benzene. Thus the band at 3000 A° in pyrazine is due to the promotion of a lone pair of electrons on the nitrogen atom into the antibonding π^* orbital of the aromatic system.

In case of compounds containing a carbonyl group in conjugation with an ethyienic group (α, β-unsaturated ketones and aldehydes) a weak $n \rightarrow \pi^*$ band is observed at 310–330 nm. This band is poorly defined.

3. $\sigma \rightarrow \sigma^*$ Transitions

These transitions can take place in those compounds in which all electrons are involved in single bonds and there are no lone pair of electrons. For example, saturated hydrocarbons.

Since $\sigma \rightarrow \sigma^*$ transitions needs a large amount of energy so the absorption band appears in the far ultraviolet region (126.135 nm). In case of methane λ_{max} occurs at 121.9 nm and in case of ethane λ_{max} occurs at 135 nm.

Most of the commercial spectrophotometers are generally non-operative for wavelengths less than 180-200 nm so $\sigma \rightarrow \sigma^*$ transition cannot be normally observed.

4. n $\rightarrow \sigma^*$ Transitions

Saturated compounds which have a lone pair (non-bonding) of electrons can undergo $n \rightarrow \sigma^*$ transition. This transition is in addition to $\sigma \rightarrow \sigma^*$ transition. This transition ($n \rightarrow \sigma^*$) needs lesser energy as compared to energy required for $\sigma \rightarrow \sigma^*$ transitions and so the absorption band for these absorptions occur in the near ultraviolet region (180 – 200 nm). We also know of a few compounds in which this type of absorption occurs at higher wavelength e.g., $(CH_3)_3N$ absorbs at 227 nm (λ_{max} = 227 nm). For this compound $\sigma \rightarrow \sigma^*$ transition is observed at 99 nm.

For making absorption measurements in u.v. region aliphatic alcohols, alkyl halides etc. are used as solvents. This is done because they start absorbing above 260 nm. But if the measurement are to be made in the region 200 – 260 nm these are not used and instead saturated hydrocarbons are used as solvents. However, these hydrocarbons are not good solvents.

The solvents commonly used in different wavelength regions are as given in Table 6.2.

Table 6.2

Solvent	*Wavelength*
Methanol	203 nm
Ethanol	204 nm
Water	191 nm
Chloroform	237 nm
Cyclohexane	195 nm
Dichloroethane	220 nm

INSTRUMENTS AND COMPONENTS

The desired parameter in spectrophotometry is absorbance, A, but it cannot be directly measured. In actual practice, values of the incident radiation power, P_0, and the transmitted power, P, or their ratio must be obtained. From these values and the relationship.

$$A = \log \frac{P_0}{P}$$

the absorbance is determined.

Fig. 6.10 is a schematic diagram of a typical single beam spectrophotometer system. Another type is a double beam spectrophotometer. The differences between single and double beam systems will be discussed at a later stage, here it is adequate to recognize that the components and their functions in the two designs are practically identical.

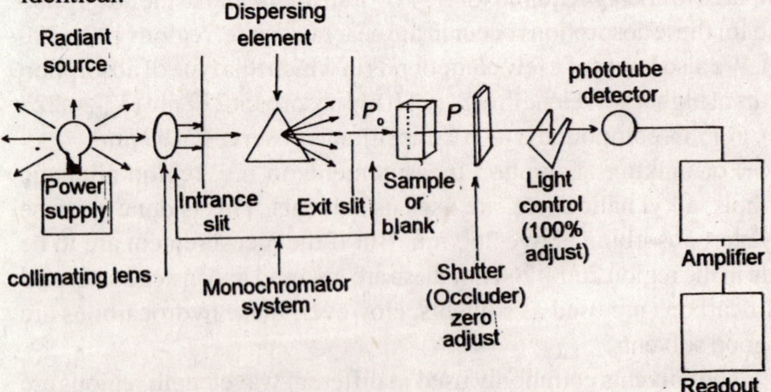

Fig. 6.10: Schematic diagram of a single beam spectrohotometer

Power Supply

The power supply serves the following functions:

1. It decreases the line voltage to the instrument's operating level with a transformer.
2. It converts alternating current to direct current with a rectifier if direct current is required by the instruement.
3. It smooths out any ripple which may occur in the line voltage so as to deliver a constant voltage to the source lamp and instrument.

Radiation Sources

The requirements of a radiations source are as under :

1. It must be stable.
2. It must be of sufficient intensity for the transmitted energy to be detected at the end of the optical path.
3. It must supply continuous radiation over the entire wavelength region in which it is used.

The tungsten filament lamp is the most common source of visible radiation, being usable in the 320 – 2500 nm region, whereas the hydrogen lamp yields a continuous spectrum from 180 to 375 nm, the ultraviolet region. Deuterium lamps are more intense than those of hydrogen and cover the same wavelengths. This extra intensity may at times it advantageous to substitute deuterium for hydrogen lamps.

It is important to recoginze that even though a particular source may emit radiation over a wide wavelength range, the power of the radiation will not be constant but will change considerably with changes in wavelength as shown in Fig. 6.11. Even though a chosen source may be adequate for a region, it may not be the best choice. The lamps should be maintained at the proper power, adjusted, and aligned with the optical path.

Radiation Sources in u.v. Spectrometers

Generally hydrogen or deutrium lamps are used as source of radiation in u.v. spectrometers. Some times xenon discharge lamps and mercury are lamps are also used as source of radiation in u.v. spectrometers. Irrespective of the source used the excitation is done by passing electrons through a gas molecule which produces electronic, vibrational and rotational excitation in the gas molecules. If the gas pressure is low only line spectra are emitted, however at higher pressures band spectra and continuous spectra will be obtained.

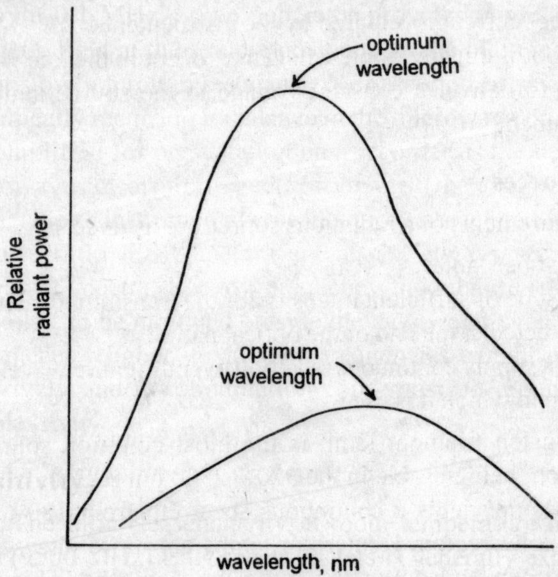

Fig. 6.11: Schematic diagram of spectral distribution curve of radiant energy source lamps.

(i) Hydrogen discharge lamps

In such lamps hydrogen is stored under high pressure and and electric discharge is passed through them. This excites hydrogen molecules which emit u.v. radiations. Since the pressure is high so we get a continuous hydrogen spectrum. Such lamps are used in the range 3500 – 1100 A° (350 – 120 nm). These lamps are found to be robust and stable so these are widely used.

(ii) Deutrium lamps

In such lamps deutrium (Heavy hydrogen) is used instead of ordinary hydrogen. The intensity of radiations emitted by such lamps is found to be 3-5 times more as compared to that in hydrogen lamps.

These are more expensive than hydrogen lamps and are used when more intensity is needed.

(iii) Xenon discharge lamps

In such lamps xenon is stored under pressure (10 – 30 atmo). They contain two electrodes separated by about 8 mm. When a low voltage is passed an electric arc is struck between these electrodes and the u.v. light is produced. The intensity of these u.v. radiations is much more than those of hydrogen lamps.

(iv) Mercury Arc

In these mercury vapours are stored under high pressure and the mercury atoms are excited by passing electric discharge. These radiations are a standard source for u.v. experiments but beacuse of presence of sharp lines or bands, this source is not suited for continuous spectral studies.

2. Monochromators

The function of monochromators is to disperse the radiations according to wavelength. It consists of an entrance slit, a dispersing element and an exit slit. The dispersing element is so adjusted by rotating it to vary the nominal wavelength passing through the exit slit. The dispersing element may be a prism or a grating which are generally made of glass, quartz or fused silica. Glass has the highest resolving power but it is not transparent to radiations between 200 – 300 nm because of strong absorption by glass in this region. The use of quartz is wide in u.v. spectrometer because it is transparent throughout the entire u.v. range.

Fused silica prisms are transparent in short wavelength region and are used it very intense radiation is to be produced.

Theoretically, a monochromator system should yield light of a single wavelength since it is only by the use of a single wavelength that maximum selectivity, minimum intereference, and maximum adherence to Beer's Law can be obtained. Even though various combinations of components are used commercially in an attempt to approach the ideal system, monochromatic radiation is seldom obtained. A monochromator system consists of a minimum of two components :

1. **a dispersion device,** such as prisms or grating, whose function is to spread out the various wavelengths of the incident beam.
2. a series of slits, mirrors, and lenses whose function (among others) is to isolate specific wavelengths of the beam.

A number of monochromator elements are shown in Fig. 6.12.

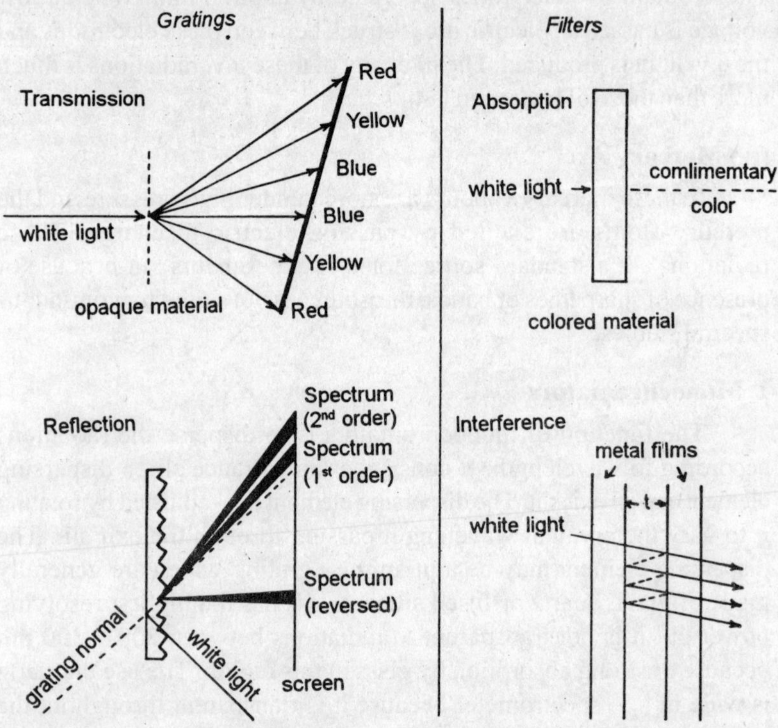

Fig. 6.12: Monochromator elements. Upper left: transmission prism; lower left; reflection prism; upper center; transmission grating; lower centre; reflection grating; upper right; absorption filter; lower right; interference filter.

Prisms may be either transmission or reflection. **Transmission prisms** (Cornu) are prepared from two pieces of triangular quartz. A beam of radiation is divided into two polarized beams as it passes through quartz. These two beams have different velocities. In order to compensate for this velocity difference the quartz is cut, rotated, and glued back together. In contrast to this, the **reflection prism**

(Littrow) is prepared from a single piece of quartz the back surface of which is coated with a thin silver mirror. The double refraction of the entering light is compensated for as it is reflected and returned through the prism.

The use of **transmission** or **reflection gratings** as dispersive elements is becoming increasingly popular. The **transmission grating** is based upon the fact that white light, when impinged upon a substance containing parallel grooves, will be separated into its component wavelengths, the degree of separation (resolution) depending upon the number of grooves per inch. The **reflection grating,** on the other hand, contains a mirrored front surface to eliminate substrate interference. In the manufacturing process the gratings are cast from a master grating. This yields very satisfactory monocrhometers at relatively inexpensive prices.

Strictly speaking, filters are not dispersive devices, but since they are frequently substituted for gratings or prisms or used in conjunction with them in order to eliminate higher orders of wavelengths and to give better resolution, a discussion of them is included. **Absorption filters** are composed of transparent colored materials. The wavelengths of light corresponding to the colour of the filter are absorbed; the complementary colour is transmitted and subsequently used in the analysis.

Intereference filters are constructed of two layers of thin semi-transparent metal films coated on the inside of transparent plates. Sandwiched between the plates is an absorbing substance. The thickness of the centre material is critical and is such that most of the wavelengths of the impinging light are subject to destructive interference. The wavelength of the desired light, on the other hand, is not subject to this interference and is transmitted. Variation of the thickness of the centre space and/or the number of layers of spacers and metal films allows the construction of continuously variable interference filters which cover wide regions of wavelengths. The spectrophotometer of Fig. 6.2. uses this type.

Prisms, gratings, and variable interference filters must be rotated in order to align the desired wavelength with the optical path of the system. This is accoplished by movement of a hand or motor-driven cam through an external wavelength adjustment knob. These may become maladjusted.

3. Detectors

Following three types of detectors are commonly used:

(a) *Barrier Layer Cell.* It is a photo voltaic cell shown in Fig. 6.13.

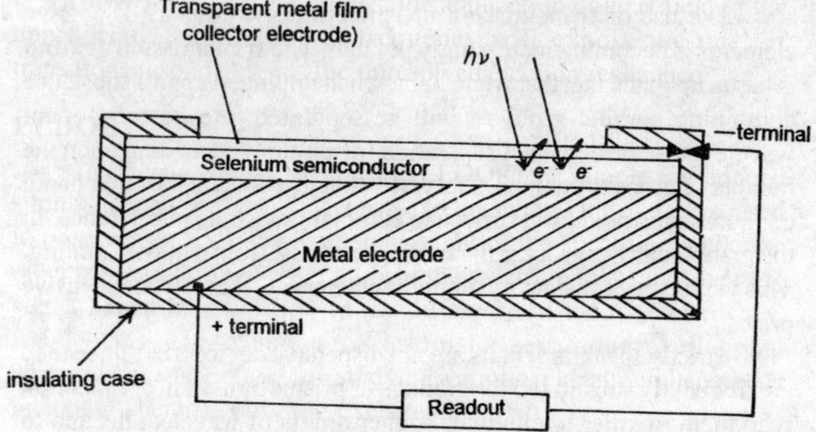

Fig. 6.13: Schematic diagram of a photovoltaic (barrier layer) cell. Light strikes the semiconductor, emitting electrons which cause a potential as they are collected at the transparent metal film.

It consists of a semi-conductor like selenium which is deposited on a strong metal base such as of iron. A thin layer of silver or gold is then sputtered over the surface of the semi-conductor which acts as a second collector electrode.

The radiations that fall on the surface produce electrons at the selenium-silver interface. Since a barrier exists between selenium and iron so the flow of electrons is prevented. It results in an accumulation of electron on the silver surface which produces an electrical voltage difference between silver surface and base of the cell. If the external circuit has a low resistance, a photo current will flow which is directly proportional to the intensity of the incident radiation. The sensitivity of a photovoltaic cell is not very high and such a cell is used in such instruments as photometers which allow wide band of radiations to strike the detector.

This cell is simple in design and needs no external supply of power. It can be hooked directly to a micrometer or glavanometer to read its output. Since it has only fair response time so it cannot cause reduction of noise. As the time passes such a cell becomes useless due to transformation of selenium layer.

(b) **Pohto Cell.** It consists of a light sensitive cathode which is in the form of a half-cylinder of metal which is contianed in an evacuated tube. The anode is also present in the tube and is fixed roughly about the axis. The inner side of the cell is coated with some light sensitive layer. (Fig. 6.14).

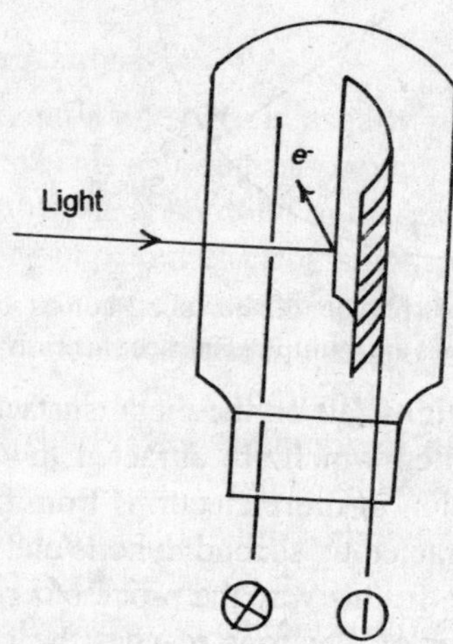

Fig. 6.14: Photocells

When the incident light falls on such a cell, electrons are emitted by the surface coating. These electrons are attracted and collected by anode. The current so produced is a measure of the incident radiations.

This type of cell is more sensitive because high degree of amplification can be used. In case quartz or silica windows are used then the range of such a cell can be increased to cover both near u.v. and far u.v. region.

(c) *Photomultiplier Tube.* This is generally used in a u.v. spectrophotometers. A photomultiplier tube is shown in Fig. 6.15. and an electron-multiplying amplifier. It consists of an evacuated tube containing one photo-cathode and 9-16 electrodes known as *dynodes.* The surface of each dynode is of Be—Cu, Cs—Sb or some such similar material.

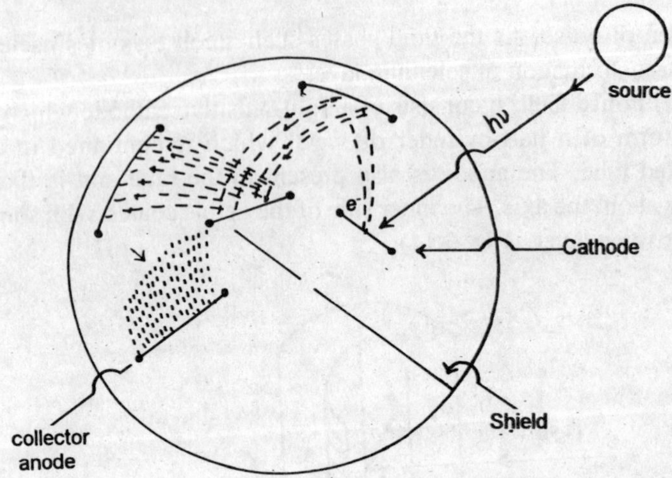

Fig. 6.15: Schematic diagrams of a photomultiplier tube. Photons strike the cathode and the ejected electrons are multiplied in each successive step

When incident radiations fall on the metal surface of a photo-cathode electrons get emitted which are attracted towards the first dynode. it results in emission of more electrons from the surface of first dynode. These are attracted by second dynode and electrons are emitted by second dynode. In this way the process is repeated at all the dynodes until a shower of electrons reaches the collector. The number of electrons that reach the collector is a measure of the incident radiations on the detector. To get a steady signal dynodes are operated at optimum voltage.

This is quite sensitive and gives an extremely fast response.

4. Recording System

Recording system receives the signal from photomultiplier tube and the recording is done by a recorder pen.

5. Sample Cells

These cells contain the sample and these must satisfy the conditions given below :

(i) They should be uniform and surfaces that face the incident light must be optically flat.

(ii) They should be made of such a material that is inert to solvents

(iii) They must transmit light of the wavelength used.

Commonly used cells are made up of quartz or fused silica. Such cells are generally available in matched pairs so that one of these is used as reference cell.

A Single beam and a Double beam Ultraviolet Spectrophotometer

Fig. 6.16 shows a single beam and a double beam spectrophotometers.

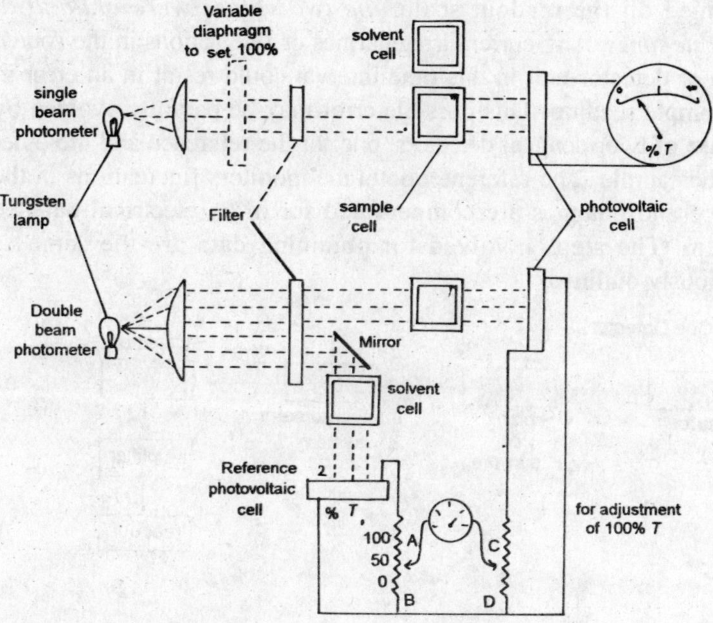

Fig. 6.16: A single-beam and a double beam photometer

The single beam spectrophotometer is a direct reading instrument but the double beam photometer is a null type photometer.

In the single beam system (Fig. 6.17) it is necessary to obtain values of the incoming power, P_0, and the transmitted power, P. This is accomplished at a given wavelength as follows:

1. With the shutter closed (no light atrikes the detector) the transmittance scale is adjusted to 0%.
2. With the shutter open and the blank in the system the transmittance scale is adjusted to 100%. This corresponds to the incoming power, P_0, in the diagram.

3. With the blank replaced by the sample, the percent transmittance scale yields a % T value. This is percent ratio ($P_1/P_0 \times 100$) of the power of the light now exiting, P, to that which was formerly exiting, P_0.

Although the ratio of the values of the two powers was obtained on the readout scale, *the two values were obtaind at separate times.* Any current irregularities or fluctuations in the source lamp or detector unit in this time interval could result in an error in the sample reading. This possible error may be partially avoided by the use of two identical detectors, one for the reference and the other for the sample. The referene phototube monitors fluctuations in the source and changes are compensated for in an electrical balance system. The steps involved for obtaining data are the same as previously outlined.

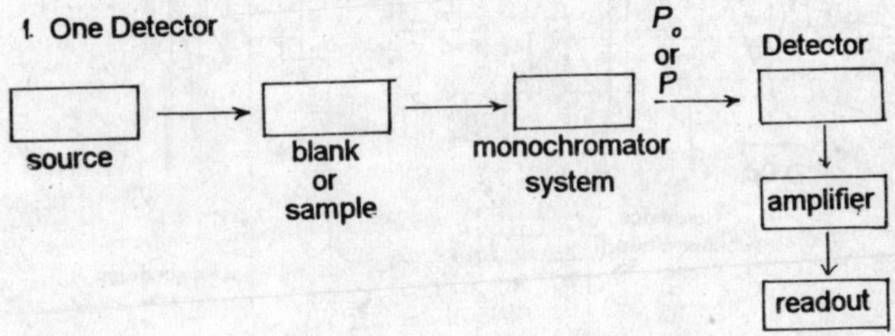

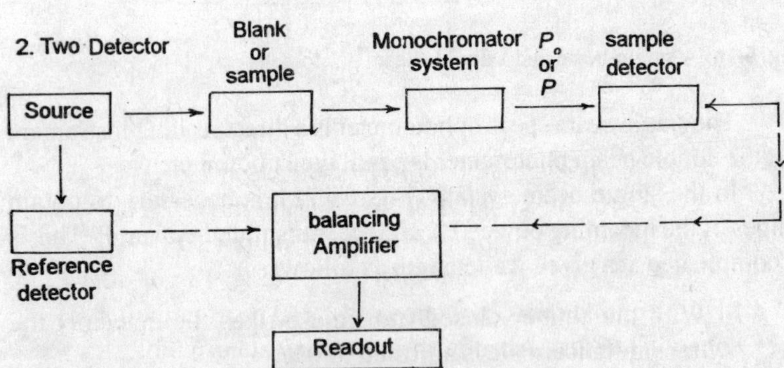

Fig. 6.17: Schematic diagrams of single beam spectrophotometers. The two detector system (bottom) partially compensates for power fluctuations with time

Some of the shortcomings of single beam systems have been overcome with the introduction of **double beam** (Fig. 6.18) instruments. In these the intensity of the source radiation is split into two separate beams: one for the sample and the other for the reference. The radiations intensity may be split in either *space* or in *time*.

In the **double beam in space** instrument the exiting beam, P_0, from the monochromator is split into two equal parts; one-half passes through the blank to the reference detector, the other through the sample to the sample detector. The output from the two detectors is electrically balanced, amplified if essential, and the ratio of the two output is fed to the readout device.

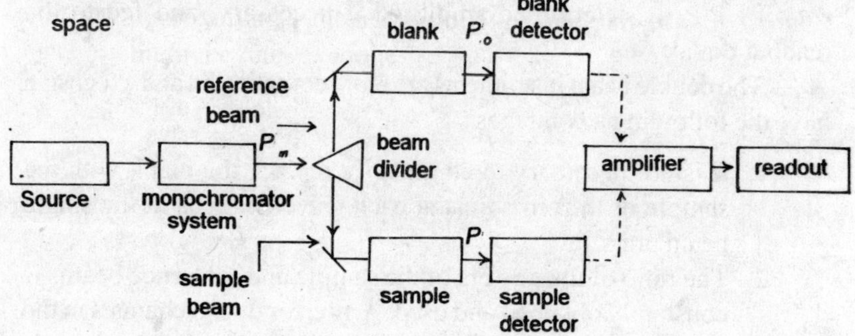

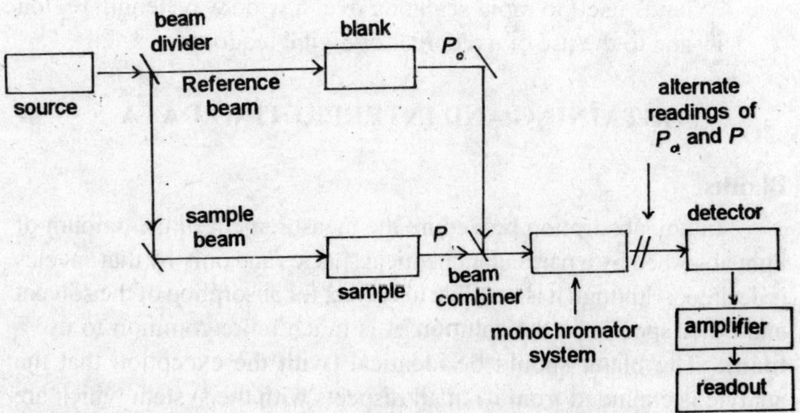

Fig. 6.18: Schematic diagrams of double beam spectrophotometers. Top: double beam in space. Bottom: Double beam in time.

The design has the disadvantages of requiring two detector units. In addition, the power of the beam from the monochromator, P_m, is out in half by the beam divider to form the respective powers of the reference, P_0, and sample, P, beams. This loss of power should be compensated for by the addition of either more sensitive detectors or more power amplifiers.

In the double beam in time design the full energy from the source lamp is alternated between the reference and sample beams at a fixed frequency. The reference power, P_0, and sample power, P are alternately read by the single detector, the differences in intensities are compared, and the required electrical corrections are made. The *ratio* of P/P_0 is determined amplified if necessary, and fed to the readout device.

The double beam instruments are more complicated and expensive have the following advantages :

1. It is not necessary to continually replace the blank with the sample or to zero adjust at each wavelength as in the single beam units.
2. The ratio of the powers of the sample and reference beams is constantly obtained and used. Any error due to changes in the intensity of the source and fluctuation in the detector is minimized.
3. Because of the previous two factors the double beam system lends itself to rapid scanning over a wide wavelength region and to the use of a recorder or digital readout.

OBTAINING AND INTERPRETING DATA

Blanks

In any absorption procedure the measurement of the amount of light absorbed by a particular chemical species and only by that species is desired. Although it is possible to correct for absorption of the solvent and other species in the solution, it is much more common to use a **blank.** The blank should be identical (with the exception that the analyte is excluded from it) in all respects with the system which are being determined. It should also be subjected to all prior processes such as extraction, pH adjustment, or heating to which the unknowns and standards are subjected.

For example, in the determination of iron with 1, 10-phenanthroline, after dissolution any remaining Fe(III) is reduced with hydroxylamine hydrochloride to Fe(II) in a solution which is buffered at approximately pH4. The resulting Fe(II) is complexed with 1, 10-*ortho*-phenantroline and the absorbance of the iron(II)—*ortho*-phenanthroline complex is obtained. The blank should contain the solvent, the hydroxylamine hydrochloride, the buffer, and the *ortho* phenanthroline, but should contain no iron. The reagents should be added in the same order and in the same concentrations, and allowed to stand the same length of time in the samples, standards and blank.

Obtaining a Spectrum

For obtaining spectrum first, check and adjust the components of the system. To obtain optimum power at the detector, the only alternative to the proper choice and adjustment of the source lamp, sample cells, monochromator elements and detectors is the use of wider slit widths. However, since wider slit widths will lead to loss of detail in the absorption spectra, this alternative is unsatisfactory.

In any new spectrophotometric work an absorption spectrum should be first obtained and the wavelength of maximum absorption chosen or verified, even if information is available from the literature. Sometimes, the wavelength of maximum absorption might not match because the prism, grating, or filter is out of adjustment in one or another of the instruments. Mechanical slack in the system might lead to differing results, depending upon the direction of approach. *i.e.*, from lower to higher or from higher to lower wavelengths. The possibilities of variation in the chemical systems must also be considered when utilizing a procedure from the literature in a specific application.

Mechanics of Recording the u.v. Spectrum

The u.v. radiations from the source are passed through monochromators which automatically selects the required wavelengths. The monochromatic beam is split into two parallel beams one of which passes through a transparent cell containing very dilute solutions of the organic compound. It is called the sample beam. The second beam is directed through an identical cell containing solvent only. This is called reference beam.

The instruments directly measure the optical density at the said wavelength. The wavelength of light is changed and new optical density is recorded. The intensities of two transmitted beams are electronically compared and a plot of absorbance against wavelength is automatically recorded over the desired range.

Solvents which are fairly transparent over the required range of wavelength are used. These instruments record the wavelength of maximum absorption of light and the intensity of absorption. The spectrum thus obtained consists of one or more **broad absorption peaks or humps.** The wavelength corresponding to the top of a peak is called λ_{max} and the intensity of absorption in that position is called molar extinction coefficient ε (epsilon).

Qualitative Procedures

(i) *Functional group analysis.* Although theoretically every chemical species has a unique absorption spectrum, practically the absorption spectra of different species are frequently difficult to distinguish from each other in the visible and ultraviolet regions. The various possible combinations of electronic and vibrational energies are so close to each other that broad absorption bands are obtained if a high resolution instrument is not used. In spite of this drawback, some functional groups can be identified from the ultraviolet absorption spectra. The most common use is for aromatic and conjugated organic compounds and for complex inorganic ions. Tables of avrious functional group absorption regions are available in the literature.

(ii) *Comparison of spectra.* Qualitative matching of spectra can be done by comparison of an unknown against a known sample in the laboratory or, alternatively, by matching the unknown spectra with those available in the literature.

Since percent transmittance is not a linear function, whereas absorbance is linear with respect to concentration, the following discussion will be limited to absorption spectra:

The absorption spectra for various concentrations of a hypothetical compound are shown in Figure 6.19 (a). In this **family of curves** it is difficult to qualitatively compare two curves because of the change in the shapes with the change in concentration. When the data are plotted as log absorbance vs. wavelength, the individual curves are displaced equal amounts and the spectra of identical solutions of the diofferent concentrations may be compared. As an alternative to calculating and susbequently plotting log absorbance values, absorbance may be plotted on a semilog absorbance values scale. The same results will be obtained. A disadvantage of the log plotting procedures is that detail will be lost [Fig. 6.19 (b) and 6.19 (c)].

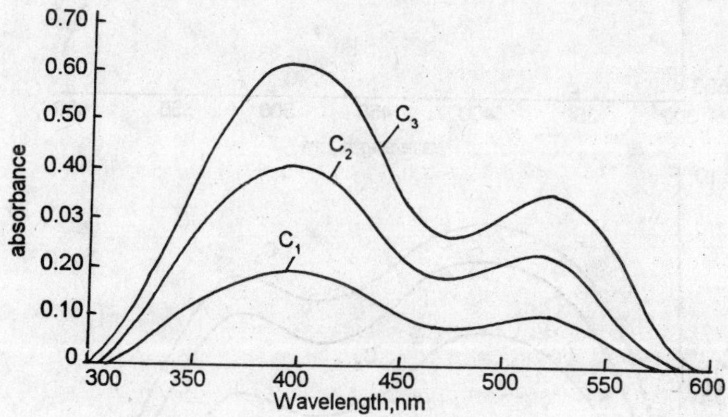

Fig. 6.19: Plotting of absorption spectra. Plot 6.19 *(a)* is absorbance vs. wavelength on a linear scale. Plot 6.19 *(b)* is log absorbance vs. wavelength on a linear scale. Plot 6.19 *c* is absorbance vs. wavelength on a semilogarithmic scale. $C_3 = 1.5$, $C_2 = 3\ C_1$.

(iii) *Proof of two equilibria species.* In a solution of any acid—base indicator, HX, the following equilibrium exists :

$$HX \rightleftharpoons H^+ + X^-$$

where HX is the acidic and X– is the basic form of the indicator. A series of solutions, all containing the same total concentration (i.e., [IIX] ı [HX⁻]) of the indicator would yield, with a change in pH, a family of absorption curves similar to those shown in Fig. 6.20 *a*, provided both species, HX and X^-, absorb over the studied wavelength region.

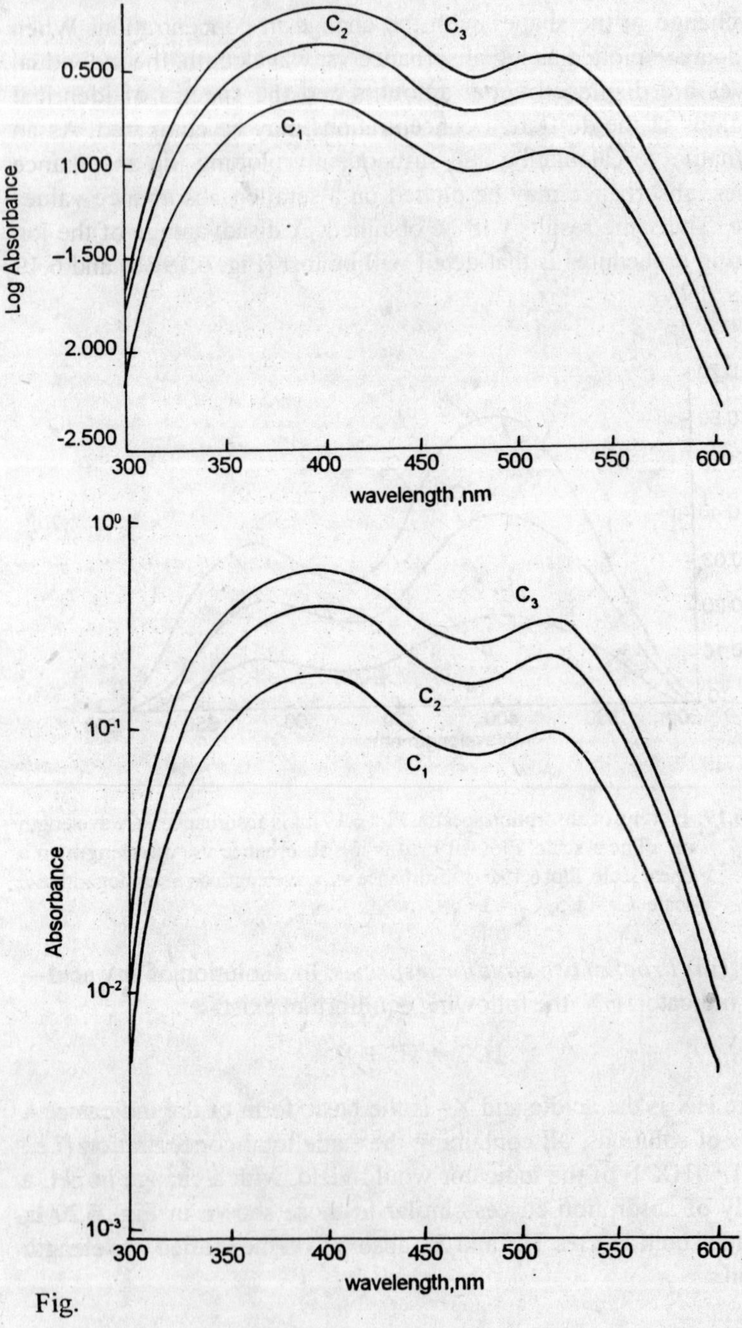

Fig.

The common point at which all of these curves cross is referred to as an **isosbestic point.** At the wavelength of this point the absorbance of the two species are equal. The presence of isosbestic points is proof of a two-absorbing component equilibrium. It will occur any time that two incterconvertible species are present in equilibrium provided they both absorb in the spectral region studied. For example, they frequently occur in oxidation—reduction and chelation, as well as in acid—base equilibria.

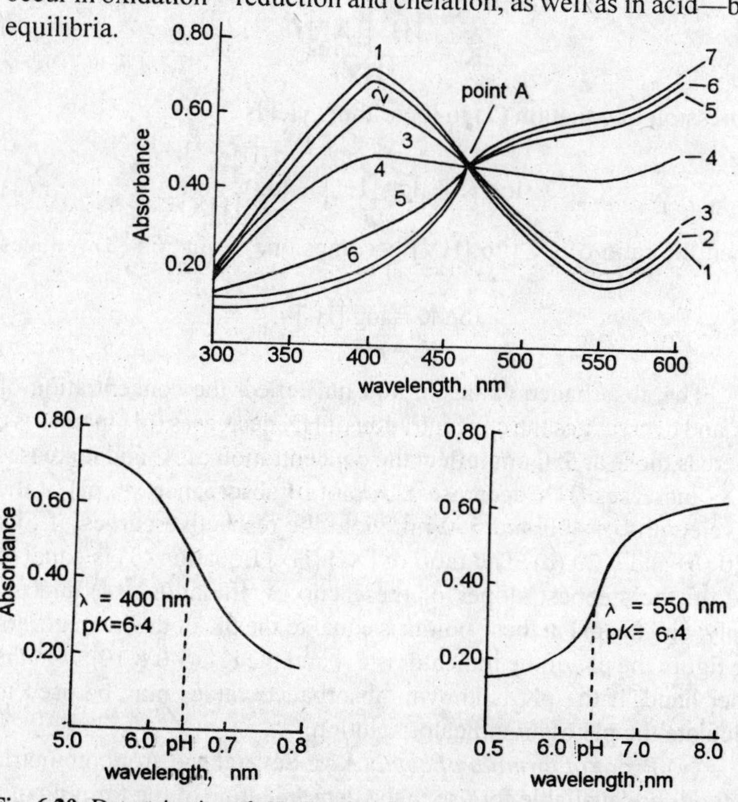

Fig. 6.20: Determination of *p*K of an indicator. (a) Family of curves of an indicator. [HX] + [X⁻] is a constant and *p*H changes. The *p*H's of the solutions are (1) = 5.0; (2) = 5.6; (3) = 5.9; (4) = 6.4; (5) = 6.9; (6) = 7.9; and (7) = 8.5. (b) Absorbance vs. *p*H at 400 nm. (c) Absorbance vs. *p*H at 550 nm.

(iv) *Dteremination of pK and pH.* Consider again the family of curves Fig. 6.20 (a) for the acid—base indicator equilibrium

$$\underset{\text{acid}}{HX} \rightleftharpoons H^+ + \underset{\text{base}}{X^-} \qquad ...(1)$$

The total concentration of the indicator is held constant and the ratio of acid to base ([HX] to $[X^-]$) changes with pH. At a relatively low pH (high acidity) the above equilibrium shifts to the left and curve 1 is, for all practical purposes, that of the acid, HX, species. On the other hand, curve 7, that of the basic form, X–, would be obtained from highly basic solutions.

The equilibrium constant for the previous equation is written as

$$K = \frac{[H^+][X^-]}{[HX]} \quad ...(2)$$

Expression of equation (2) in logarithms yields

$$\log K = \log[H^+] + \log\frac{[X^-]}{[HX]} \quad ...(3)$$

When the ratio of $[X^-]$ to [HX] becomes one, Equation (3) reduces to

$$\log K = \log [H^+] \quad ...(4)$$

$$pK = pH. \quad ...(5)$$

The absorbance values at 400 nm reflect the concentration of HX and decrease as the concentration of HX decreases (pH increases), whereas those at 550 nm reflect the concentration of X^- and increases as X^- increases (HX decreases). A plot of absorbance vs. pH at the wavelengths of 400 and 550 nm yield the respective curves of Fig. 6.20 (*b*) and 6.20 (*c*). The ratio of $[X^-]/[Hx]$ Equation (3) is equal to one on the steepest slopes of these curves, Euqations (4) and (5) apply, and the pH at these points is equal to the pK of the indicator. In the figure the pK of the indicator is 6.4 and the K is 4.0×10^{-7}. On the other hand, if the pK is known, absorbance values may be used to calculate the pH of an indicator solution.

(v) *Proof of formula of complexes.* Several spectrophotometric methods are available for use in the determination of the formula of a complex ion or metal chelate. The two primary procedures are the mole ratio method and the method of continuous variations.

Quantitative Procedures

Before obtaining quantitative data it is essential to consider the absorption spectra of the different compounds and of possible intereferences which are involved.

Choice of Wavelength

A number of criteria are considered in the choice of wavelength. It may be necessary to balance one against another.

1. The highest absorption peak will yield the maximum sensitivity.
2. A chosen wavelength should yield a minimum intereference from other species which may be in the system.
3. Because of the previously discussed errors which arise from sharply rising or sharply falling portions of the absorption curves, these sections should be avoided.
4. Care should be used to avoide a sharp band if the slit width exceeds the width of the band.

This is demonstrated in Fig. 6.21. The solid line represents the absorption curve for the sample species and the dasbed line represents that of an interference. This interference may be caused by either the necessary excess of one of the colour-forming reagents or may be naturally occurring in the studied system. When all factors one considered, point B would be the chosen wavelength when using a 20 nm band width. Whereas the compound possesses a greater absorptivity, and thus a greater potential sensitivity at both A and F than at B, this advantages is lost when the potential for interference (dashed line) and the slit width are both considered. Points C and E are avoided because of the error associated with absorbance measurements on sharply rising or falling curves. Point D would yield an accurate reading but its sensitivity is less than one-third of that at point B.

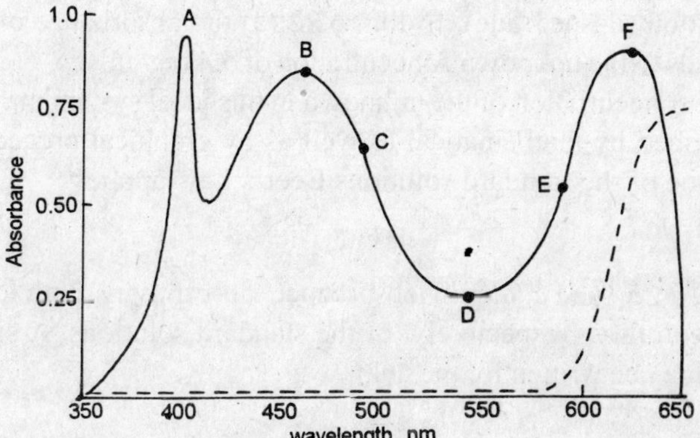

Fig. 6.21: Choice of experimental wavelength. Solid line: measured species. Dashed line: interfering species.

Preparation of Calibration Curves

After the instrument has been adjusted and the wavelength has been chosen, the proposed system is checked for adherence to Beer's Law. The following steps are involved in checking for this adherence:

1. Prepare a series of standard solutions in the concentration range desired. The composition of the solutions should be as close as possible to that of the system being studied and should yield absorbance readings in the lower section of error curve. If the contaminants and/or possible interferences in the samples are known they should be reproduced in the standards.
2. A blank must be prepared. Its composition was previously discussed.
3. Set the chosen wavelength. Calibrate the 0% transmittance with the shutter closed and the 100% transmittance with the shutter open and the blank in the optical path.
4. Obtain the absorbances at various concentrations of the standards.
5. Plot absorbance vs. concentration to check for adherence to Beer's Law.

Experimentally, the ideal calibration curve is a straight line which intersects the origin Fig. 6.22 (a). The concentration of the unknown is obtained by reading its absorbance and determining the corresponding concentration (dashed line). In Fig. 6.22 (a) the absorbance of 0.53 corresponds to the unknown concentration of 5.4 mg/ ml.

The concentration of the unknown in this ideal system may also be established by mathematical as well as by graphical procedures. For any one of the standard solutions Beer's Law applies :

$$A_s = a_s\, b_s\, c_s \qquad ...(1)$$

where A_s, a_s, b_s, and c_s are the absorbance, absorptivity, path length, and concentration, respectively, of the standard solutions. A similar equation may be written for the unknown

$$A_x = a_x\, b_x\, c_x \qquad ...(2)$$

where x subscript refers to parameters of the unknown. Combination of the two previous equations yields

$$\frac{A_s}{A_x} = \frac{a_s\, b_s\, c_s}{a_x\, b_x\, c_x} \qquad ...(3)$$

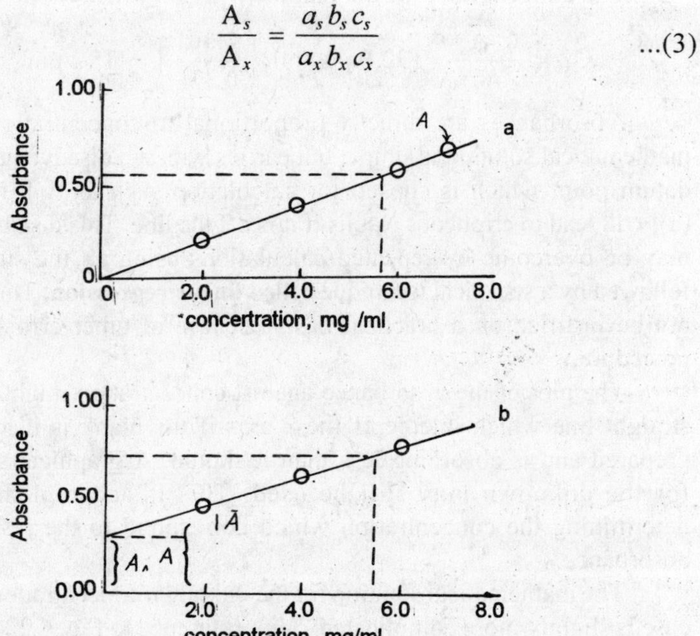

Fig. 6.22: Beer's Law calibration curves

Since the absorptivities of the standard and unknown are the same and the cell depths, b_s and b_x, are equal, Equation (11) reduces to

$$\frac{A_s}{A_x} = \frac{c_s}{c_x} \qquad ...(4)$$

Example

The absorption spectrum for the titanium-peroxide complex ion in perchloric acid exhibits a maximum at 400 nm. The absorbance of a 32.0 μg/ml solution of Ti gave an absorbance of 0.456. An unknown solution treated in an identical fashion gave an absorbance of 0.501. What was the concentration of the unknown? Assume identical cells.

$$\frac{A_s}{A_x} = \frac{C_s}{A_x}$$

The concentration of the unknown is

$$C_x = \frac{C_s A_x}{A_s} = (32.0\ \mu g/ml)\left(\frac{0.501}{0.456}\right) = 35.2\ \mu m/ml.$$

Absorbances are directly proportional to concentration. This mathematical solution is simple but carries with it a disadvantage: the datum point which is chosen for calculation such as A [Fig. 6.22 (a)] will lead to erroneous results if it is off the line. This disadvantage may be overcome by repeated calculations using all the standards followed by a statistical technique called **linear regression.** This might not be justified on a practical basis because of other errors in the procedure.

The plot of the absorbance against concentration might yield a straight line which intercepts the *y* axis if the blank is incorrectly prepared and is absorbing less than it should. A graphical solution for the unknown may still be used. This is accomplished, by determining the concentration which correspond to the respective absorbance.

The mathematical solution for the unknown concentration in this case is slightly more complicated. With reference to Fig. 6.22 (b)

$$\overline{A} = (A - A_b) \qquad \text{... (1)}$$

where A_b is the absorbance from the origin to the intercept of the *y* axis, and A is the measured absorbance. The corrected absorbance, *i.e.,* that which is due only to the species of interest, is $\overline{A}$. Application of the same assumptions and nomenclature as in the previous case yields the following equations :

$$(A - A_b) = \overline{A} = abc \qquad \text{...(2)}$$

for both the standards and unknowns. The equation

$$\frac{\overline{A}_s}{A_x} = \frac{a_s\, b_s\, c_s}{a_x\, b_x\, c_x} \qquad \text{...(3)}$$

now applies since the absorptivities of the standard and unknown are the same and since the path lengths of the cells for the standard and unknowns are equal, Equation (3) reduces to

$$\frac{\overline{A}_s}{A_x} = \frac{c_s}{c_x} \qquad \text{...(4)}$$

The corrected absorbance, $\overline{A}$, not the measured absorbance, A, is now directly proportional to concentration. The mathematical solution suffers from the same disadvantage as in the previous case, namely, the chosen data for calculations might not be representative.

Example

In the determination of acetone in bioloigical fluids, a water blank instead of a reagent blank was used. A 2.5 mg% standard gave an absorbance of 0.81 and an unknown gave 0.62. In a separate determination it was found that the reagent blank gave an absorbance of 0.05 when read against the water blank.

Calculate the concentration of the acetone in the unknown.

$$\frac{\overline{A}_s}{A_x} = \frac{(A_s - A_b)}{(A_x - A_b)} = \frac{c_s}{c_x}$$

therefore

$$C_x = (2.5\text{ mg\%})\left(\frac{0.62 - 0.05}{0.81 - 0.05}\right) = 1.9\text{ mg\% acetone.}$$

If the absorbance vs. concentration plot is non-linear, *i.e.*, shows a deviation from Beer's Law, a graphical but not a mathematical solution may be used.

Photometric Titrations

A photometric titration follows the change in the absorbance of a system with respect to the change in concentration of an absorbing species, the change in the concentration of the absorbing species being brought about by a chemical reaction with the titrant. The absorbance after each added increment is plotted against the added volume of titrant. The equivalence point is subsequently determined by the intercept of two straight lines. The titrations may be performed in the cell or in an external unit.

An example is shown in Fig. 6.23a, 6.23 b, and 6.23 c. Fig. 6.23 (a) shows the absorption spectra for the metal ion, M^{2+}, and the chelating agent HKE, by the respective solid and dashed lines. The equation for the reaction is

$$M^{2+} + HKE \rightleftharpoons MKE^{+} + H^{+} \qquad ...(1)$$

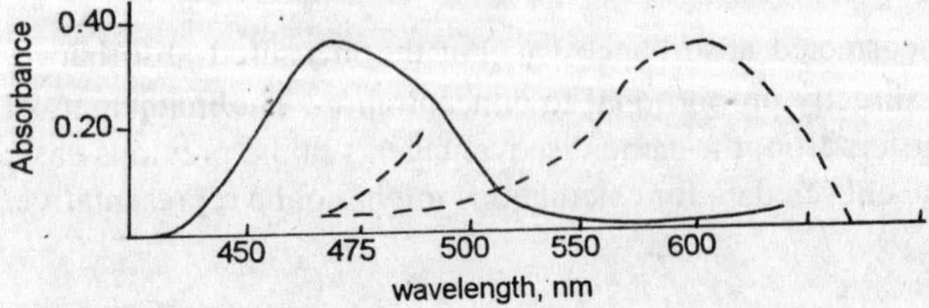

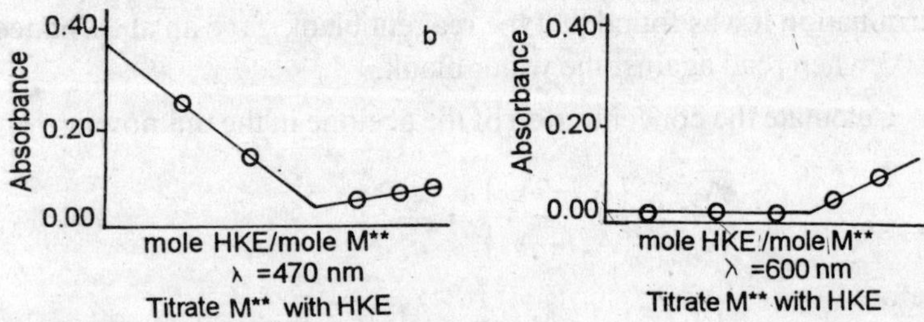

Fig. 6.23: Spoectrophotometric titration. (a) Absorption spectra of M^{2+} (solid line) and HKE (dashed line). (b) M^{2+} titrated with HKE at 470 nm. (c) M^{2+} titrated with HKE at 600 nm

The resulting chelate, MKE^+, does not absorb in the wavelength region studied. Fig. 6.23 shows the titration curve obtained at 470 nm as the absorbing species, M^{2+}, is being titrated with HKE. The declining portion on the left represents the decrease in concentration of the unreached M^{2+} in the solution, whereas past equivalnece the right side of the curve reflects the addition of an excess of the chelating agent, HKE. The same titration may be prrformed at 600 nm. Since the species in excess, M^{2+}, is not an absorbiung species at this wavelength, the line prior to eqauivalence is flat. Past equivalence, an excess of HKE is being introduced and an increase in absorbance occurs.

For best results, appropriate dilution correction must be made. The following equation

$$A_c = \left(\frac{V_i + V_a}{V_i}\right)(A_m) \qquad ...(2)$$

where A_c is the corrected absorbance; V_i and V_a are the respective initial and added volumes, and A_m is the measured absorbance may be used. However, a negligible error will result from ignoring dilution corrections if the concentration of the titrant is much greater than the concentration of the solution which is being titrated.

Example

An absorbance of 0.436 was obtained after 11.5 ml of titrating agent was added to 68.0 ml of an initial solution. What was the corrected absorbance of the solution? What would the % age error have been if the correction was not made?

$$\text{Corrected absorbance} = \frac{(11.5\text{ ml} + 68.0\text{ ml})}{68.0\text{ ml}} \times 0.436 = 0.510$$

$$\%\text{ age error} = \frac{11.5\text{ ml}}{79.5\text{ ml}} \times 100 = \frac{(0.510 - 0.436)}{0.510} \times 100 = 14.5\%$$

If the titration is performed in an external unit it is necessary to continually remove known aliquots for absorbance measurements. These aliquots must be returned to the original solution after each measurement, a difficult process at best; or alternatively, the solution must be corrected for the change in concentration of both reagents. It is much more convenient to perform the titration directly in a cell if the equipment is available.

The advantages of spectrophotometric titrations are many :

1. Only one reacting or product species need absorb. This offers the ability to titrate non-absorbing or non-coloured species in both the visible and ultraviolet regions.
2. In contrast to normal absorbance procedures, it is not essential that the titration be carried out in the region or wavelength of maximum absorbance. Thus, a greater choice of wavelengths is available.
3. Since the end point is determined by the intercept of the straight lines, it is essential to obtain only a few points prior to and after the end point. As a matter of fact, due to dissociation, the curves will frequently be rounded at the end point of the titrations. The difference between a true curve and the extrapolated straight lines is a measure of the degree of dissociation of the resulting product. Equilibrium constants may be calculated from these differences.

4. Relatively large quantities of other absorbing species at the chosen wavelength are necessary before noticeable interference results.
5. Turbid solutions may be titrated in selected circumstances.
6. More dilute solutions may be used than in many other types of titrations.
7. A variety of equipment may be used.

In summary, spectrophotometric titrations are convenient, rapid, and accurate methods of obtaining data once the original setup is acquired. They offer the possibility of the use of a wide range of concentrations, reagents, wavelengths, procedures, and apparatus.

Analysis of a multicomponent mixture. Absorption measurements may be used for analysis, without prior separation, of the individual components of a multicomponent system provided certain conditions are met :

1. There must be no chemical interaction between the components. Each must react in the mixture just as they would react in their respective individual solutions.
2. Absorption spectra of the individual components must be different from each other over the wavelenth range being studied. It is desired that the ratio of absorptivities of the two components at the wavelength studied be as large as possible.
3. Each components must adhere to Beer's Law at each wavelength utilized and through the concentration range being studied.

For a single component system the equation

$$A = \varepsilon bc \qquad ...(1)$$

applies. Since absorbances are additive, it follows that

$$A_{total} = A_x + A_y \qquad ...(2)$$

where A_x and A_y are the absorbances of the two unknows, x and y. Substitution of Equation (1) into (2) for each of the two wavelengths yields

$$A_{\lambda_1} = \varepsilon_{x_{\lambda_1}} bc_x + \varepsilon_{x_{\lambda_1}} bc_y \qquad ...(3)$$

at the first wavelength, and

$$A_{\lambda_2} = \varepsilon_{x_{\lambda_2}} b c_x + \varepsilon_{x_{\lambda_2}} b c_y \qquad ...(4)$$

at the second wavelength, where A_{λ_1} and A_{λ_2} are the respective absorbances at the first and second wavelengths, $\varepsilon_{x_{\lambda_1}}$ and $\varepsilon_{x_{\lambda_2}}$ are the molar absorbancies of the first unknown at each of the two wavelengths, $\varepsilon_{x_{\lambda_1}}$ and $\varepsilon_{x_{\lambda_2}}$ are the molar absorbancies of the second unknown at each of the two wavelengths and c_x and c_y ar the molar concentrations of the two unknowns.

If the cell lengths are known, and the absorptivities of each species at each chosen wavelength are known or can be determined, the two equations may be solved simultaneously. Theoretically, any number of components may be simultaneously determined provided sets of data are obtained at as many wavelengths as there are components. A three component system, for example, would require data at three wavelengths. On a practical basis, the instruemental problems as well as the mathematical solutions set a limit. Automated equipment for obtaining the data and the use of computers for solving the equations are extending the scope of applications.

APPLICATIONS

Ultraviolet and visible spectrophotometric applications are numerous. The techniques may be applied to many of the organic chromophoric groups as well as many inorganic species. For example, ultraviolet spectrophotometry has been used for monitoring more than 60 metal ions in HCl which have been previously separated by liquid chromatography. It has been applied to the analysis of drugs and pharmaceutical products such as antihistamines, to steroid phosphates in cosmetic creams and emulsions as well as to cholesterol in milk fat and blood. Automated procedures are now utilized for analysis in clinical samples, food products, vitamins, cast iron, steels, and alloys. Many non-absorbing species may be indirectly determined by first converting to an absoring species.

The procedures are sensitive in the 10^{-4} to 10^{-7} M concentration range, are moderately selective, and have a relative error of 1% to 4%. This relative error may be reduced by special techniques. In many cases the sample may be recovered for subsequent study.

(i) Detection of Conjugation

The ultraviolet spectrum is not used primarily to show the presence of individual functional groups but rather to show their mutual relationship. The chief relationship is regarding conjugation such as

(a) Conjugation between two or more carbon-carbon double (or triple) bonds.

$$—C=C—C=C—$$

(b) Conjugation between carbon-carbon and carbon-oxygen double bonds

$$—C=C—C=O$$

(c) Conjugation between double bonds and an aromatic ring and even the presence of aromatic ring itself.

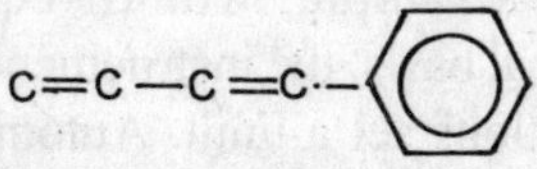

$$C=C-C=C-C_6H_5$$

U.V. spectrum can also reveal the number and location of substituents attached to the carbons of the conjugated system.

(ii) Detection of the Geometrical Isomers

The geometric isomers can be identified by u.v. spectroscopy as the trans isomers exhibit λ_{max} at slightly lower wavelengths and have larger extinction coefficients than the cis-isomers. For example trans stilbene $C_6H_5-CH=CH-C_6H_5$ show λ_{max} 294. ($\varepsilon = 24{,}000$) and cis-isomers has λ_{max} 278 ($\varepsilon = 9350$).

(iii) Qualitative Analysis and Purification Control

It is also used in qualitative analysis and purification control in dye stuffs and pharmaceutical industries.

For identification of compounds containing unbonded electrons (n-electrons) or those with conjugated double bond system (π-electron) a comparison is done by comparing the absorption spectrum with the spectra of known compounds.

In u.v. absorption spectrum the absorption curve is influenced by the whole molecule as well as by particular groups containing the absorbing electrons. This makes it difficult to identify a particular group so this method is not as useful as the N.M.R. method.

For purification control the presence of impurities can be done by u.v. spectroscopy which is considered as the best method for such detection. This method is considered superior to other methods because the bands due to impurities are very intense.

In manufacture of nylon it is desirable that the starting materials are very pure and the purity of these materials could be checked by u.v. spectroscopy. This helps in getting a product of high quality.

(iv) Quantitative Analysis

This method could be used for quantitative determination of those compounds which absorb u.v. radiations. Quantitative determination is based on Beer's Law and Lambert's Law.

Beer's Law

The fraction of the incident monochromatic light absorbed by a homogeneous medium in proportional to the quantity of absorbing medium.

Lambert's Law

The proportion of monochromatic light absorbed by a homogeneous medium is independent of the intensity of the incident light and each successive layer absorbs an equal fraction of the light incident upon it.

In solutions the amount of absorption depends upon the concentration of the solution and thickness of the layer.

By combining Beer's Law and Lambert's Law, we get

$$\log\frac{I_0}{I} = \varepsilon \times c \times l$$

$$= A$$

where I_0 is the intensity of the incident light I is the intensity of transmitted light or emergent light. ε (epsilon) is the molar extinction coefficient or molar absorptivity and it is a measure of absorption efficieny, *c* is the molar concentration (moles/litre) of absorbing solutes in non-absorbing solvents, *l* is the thickness of solution in centimetres (path length) and $\log\frac{I_0}{I}$ or A is the absorption or optical density.

Rearranging the above equation, we have

$$\varepsilon = \frac{A}{c \times l} = \frac{1}{\frac{\text{moles}}{\text{litre}} \times \text{cm}} = \frac{1}{\frac{\text{moles}}{\text{cm}^3} \times \text{cm}} = \frac{\text{cm}^3}{\text{moles}}$$

Thus the unit of ε is $\text{cm}^2\ \text{mole}^{-1}$.

Let T $(\text{Transmittance}) = \frac{I}{I_0}$

$$\therefore \qquad A = \log \frac{I_0}{I}$$

$$= \log \frac{I}{T}$$

If $\quad T = 0 \rightarrow 1 \quad$ or $\quad \% T = 0 \rightarrow 100$

$\therefore \quad A = a \rightarrow 0$

When working in terms of absorption instead of transmittance we have the following relations

$$\text{Absorption} = 1 - T$$

$$\% \text{ Absorption} = 100(1 - T)$$

$$= 100\left(1 - \frac{I}{I_0}\right)$$

or $$\frac{\% \text{ absorption}}{100} = 1 - \frac{I}{I_0}$$

or $$\frac{I}{I_0} = 1 - \frac{\% \text{ absorption}}{100}$$

or $$T = 1 - \frac{\% \text{ absorption}}{100}$$

$$A = \log \frac{I_0}{I}$$

$$= \log \frac{1}{T}$$

$$= -\log T$$

$$A = -\log\left[1 - \frac{\% \text{ absorption}}{100}\right]$$

For the determination of concentration of a substance wavelength of maximum absorption for the compound is selected. If several absorbing species are present then the absorbance of solution will be the sum of absorption of all the species present in solution provided there is no interaction amongst the species. In case a solution contains *n* components having absorbance at wavelengths $\lambda_1, \lambda_2, \lambda_3 \ldots \ldots \lambda_n$, then absorbance will be given by

$$\lambda_1 A_1 = \varepsilon_l^{\,a} . c^a . l + \varepsilon_l^{\,b} . c^b . l + \ldots \ldots + \varepsilon_l^{\,n} . c^n . l$$

$$\lambda_2 A_2 = \varepsilon_2^{\,a} . c^a . l + \varepsilon_2^{\,b} . c^b . l + \ldots \ldots + \varepsilon_2^{\,n} . c^n . l$$

$$\ldots \ldots \ldots \ldots \ldots \ldots \ldots \ldots \ldots \ldots \ldots \ldots \ldots \ldots \ldots \ldots \ldots \ldots \ldots \ldots \ldots$$

$$\ldots \ldots \ldots \ldots \ldots \ldots \ldots \ldots \ldots \ldots \ldots \ldots \ldots \ldots \ldots \ldots \ldots \ldots \ldots \ldots \ldots$$

$$\lambda_n A_n = \varepsilon_n^{\,a} . c^a . l + \varepsilon_n^{\,b} . c^b . l + \ldots \ldots + \varepsilon_n^{\,n} . c^n . l$$

Where $A_1, A_2, \ldots \ldots A_n$ are the absorbance of the system at wavelengths $\lambda_1, \lambda_2 \ldots \lambda_n$ respectively $\varepsilon_l^{\,a}, \varepsilon_2^{\,a}, \ldots \varepsilon_n^{\,a}$ are the extinction coefficients of component A at wavelengths $\lambda_1, \lambda_2, \lambda_n$ respectively $c^a c^b c^n$ are the concentrations of the components and l is the thickness of the layer (absorbing).

The absorbance of the system is measured at *n* different wavelengths and extinction coefficients of various components are also determined at each of these wavelengths. Using this data concentrations of any number of absorbing species in a mixture can be found out. This method has been used to quantitatively analyse a mixture of *o*–, *m*–, and *p*–xylenes etc.

(v) Molecular Weight Determination

Determination of molecular weight spectrophotometrically is prossible only for those compounds whose derivatives could be prepared. For example to find out the molecular weight of an amine first of all its derivative (generally a picrate) is prepared. A solution of this derivative is then prepared and the optical density of the solution is measured at $\lambda_{max} = 380$ nm. To calculate the concentration of solution the following formula is used

$$c = \frac{\log \dfrac{I_0}{It}}{\varepsilon_{max} . x l}$$

Knowing the value of c and the weight of amine picrate (ω) it is possible to calculate the molecular weight of amine picrate and so the molecular weight of amine can now be calculated.

(vi) Calculations of Dissociation Constants of Acids and Bases

To make such calculation, let us consider an acid HA.

The acid HA on dissolution in water dissociates to form H_2O^+ and A^- ions.

$$HA + H_2O \dashrightarrow H_2O^+ + A^-$$

Then by definition dissociation constant (K_a) for this acid is given by

$$K_a = \frac{{}^aH_3O + {}^aA^-}{{}^aHA}$$

For every dilute solution, it become

$$K_a = \frac{\left[H_3O^+\right]\left[A^-\right]}{[HA]}$$

Taking logarithm of both sides, we get

$$-\log K_a = -\log\left[H_3O^+\right] - \log\left[A^-\right] + [AH]$$

$$pK_a = pH - \log\frac{\left[A^-\right]}{[HA]}$$

$$= pH + \log\frac{[HA]}{\left[A^-\right]}$$

Thus if the ratio $\frac{[HA]}{[A^-]}$ is known at any particular pH then we can calculate pK_a value.

The find the ratio $\frac{[HA]}{[A^-]}$ spectrophotometrically, a graph is plotted between absorbance and wavelength at different pH value (Fig. 6.24).

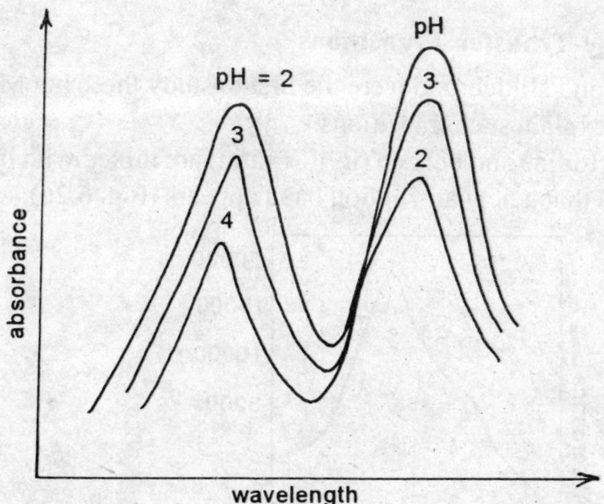

Fig. 6.24: Change in absorbance of the system (HA + A) as the pH of the system is increased from 2 to 4.

(vii) Chemical Kinetics

It is possible to study the kinetics of a chemical reaction using u.v. spectroscopy. For such a study the change in concentration of either a reactant or product with time is measured. This can be easily measured because absorbance is directly proportional to concentration. This method can be used for relatively slow reactions. To study fast reactions, spectroscopically another method known as *stop-flow* method is used. In stop flow method two solutions are allowed to enter through two different openings *x* and *y* (Fig. 6.25) and then they both pass through the reaction chamber B. The piston D is used to stop the flow of the mixed solution. The absorbance of any species that absorbs in u.v. region is measured at *c* with the help of a u.v. spectrophotometer.

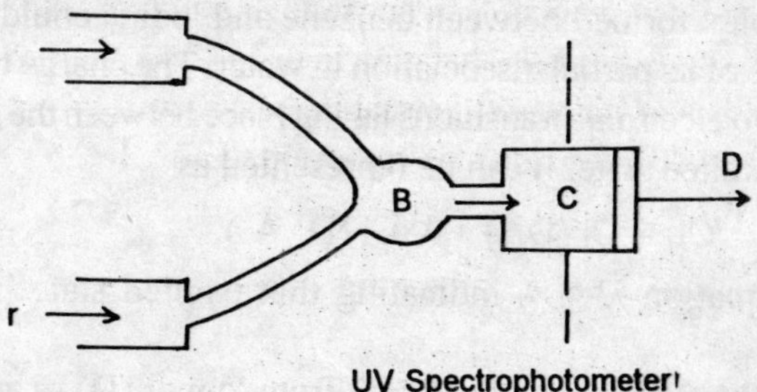

Fig. 6.25: Rapid stop-flow appartus for kinetic studies

(viii) Charge-Transfer Transitions

Bensi and Hildebrand were the first to study these but Mulliken was first to explain such transitions.

When Iodine and benzen (in 1 : 1 ratio) are mixed with heptane, it was found that a new absrоption band appears (Fig. 6.26).

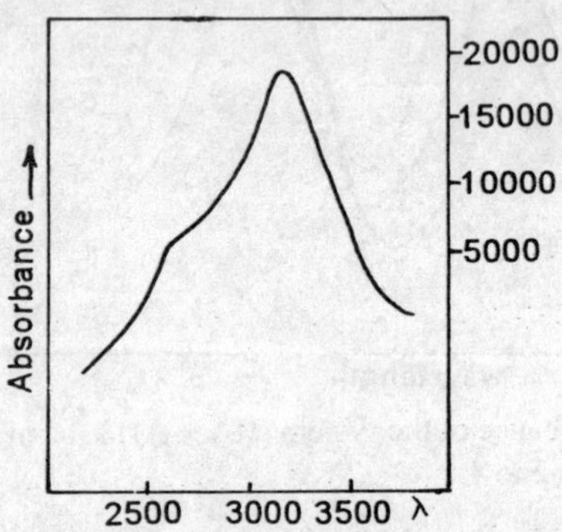

Fig. 6.26: The charge transfer bands of benzene and iodine

Mulliken explained that this new band arises because of a radiation by molecular complex that has been formed between benzene and iodine.

$$\underset{\text{Bezene iodine}}{D + A} \rightleftharpoons (DA) \text{ complex}$$

$$\uparrow$$

$$D^{\oplus} A^{-}$$

[Resonance in (DA) complex]

Now the total wave function in ground state is

$$Cl_0 = Cl\,(DA) + \lambda\, Cl\,(D^+ A^-)$$

λ^2 is a measure of charge transfer

The complex formed between benzene and iodine could not be isolated because of its partial dissociation in water. The charge transfer band arises due to electronic transitions taking place between the ground state and first excited state. It can be represented as

$$Cl_1 = Cl\,(DA) + \lambda'\, Cl\,(D^+ A^-)$$

In this equation $\lambda' > \lambda$ indicating that excited state is more ionic.

Because the electron is transferred from Donor (D) to acceptor (A) so the energy of charge transfer depends upon the ionisation potential of D and electron affinity of A. The following empirical relationship is found to be correct in a number of cases:

$$(hv)_{CT} = I_D - E_A + \Delta$$

where

$(hv)_{CT}$ = Energy of charge transfer

I_D = Ionisation potential of Donor

E_A = Electron affinity of acceptor

Δ = Some constant

From the equation it is quite clear that for a series of different donors, with the same acceptor a plot of $(hv)_{CT}$ against I_D must be a straight line.

(ix) Tautomeric Equilibrium

The % age of keto and enol-forms present in a tautomeric equilibrium can also be determined using u.v. spectroscopy. For example in case ethylacetoacetate, we have

$$CH_3-\overset{\overset{\displaystyle O}{\|}}{C}-CH_2COOC_2H_5 \qquad CH_3\underset{\underset{\displaystyle OH}{|}}{C}=CHCOOC_2H_5$$

In this the λ_{max} for keto form is all 275 nm and λ_{max} for enol form is 244 nm. ε values for keto and enol forms are 16 and 160000 respectively. The keto form shows only a weak $n \rightarrow \pi$ band but enol form shows a strong band. From the strength of 244 nm band it is possible to measure the proportion of tautomers present in equilibrium mixture. Similarly % ages of enols in ethers and alcohols can be found.

(x) Structure of Chloral

The possible structures for chloral are

$$CCl_3-\overset{\overset{\displaystyle OH}{|}}{\underset{\underset{\displaystyle OH}{|}}{C}}-OH \qquad\qquad CCl_3-CHO$$

(I) (II)

When the u.v. spectrum of chloral was recorded in hexane and an absorption band was observed at 298 nm ($\varepsilon = 33$), but when its u.v. spectrum was recorded in aqueous solution no band was observed. From this it was confirmed that the actual structure of chloral is I and not II.

SOLVED EXAMPLES

Example 1

Which of the following compounds absorbs u.v. radiation?

Heptane, Benzene, Butadiene, water, heptene, chlorohexane, ethanol, *n*-butylamine, acetone, ethylene, nitrobenzene, benzoic acid.

Solution

Heptane	:	No absorption occurs because $\sigma \rightarrow \sigma^*$ transition needs very high energy.
Benzene	:	Absorption occurs because of $\pi \rightarrow \pi^*$ transition.
Butadiene	:	Strong absorption because of $\pi \rightarrow \pi^*$ transition
Water	:	No absorption occurs.
Heptene	:	Absorption occurs due to $\pi \rightarrow \pi^*$ transition.
Chlorohexane	:	No absorption observed in u.v. region.
Ethanol	:	No absorption occurs.
n-Butylamine	:	No absorption occurs.
Actene	:	It absorbs u.v. radiations due to $\pi \rightarrow \pi^*$ and $n \rightarrow \pi^*$ transition.
Ethylene	:	Absorbs because of $\pi \rightarrow \pi^*$ transition.
Nitrobenzene	:	Absorbs because of $\pi \rightarrow \pi^*$ and $n \rightarrow \pi^*$ transitions.
Benzoic acid	:	Absorbs because of $\pi \rightarrow \pi^*$ and $n \rightarrow \pi^*$ transitions.

Example 2

Which of the following do not absorb above 2000°?

n-propyl alcohol, Benzene, Diethyl ether, Methyl vinyl ketone, Methyl alcohol.

Solution

In u.v. region *i.e.*, above 2000 A° or 200 nm no absorption occurs in case of *n*-propyl alcohol, Diethyl ether, and methyl alcohol.

Example 3

Which of the following needs maximum energy.

$$\sigma \rightarrow \sigma^*, n \rightarrow \sigma^*, \pi \rightarrow \pi^*$$

Solution

$\sigma \rightarrow \sigma^*$ needs maximum energy.

Example 4

Which will have greater λ_{max}?

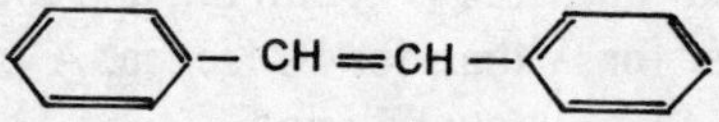

— CH = CH —

or

— CH = CH — CH = CH —

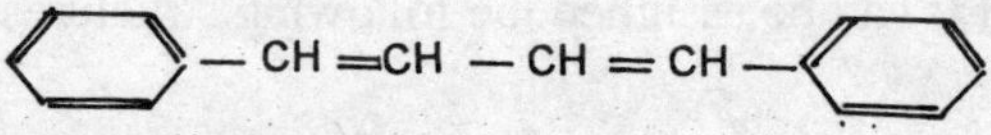

Solution

More the extent of conjugation larger will be the value of λ_{max} in u.v. spectrum. In this case the value of λ_{max} will be more for

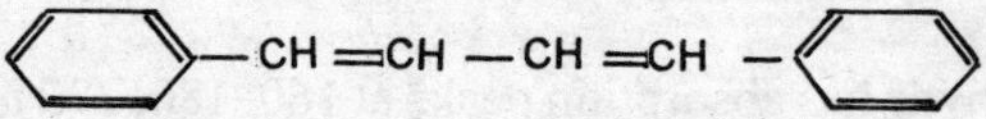

— CH = CH — CH = CH —

Example 5

Which will have larger λ_{max}?

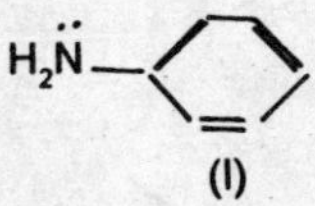

$H_2\ddot{N}$— (I) or. $Cl^- H_2N^+$— (II)

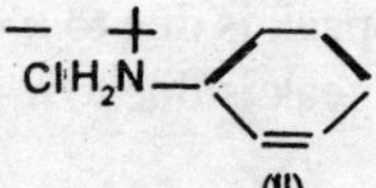

Solution

Since in (I) a lane pair of electrons is available on nitrogen for conjugation with the aromatic ring so λ_{max} will be larger for it as compared to that for (II) in which no such pair of electrons is available for conjugation with the ring.

Example 6

Two isomeric compounds A and B have the molecular formula C_6H_8. Both of them can decolourise bromine solution in CCl_4 and also alkaline $KMnO_4$. On catalytic hydrogenation both yield cyclohexane (C_6H_{12}). A shows an absorption maxima at 256 nm and B shows no absorption beyond 200 nm. Assign structures to A and B.

Solution

Both A and B add four hydrogen atoms on catalytic hydrogenation and form C_6H_{12} thus both A and B are cyclic hydrocarbons containing two double bonds. Thus A and B are cyclohexa dienes.

This fact is furthur supported by their action of bromine solution in CCl_4 and alkaline $KMnO_4$.

The absorption maxima of A is at 256 nm and for B is less than 200 nm so λ_{max} is larger for A than for B. Hence, A is a conjugated diene where is B is not a conjugated diene.

Thus A and B can be assigned the following structures:

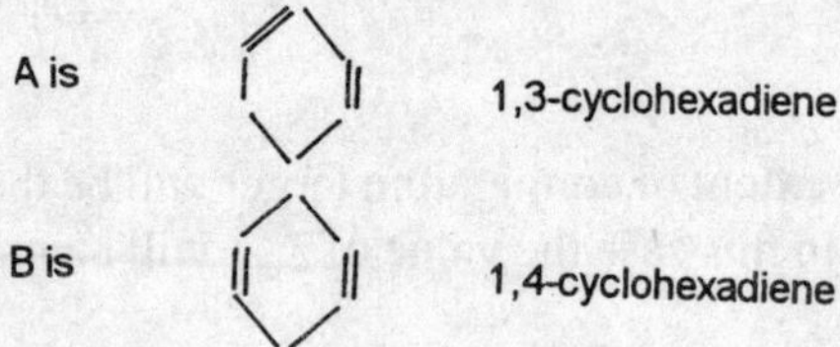

Example 7

Acetaldehyde has absorption peaks at 160, 180, 299 nm. Which type of absorption is responsible for each of these absorptions?

Solution :

160 nm peak is due to $\pi \rightarrow \pi^*$ transition
180 nm peak is due to $n \rightarrow \sigma^*$ transition
299 nm peak is due to $n \rightarrow \pi^*$ transition.

Unsolved Problems

1. A compound exhibits absorption peaks at 190 nm and 300 nm. What type of absorption is associated with each absorption?

(Ans. $n \rightarrow \pi^*$ for 300 nm $n \rightarrow \sigma^*$ for 190 nm)

2. Making use of Woodward's rules, predict the wavelength of maximum absorption for each of the following compounds.

(a)

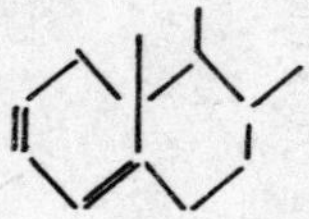

(b) C_8H_{17}

ACO H

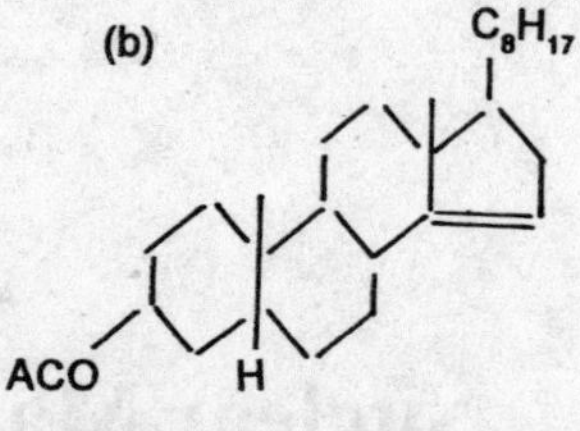

[**Ans.** (a) 273 nm (b) 244 nm]

3. At a wavelength of 356 nm, the molar absorptivity of phenolic alkaloid numismine is 0.1 μHCl is 400; in 0.20 μNaOH is 17,100 and in pH 9.50 buffer is 9800. What is the pKa?

(**Ans.** 9.39)

7

Nuclear Magnetic Resonance

Introduction

As the name implies, this form of spectroscopy depends upon the magnetic property of an element, thus, it delas with the transitions between magnetic energy levels of nuclei of a molecule. To create magnetic energy levels the nuclei is kept in a magnetic field. Without the magnetic field the spin state of the nuclei are degenerate and so no energy level transitions are possible. The application of magnetic field brings about the separation of these levels and then radio frequency radiations can cause transitions between these energy levels.

When a sample is subjected simultaneously to two magnetic fields, one stationary and the other varying at some radio frequency then a different type of interaction between matter and electromagnetic radiations can be observed. At some particular combination of magnetic fields, energy is absorbed by the sample, and the absorbance can be observed as a change in signal developed by radio frequency detector and amplifier. This energy absorption can be related to the magnetic dipolar nature of the spinning nuclei. This technique is known as **Nuclear Magnetic Resonance (NMR).** It involves the magnetic energy of the nuclei when they are placed in magnetic field. A strong magnetic field causes the energies of certain nuclei to be split into two or more quantized levels, owing to the magnetic properties of these particles. The energy difference between magnetic quantum levels of some atomic nuclei correspond to radiation energy in the frequency range of 0.1 to 100 mHz (wavelengths between 3000 and 3m), which in the radio-frequency (rf) portion of the electromagnetic spectrum and this constitutes the most fundamental part of NMR spectroscopy.

The three molecular parameters on which NMR depends are:

(i) *shielding constant of the nuclei*

(ii) *coupling constant of the nuclei*

(iii) *life-time of energy levels.*

For an organic chemist the nuclear magnetic resonance, spectroscopy is more powerful tool that ultraviolet and Infrared spectroscopy to obtain structural information from the spectrum.

The NMR (or *pmr*) spectrum provides information about the chemical environement of hydrogen atoms, the number of hydrogen atoms in each different kind of environment and the structure of groups adjacent to each hydrogen atom. Simplest atomic nucleus is a proton. This spectrum is sometimes called **Proton Magnetic.**

Resonance (PMR) Spectrum

The value of NMR lies in the fact that combined hydrogens in different types of bonding situations will yield different radio-frequency absorption spectra. The change in absorption frequency with a change in the character of the hydrogen atom is the foundation of NMR spectroscopy structure elucidation. The procedure offers information regarding the detailed structure and configuration of organic molecules. It yields no information regarding elemental analysis. Micro samples may be used and, since the analysis is non-destructive, the samples may be recovered. Initially, NMR was a supplement to infrared and mass spectroscopy; however, it is now a more valuable tool for organic structure elucidation than either of the former. In practice all three are frequently used to support each other.

Its primary function is as a qaualitative tool; however, current developments and improvements in the instruments are greatly extending the quantitative capabilities.

Like ultraviolet, visible, and infrared spectroscopy, nuclear magnetic resonance depends upon the absorption of certain wavelengths or radiation. The radiation in this case is in the **radio-frequency, rf,** range of approximately 5×10^8 cycles/sec. Substitution of this value and the value of Planck's constant, $h = 6.6 \times 10^{-27}$ erg sec, into the relationship

$$E = hv$$

shows the energy to be approximately 3.3×10^{-18} erg. This is considerably less than that required for any of the previous absorption, excitation, or fluorescence procedures; however, it is adequate to cause a change in the angular momentum (**spin**) of some atomic nuclei as they absorb this radiation. The principle of nuclear magnetic resonance is based upon this interaction between radio-frequency radiation and nuclei. The primary applications are with respect to *rf* interactions with the ^1_1H nuclei. Our discussions will be exclusively limited to this species.

The first observations of NMR signals were observe independently by Purcelle at Harvard and Block at Stanford in 1945. The first application to study the structure of ethyl alcohol was made in 1951. Purcelle and Block were awarded Noble prize in physics in 1952 for their discovery.

The radiowaves have a frequency between 10^7 and 10^8 Hz. The energy associated with such waves is found to be between 10^{-18} to 6.6×10^{-19} ergs. So we find that the amount of energy involved in radio-frequency waves is very small. But even this small amount of energy can affect the nuclear spin of the atom of a molecules.

Principles of NMR Spectroscopy

Just like electrons, the nuclei of certain atoms behave as if they are spinning around their axes. When the particles in a nucleus do not have their spin paired, a resultant spin appears. The spinning nucleus produces an electric current and has a magnetic field associated with it and it behaves like a tiny bar magnet.

Only some nuclei have magnetic properties and so NMR spectrum is obtained only from such molecules which contain such nuclei. NMR spectroscopy is often concerned with nuclei with $I = \frac{1}{2}$ *e.g.*, ^1H, ^{31}P, ^{19}F etc. No spectra is obtained for $I = 0$. However, in special cases spectra can be obtained from nuclei where $I \geq 1$. The nuclei with odd mass numbers such a ^1H, ^{31}P, ^{19}F etc. and nuclei of odd atomic numbers and even mass numbers like ^2H, $^{10}B_5$, $^{14}N_7$ etc. show magnetic properties. Nuclei with even atomic number and even mass number such as $^{12}C_6$, $^{16}O_8$ etc. do not show magnetic properties.

PRINCIPLES

Properties of the Hydrogens Nucleus

The nuclei of atom possess electrical charges. As the nucleus is spinnings about its own axis the motion sets up a magnetic field and produces a **magnetic energy.** In addition, the nucleus has a momentum and **a mechanical energy.** Both the mechanical and magnetic energies must be taken into consideration so as to interpret and understand nmr.

The Application of a Magnetic Field

Imagine a hydrogen nucleus spinning around its own axis as in Fig. 7.1. When this nucleus is placed in an external magnetic field, the magnetic energy of the nucleus will cause its magnetic field to align itself with the external magnetic field. The external magnetic field exerts a torque upon the nuclei and they assume a permitted orientation with respect to this external field. The spin axis of the nucleus tilts with respect to the external field and processes in a circular path about the filed direction. This is similar to a spinning tilted top. The axis of the tilted top is not perpendicular to the earth's magnetic field.

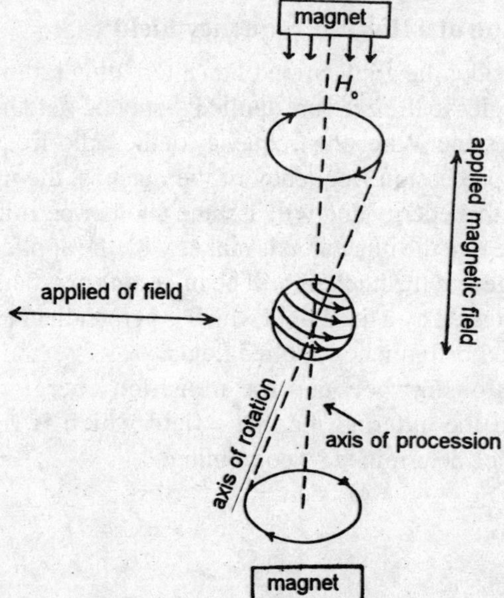

Fig. 7.1: Schematic diagram representing a spinning nucleus in a magnetic field.

When the outside magnetic field is applied to the magnetic field of the nucleus, a discrete set of orientations is possible; that is, the system is **quantized.** This number of possible, orientations is determined by I, the spin number of the nucleus. Any one of 2I+1 orientations is possible. Since $I = \frac{1}{2}$ for the $^{1}_{1}H$ nucleus, this nucleus will be able to assume only one of two, $[2(\frac{1}{2}) + 1]$, possible orientations, each of which corresponds to a permitted energy level. The more complex the nucleus with respect of spin possibilities the more energy levels are possible.

Whereas the number of possible orientations which a nucleus may assume is determined by the magnetic moment, μ.

The difference, ΔE, between the two quantum levels of the hydrogen nucleus depends upon the applied magnetic fields and the magnetic moment of the nucleus. It is expressed as

$$\Delta E = \frac{\mu H_0}{I}$$

where μ is the magnetic moment of the nucleus, H_0 is applied magnetic field, and I is the spin quantum number.

The Application of a Radio-Frequency Field

Now consider the application of an alternating radio-frequency field at right angles to the external applied magnetic field in which the nucleus is processing. When the frequency of the radio-frequency field is equal to the precession frequency of the nucleus, the nucleus will absorb or emit this energy and will assume another permitted energy level. The angle that the nuclear axis makes with the applied magnetic field will change and the nucleus will be **in resonance.** The change in frequency is detected by a third field which is perpendicular to both the magnetic and radio-frequency applied fields.

The relationship between the transition energy, the radio-frequency, and the external magnetic field which is required for resonance may be determined by combining.

$$\Delta E = h\nu$$

with the equation

$$\Delta E = \frac{\mu H_0}{I}$$

to yield

$$\nu = \frac{\mu H_0}{hI}.$$

Magnetic properties of proton. The hydrogen nucleus or proton may be considered as a positively charged spinning unit. Hence like any other rotating electric charge, it will produce a small magnetic field H, lying along its axis of spin. If this nucleus (Proton) is placed in an external magnetic field H_0 it will be aligned either *with* or against the external field (Fig. 7.2). The former position is more stable and is of lower energy. Energy must be absorb to "flip" the proton magnet to the less stable alignment against the field.

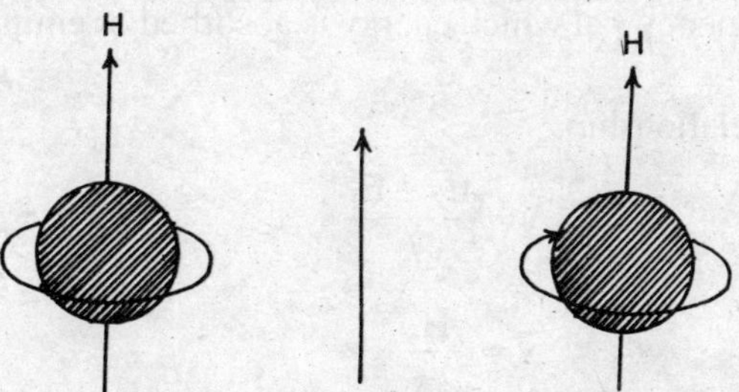

Fig. 7.2: Two spin States of the Hydrogen nucleus in a magnet filed, H_0

Theory of Nuclear Resonance

The amount of energy (ΔE) that is required to flip the proton over depends on the strength of the external field.

For any nucleus the possible values of spin angular momentum are given by the equation

$$\text{spin angular momentum} = [I(I+1)^{\frac{1}{2}}\frac{h}{2\pi}]$$

where I is the spin quantum number of the nucleus.

Since $\mu = \gamma \times$ spin angular momentum

so $$\mu = \gamma[I(I+1)^{\frac{1}{2}}\frac{h}{2\pi}]$$

Where μ is the magnetic moment of the nucleus and γ is the *gyromagnetic ratio.*

When a nucleus with magnetic moment is placed in a magnetic field H_0, the two energy levels become separate corresponding to $m_e = -\frac{1}{2}$ and $m_e = +\frac{1}{2}$. Thus corresponding energies E_1 and E_2 are given by

$$E_1 = -\tfrac{1}{2}[\frac{\gamma h}{2\pi}]h_0$$

$$E_2 = +\tfrac{1}{2}[\frac{\gamma h}{2\pi}]h_0$$

$E_1 < E_2$.

The nucleus absorbs energy while going from E_1 to E_2 and emits energy in coming from E_2 to E_1. The difference in energy between the two states

$$\Delta E \text{ is equal to } E_2 - E_1$$

so $$\Delta E = E_2 - E_1$$

The frequency γ at which energy is absorbed or emitted is given by

Bohr's relationship.

$$\nu = \frac{E_2 - E_1}{h}$$

or $$\text{v} = \frac{\gamma H_0}{2\pi}$$

The stronger the field, the greater the tendency to remain lined up with it and the higher the frequency ($\Delta E = h\nu$) of the radiation needed to do the job.

$$\nu = \frac{\gamma H_0}{2\pi} \qquad = \Delta E = h\nu,$$

Where ν = Frequency h = Planck's constant

H_0 = Strength of applied magnetic field in gauss.

γ = a nuclear constant, the **gyro-magnetic ratio,** 26, 750 for the proton.

From this it is clear that higher the applies field strength, higher is the frequency of radiation needed to flip to proton to a less stable state.

In a field of 14, 092 gauss, the energy needed corresponds to electromagnetic radiation of frequency 60 million cycles or mega cycles (Mc) per second. (1 mega cycle = 10^6 cycles). The radiation lies in the radio-frequency range and is of much lower energy than even infrared light.

In principle, we could place a substance in a magnetic field of constant strength and then obtain a spectrum in the same way we obtain infrared or an ultra-violet spectrum, *i.e.,* pass a radiation of steadily changing frequency through the substance and observe the frequency at which radiation is absorbed. In practice, however, it has been found more convenient to keep the radiation frequency constant and to vary the strength of the magnetic field. At some value of the field strength, the energy needed to flip the proton matches the energy of the radiation, absorption takes place and a signal is observed. Such a spectrum is known as a *nuclear magnetic resonance (NMR) spectrum* (Fig. 7.3). Please note that the frequency at which absorption occurs depends on the magnetic field which the proton actually *feels.* The *effective* magnetic field is not exactly the same as the *applied* field strength. The effective field strength of each proton depends upon the environment of that proton, such as *electron density* of the proton, and the *presence of other nearby protons.* Each proton of each set of equivalent protons will have a slightly different environment from every other set of protons and will need a slightly *different applied* field strength to produce the same effective field strength at which absorption occurs.

At a given radio frequency then, *all protons absorb at the same effective field strength but they obserb at different applied field strengths.* It is thus applied field strength that is measured and against which the absorption is plotted field strength that is measured and against which the absorption is plotted. The result is a spectrum showing many absorption peaks whose relative positions (due to different environments of protons) can given almost detailed information about molecular structure.

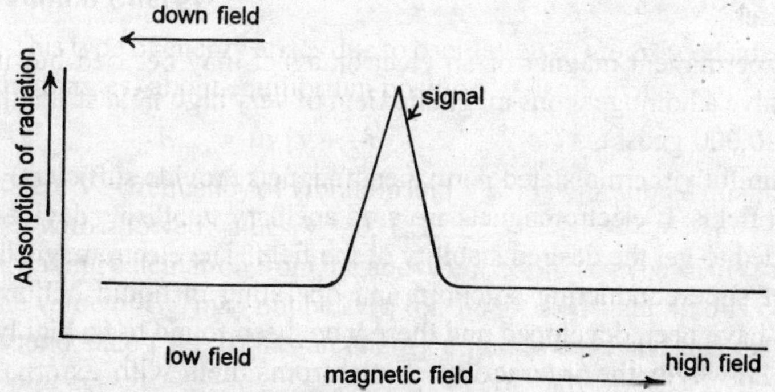

Fig. 7.3: The NMR spectrum

Apparatus used in NMR Spectrum

An idea of the apparatus used in NMR spectra can be had from

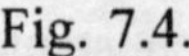

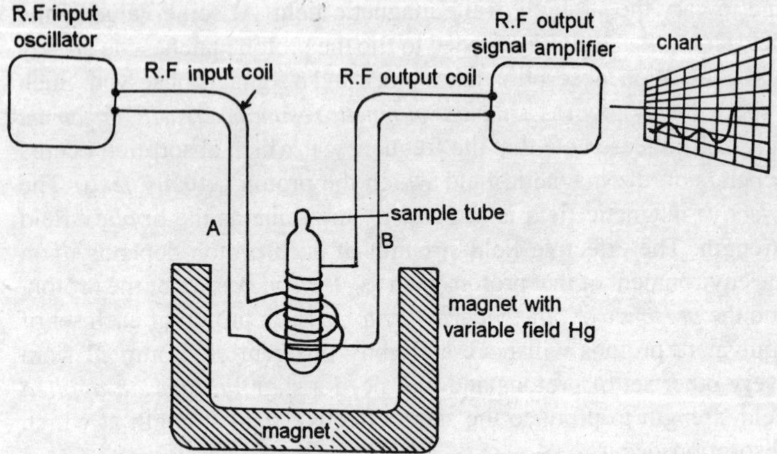

Fig. 7.4: Nuclear Magnetic Resonance Sepctrometer

The apparatus consists of the following parts:

1. Sample Tube

It is made of a chemically inert material which is durable and also transparent of γf radiations. Glass tubes are generally used because these are cheap and satisfy the above requirements. Tubes generally used are about 7 cm in length and 0.3 cm in diameter.

2. Magnet

A permanent magnet or an electromagnet may be used but it should give a homogeneous magnetic field of very high field strength (about 20,000 gauss).

Carefully thermo stated permanent magnets provide sufficiently constant fields. If electromagnets are used ancillary stabilising devices are needed to get the desired stability of the field. The electromagnets made of superconducting solenoid and operating in liquid helium crystals have been developed and these have been found to be highly stable. However the interaction of electromagnets with external magnetic fields is much less in comparison to permanent magnets.

The spectrometers operating upto 100 MHz generally use permanent magnets or conventional electromagnets. Super-conducting magnets can realise fields that correspond to 230 MHz.

3. Sweep Generator

To make a nucleus resonate, the procession frequency of nucleus must become the same as that of applied *rf* radiation. When H_o is kept constant, the precession frequency is fixed. To bring about resonance of nucleus the frequency of *rf* field has to be changed so that it becomes equal to the resonance frequency (frequency sweep method). When *rf* radiation is kept constant the resonance frequency of nucleus has to varied by changing H_0 (field sweep method). In most cases field sweep method is used.

4. Radio Frequency Generator

Radio frequency oscillator is used to produce radio frequency radiations.

For maximum interaction of *rf* radiation with the sample, the coil of oscillator is wound around the sample tube. The oscillator irradiates the sample with *rf* radiations. The oscillator coil is wound perpendicular to the applied magnetic field. Such an arrangement does not allow *rf* field to affect the effective magnetic field in the process of irradiation.

5. Radio Frequency Receiver

Radio frequency radiations while passing through the magnetised sample undergo absorption and dispersion. The resonance frequency can be determined by observing either of the two, but it is easier to interpret an absorption spectrum than a dispersion spectrum.

Following methods are used to detect the signals:

(a) The first method makes use of radio frequency bridge. Its network balances out transmitter signal and allows the absorption and dispersion signals to appear as an out of *e.m.f.* across the bridge. The coil used for surrounding the sample tube serves both as a transmitter and receiver coil in this method.

(b) The second method makes use of a separate receiver coil. This is known as *crossed coil method* or *nuclear induction method.* When two coils are fixed at right angles to each other as well as to the direction of static magnetic field they will not be effectively coupled. This separates the transmitter signal from the absorption and dispersion signals.

6. Read out system

Since the absorption signal received from radio frequency receiver is very weak it has to be considerably amplified before feeding it to a chart recorder.

THE ABSORPTION SPECTRUM

The $^{1}_{1}$H nuclei may possess only one of two possible energy states. The transition from the low to high energy corresponds to the absorption process, while that from the high to the low to an emission. If an equal population of the nuclei existed at the two energy levels, the energy lost by emission would equal the energy gained by absorption and nothing could be measured. This is not the case. A very slight (approximately 0.0001%) excess of nuclei exists in the lower energy spin state. It is this slight excess upon which nuclear magnetic resonance depends.

Width of Absorption Peaks

The width of the absorption lines in an NMR spectrum is determined by the magnetic field strength and the length of time that the excited nucleus remains in this excited state. A narrow absorption line is desirable, otherwise resolution and fine structure detail are lost.

At first glance it seems that the very slight excess of lower energy spin state nuclei would be used up as the absorption process continues and that a decrease in the intensity or even a complete loss of absorption would take place as the nuclei population in the lower energy state becomes equal to that of the higher energy state. However, after absorption of radiation the nucleus remains excited for only a limited period of time and returns to its stable energy level.

The width of an absorption line is inversely proportional to the average time, the **relaxation time,** the nuclei remain in the excited state. The excited nuclei may lose their energy, **relax**, and return to the ground state by one of several processes:

(a) In **transverse relaxation** the excited nucleus gives its energy to an unexcited nucleus of a similar neighbouring molecule. This neighbouring nucleus, in turn, becomes excited. There is no net change in the energy of the system, but the relaxation time of any single nucleus is decreased. This process does not assist in maintaining the required excess of lower spin state nuclei. Since transverse relaxation time is quite short for solid samples and brad absorption lines are obtained, *solid samples are rearely utilized in NMR spectroscopy.*

(b) In **longitudinal relaxation** the energy of the excited nuclei is changed to heat instead of to other unexcited nuclei. As a result the temperature of the system rises slightly, previously excited nuclei become unexcited, the population of unexcited molecules increases, and more molecules become available for the absorption process.

Since the applied field must be constant over the entire sample, it is necessary that the samples be homogeneous and the sample holders be consistent throughout their entire length. Any change in either will lead to a variation of absorption frequency and a broadening of the lines. Other factors, like the presence of paramagnetic molecules or ions in the sample, can also lead to line broadening.

Number of Major Absorption Peaks

The number of major absorption peaks in an NMR spectrum is equal to the number of *different types* of protons which exist in the molecular compound. The protons within the molecule which possess the same magnetic environment are said to be **equivalent** and will absorb at the same applied field strength. Protons with different magnetic environments will absorb at different applied field strengths and are **non-equivalent.**

The equivalency or non-equivalency of protons in any given molecule is determined by the type and character of its bonding to adjacent atoms and in some cases to a group of atoms. Some general considerations are given below:

(a) Hydrogens bonded to different atoms are non-equivalent. For example, in methyl alcohol, CH_3OH, the hydrogens bonded to the carbon are not equivalent to that on which is bonded to the oxygen and two major absorption peaks will be obtained. Because of the tetrahedral nature of the carbon atom, the three hydrogens in the methyl group are equivalent to each other and appear as a single major absorption peak in an nmr spectrum. The other major absorption peak arises from the OH bonding (Fig. 7.5).

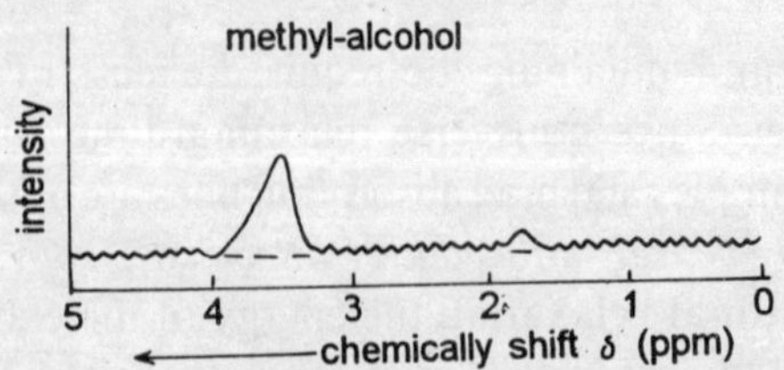

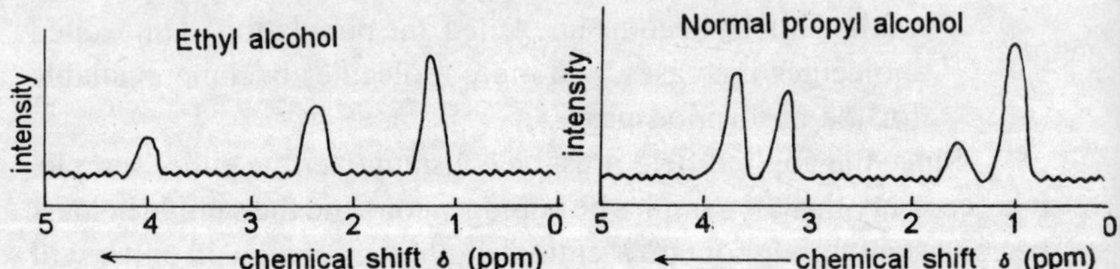

Fig. 7.5: Low resolution NMR spectra of methyl, ethyl, and normal propyl alcohol

(b) The two hydrogens in a CH_2 group are equivalent to each other. Three major absorption peaks would be expected from CH_3—CH_2—OH; one from the CH_3 group, one from the CH_2 group, and one from the OH group.

(c) At times it is essential to consider not only the atom to which the hydrogen is bonded but also any bonds which are once or twice removed from this primary bond. For example, the bonding of the CH_3 hydrogens in a CH_3—CH_2—group will not be equivalent to the bonding of the CH_3 hydrogens in a CH_3—O—group. The compound methyl ethyl ether, CH_3—O—CH_2—CH_3 will show three major absorption bands—one for the hydrogens of the methyl group with the CH_2—O bond, one for hydrogens of the methyl group with the CH_3—CH_2—bond, and one from the hydrogens of the CH_2 group. On the other hand, diethyl ether, CH_3—CH_2—O—CH_2—CH_3, shows only two major bands.

(d) Bonding to atoms which are twice removed must also be considered. For example, all of the groups of hydrogen atoms in CH_3—CH_2—CH_2—OH are non-equivalent and four major absorption peaks will be observed.

(e) In the benzene molecule all six hydrogens are equivalent since the bonding of all six to the carbon atoms is identical. Only one peak will be observed. A substitution on one or more of the positions of the benzene ring will upset this equivalency—the extent of change depends upon the number and/or nature of the substituted group(s).

(f) The presence of double or triple bonds may cause a non-equivalency.

Area Under Absorption Peaks

The area under an NMR absorption peak is directly proportional to the number of equivalent protons which are causing the peak.

Area determinations may be done either by making use of a planimeter or by counting the number of squares on the graph paper. These procedures are, however, generally unnecessary with the present NMR instruments. Instead, the instruments are equipped with an **electronic integrator.** The integral curve is recorded on the vertical axis of the recorder chart simultaneously with the absorption curves. The relative areas of the peak are determined by the heights of the integral curves at the respective frequencies. These data are then reduced to a whole number ratio and may be used for quantitative determination or for the elucidation of structure (Fig. 7.6).

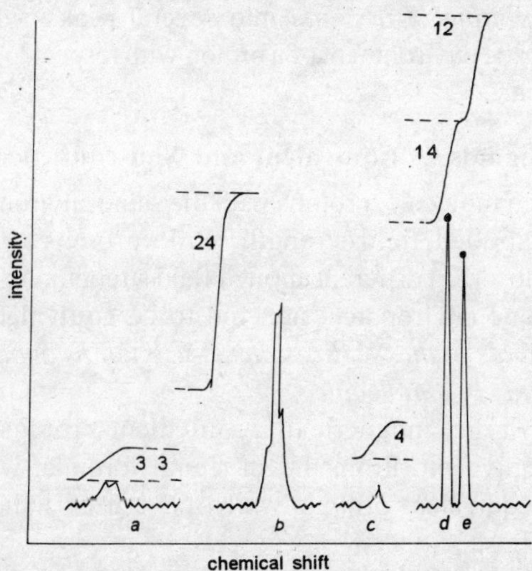

Fig. 7.6: Schematic diagram of NMR absorption curve fractions showing the use of an integral recorder. The heights of the integral curves show that the ratios of the areas of *a* : *b* : *c* : *d* : *e* are 3.3 : 24 : 4 : 14 : 12.

Example 1

Consider a spectrum of ethyl alcohol

$$\begin{array}{ccc} \text{H} & \text{H} & \\ \text{HC} - & \text{C} - & \text{OH} \\ \text{H} & \text{H} & \\ a & b & c \end{array}$$

in which the height of the integration curves shows a ratio of $a : b : c = 7.71 : 5.24 : 2.58$. This reduces to a 3 : 2 : 1 ratio.

Study of NMR Spectra

The following aspects of NMR spectrum will be studied in following discussion:

(a) *The number of signals,* These give us an idea about the number of different 'kinds' of protons present in a molecule.

(b) *the positions of the signals,* which tells us something about the electronic environment of each kind of proton.

(c) *the intensities of the signals,* which tell us how many protons of each kind.

(d) *the splitting of a signal* into several peaks, which tells us about the environment of a proton with respect to other nearby protons.

Number of Signals — Equivalent and Non-equivalent Protons

In a given molecule, protons with the same environment absorb at the same applied field strength while protons with different environment absorb at different applied field strength. A set of protons having the same environment are said to be **equivalent.** Thus the *number of signals in the NMR spectrum will tell us about "kinds" of protons presents in a molecule.*

By and large, magnetically equivalent protons are simply chemically equivalent. From the structural formula, we can easily pick out the equivalent protons which have been indicated by the same letter. Thus:

(i) CH_3—CH_2—Cl : It has two different kinds
a *b* of protons *'a'* and *'b'* and
Ethyl chloride we get NMR signals.

(ii) $CH_3—CHCl—CH_3$: It has also two non-equivalent protons and gives 2 NMR signals.
a *b* *c*
Iso-propyl chloride

(iii) $CH_3—CH_2—CH_2—Cl$: It has three non-equivalent protons and gives 3 NMR signals.
a *b* *c*
n-Propyl chloride

(iv) $CH_3—CBr_2—CH_3$: All the protons are equivalent *i.e.,* it has one kind of protons and we get one NMR signal.

2. 2-Dibromopropane

Knowing that chemically equivalent protons must be stereochemically equivalent, we can analyse the following formulae too.

(a) CH_3, (a) CH_3 > C=C < H (b), H (b) ; Gives 2 NMR signals

Iso-butylene

(a) CH_3, Br > C=C < H (b), H (c) ; Gives 3 NMR signals

2-Bromorpropene

H, H > C=C < H, Cl ; Gives 3 NMR signals

Vinyl chloride

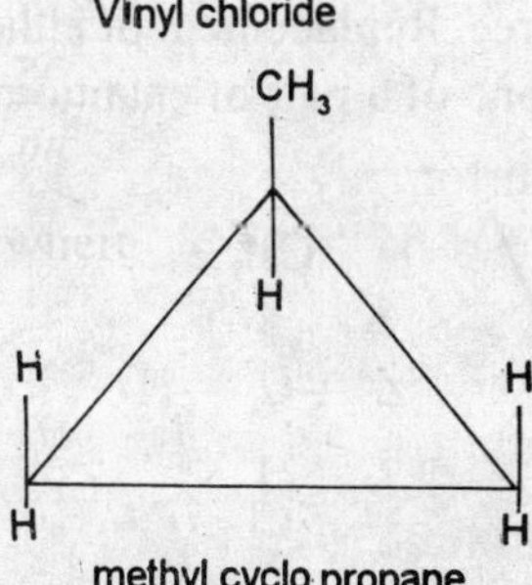

methyl cyclo propane

Gives 4 NMR signals

1, 2-Dichloropropane (optically active or inactive) give four NMR signals. It will be clear from the models or stereochemical formulae that this should be so.

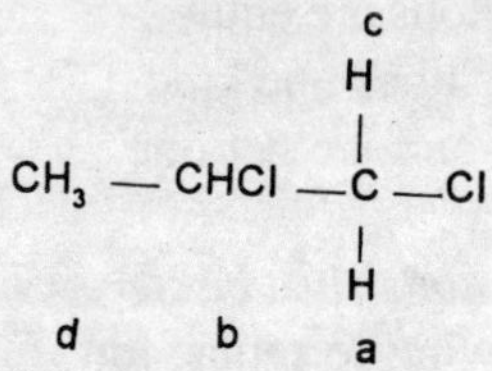

```
                 c
                 H
                 |
CH3 — CHCl — C — Cl
                 |
                 H
d        b       a
```

1,2- Dichloropropane

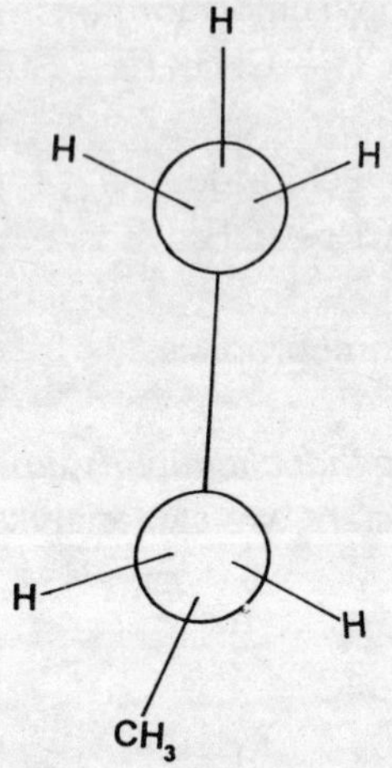

H H H H H CH_3

The equivalence of protons can be determined by the following isomer number method.

We can decide about the equivalence of protons by the process of replacement by some other atom, say Z. If on replacing either of two protons by Z, we get either the *same product* or the *enantiomeric* products (mirror images) then the two protons are chemically and magnetically equivalent, otherwise not. Thus:

(i) In ethyl chloride, for example, replacement of a methyl proton would give $CH_2Z—CH_2Cl$ while replacement of a methylene proton would give $CH_3—CHZCl$. These are two different products and methyl protons are thus non-equivalent to methylene protons. The product $CH_2Z—CH_2Cl$ is the same regardless of *which one* of the three methyl protons is replaced. The environment of three protons is the same and we expect one NMR signal for all the three. Replacement of either of the two methylene protons would give one of a pair of enantiomers:

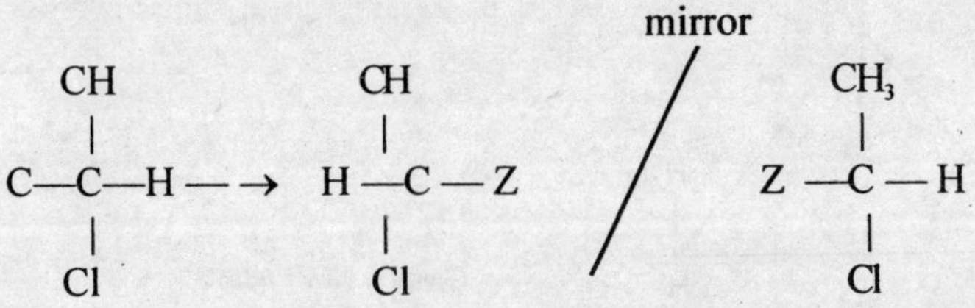

```
                                mirror
 CH                 CH            /          CH3
  |                  |           /            |
C—C—H  ——→     H — C — Z        /       Z — C — H
  |                  |         /              |
 Cl                 Cl        /              Cl
```

These two enantiotropic protons are equivalent and we get one NMR signal for the par.

Such pairs of protons are known as **enantiotropic protons**. The environments of these two protons are mirror images of each other; these protons are equivalent and there is only on NMR signal for the pair.

(ii) In 2-bromopropene, the replacement of either of the winylic protons gives a pair of diastereomers (geometric isomers).

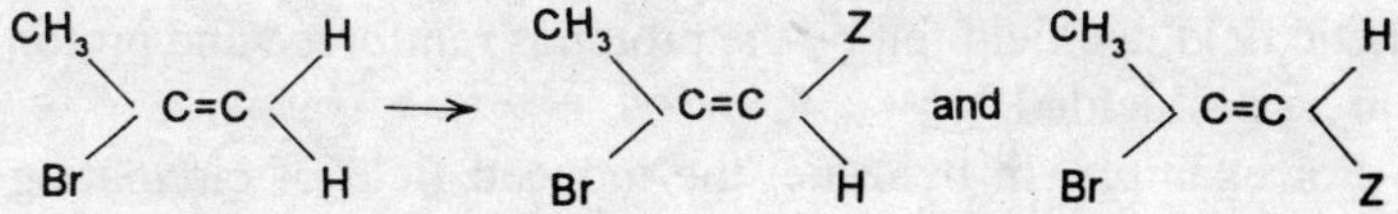

$$\begin{matrix} CH_3 & & H \\ & C{=}C & \\ Br & & H \end{matrix} \longrightarrow \begin{matrix} CH_3 & & Z \\ & C{=}C & \\ Br & & H \end{matrix} \quad \text{and} \quad \begin{matrix} CH_3 & & H \\ & C{=}C & \\ Br & & Z \end{matrix}$$

The environments of these protons are neither identical nor mirror images of each other.

These protons called **Diastereotropic protons** are non-equivalent and an NMR signal is expected from each one.

(iii) In 1, 2-dichloropropane the two protons on C—1 are diastereotropic, non equivalent and so give separate, NMR signals.

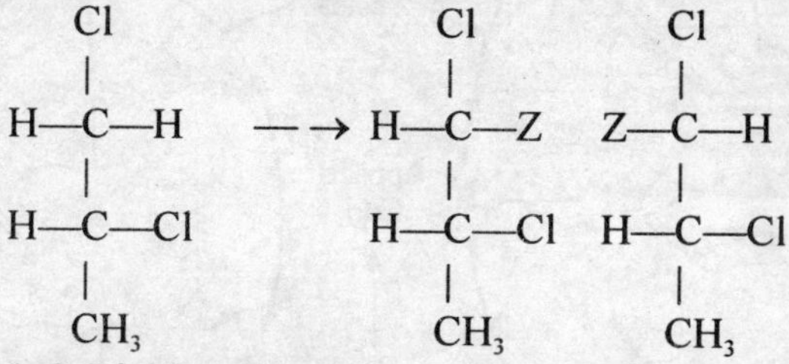

$$\begin{matrix} & Cl & \\ & | & \\ H— & C & —H \\ & | & \\ H— & C & —Cl \\ & | & \\ & CH_3 & \end{matrix} \longrightarrow \begin{matrix} & Cl & \\ & | & \\ H— & C & —Z \\ & | & \\ H— & C & —Cl \\ & | & \\ & CH_3 & \end{matrix} \quad \begin{matrix} & Cl & \\ & | & \\ Z— & C & —H \\ & | & \\ H— & C & —Cl \\ & | & \\ & CH_3 & \end{matrix}$$

1, 2-Dichloro Propane

Position of Signals—Chemical Shift

We have discussed above that the number of signals in NMR spectrum provides us information about kinds of protons a molecule has. Similarly the position helps us to know the kinds of protons, whether they are aromatic, aliphatic, primary, secondary, tertiary, benylic, vinylic, acetylenic, adjacent to halogen or to other atoms or groups.

When a molecule is placed in a magnetic field, its electrons circulate and produce magnetic fields known as **Induced magnetic fields.**

These induced magnetic fields of electrons at the proton get oriented in such a way so as to oppose the applied magnetic field. The field felt by the proton is diminished in this way and the proton gets **shielded.**

In the case of the induced magnetic field of circulating π electrons *about nearby nuclei,* these fields may oppose or reinforce the applied magnetic fields on the protons, depending upon the location of the proton. If the induced field opposes the applied field the proton is **shielded** as before and if the induced field reinforces the applied magnetic field, the field felt by the proton is reinforced and proton is said to be **deshielded.**

For example, in benzene, the induced field of circulating π electrons (ring currents) reinforces applied field at the aromatic protons and these protons get **deshielded.** In the case of acetylene, the induced field of π electrons opposes applied field at the acetylenic protons and these protons thus are **shielded** (Fig. 7.7).

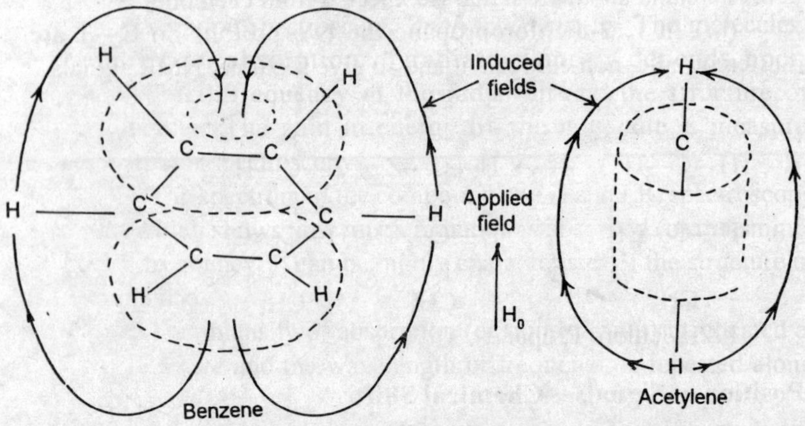

Fig. 7.7: Deshielding and the Shielding in benzene and acetylene

In the case of methyl chloride, the chlorine atom, being strongly electronegative, withdraws electrons in the vicinity of the protons (inductive effect). This lowers the electron density, in the vicinity of the proton, decreases the magnetic field and hence causes **deshielding.**

Compared with a naked proton, the shielded proton will, therefore need a higher or stronger applied field to give the particular effective field for absorption to occur to produce an absorption signal or in other words it absorbs **up field.** The deshielded proton needs only a smaller applied field, *i.e.*, lower field applied to give an absorption signal, *i.e.*, it absorbs **down field.** Shielding shifts the absorption to the right (up field) and deshielding shifts the absorption to the left (down field).

Such shifts in the position of NMR absorptions due to shielding and deshielding of protons by induced magnetic field of electrons are known as **Chemical shifts.**

Measurement of Chemical Shifts

Chemical shifts can be measured with reference to PMR or NMR signal of some, standard compound. The reference or the standard compound used is tetramethyl-silane (TMS) $(CH_3)_4Si$. A small quantity of TMS is added to the sample while getting its NMR spectrum.

Why TMS is Used as a Standard Compound?

TMS is used as a reference compound because of the following reasons:

(i) *It has 12 equivalent protons $(CH_3)_4Si$ which given an intense signal. Therefore even a small quantity of it gives a measurable signal.*

(ii) *Silicon is less electronegative than carbon, protons of TMS are, therefore, highly shielded. They give a NMR signal in a region where few other protons absorb. Hence, signal of the reference compound seldom interferes with the signals from other protons.*

(iii) *It is inert and volatile. It can, therefore, be easily removed by evaporation after the experiment is over.* Thus, advantage of using TMS as reference is that it gives a strong and sharp NMR signal in a region where only a very few kinds of other protons absorb.

Scale of Measuring Chemical Shifts

Chemical shifts are measured in Hertz (cycles per second, *cps*),on the assumption that the frequency of electromagnetic radiation is being changed whereas is actual practice magnetic field is changed. The two, however, being mathematically related, this amounts to the same thing.

$$(V = \frac{\lambda H0}{2\pi})$$

The chemical shift is generally expressed as the frequency difference (Δv) in cps between the absorption frequencies of the sample and the reference compound *i.e.,* TMS.

Magnitude of chemical shifts in Hertz *i.e.,* cycles per second (*cps*) is proportional to the strength of the applied magnetic field or, proportional to the radio frequency used in the spectrophotometer. For example, when the radio frequency used is 60 mega cycles per second (*i.e.,* 60 M*cps*) the signals will be 100 cps apart whereas the signals obtain will be 166.6 cps apart when radio frequency used is 100 M*cps*.

In view of the fact that different NMR, spectrometers use different magnetic field (usually 40, 60 or 100 M*cps*), it has been thought desirable that chemical shifts be expressed in the form which is independent of the strength (radio frequency) of the applied field (1 M*cps* = 1×10^6 *cps*).

For this, these are expressed as a fraction of the total field *i.e.,*

$$\left(\frac{\text{Shift in } cps}{\text{Radio frequency of the applied field in } cps}\right)$$

multiplied by 10^6 and reported as **parts per million** (*ppm*).

For example an NMR signal comes at 100 *cps* down field with reference to TMS in a spectrophotometer using 60 M*cps*.

$$\text{Chemical shift in } ppm = \frac{100 \times 10^6}{60 \times 10^6} = 1.66$$

In another instrument using 100 Mcps, the same signal will be 166.6 *cps* down field.

$$\text{Its chemical shifts in } ppm = \frac{166.6 \times 10^6}{100 \times 10^6} = 1.66$$

This shows that *ppm* units are dimensionless and independent of the radio frequency of the magnetic field employed in the NMR spectrometer. Further the numerator is multiplied by 10^6 only to avoid fractional values.

Two scale employed to express chemical shifts in terms of ppm are,

(i) The δ (delta) scale, wherein

$$\delta = \frac{\Delta v \text{ (in } cps)}{\text{Radio frequency used in Mcps}} \times 10^6$$

(ii) The τ **taw scale** where in

$$\tau = 10 - \delta$$

on the δ scale, position of TMS signal is taken as 0.0 *ppm* whereas on τ scale it is assigned the value 10.0 *ppm.* Most of the chemical shifts on the δ scale have value between 0 to –10 and on τ scale between 0 and 10.

A small numerical value of δ means a small down field-shift whereas a large value means a large down field shift.

On the τ scale, a small value of τ means a low field absorption and a high value means a high field absorption.

Position of NMR Signal and Effect of Electronic Environment on Chemical Shift

The chemical shift for a proton depends upon the electronic environment of a proton. Proton with the same environments (equivalent protons) have the same chemical shift. Protons which are identical are of two kinds:

(1) All hydrogens on a single carbon are identical if

(a) They are the two protons of a methyl, —CH_3

(b) They are the two protons of a methylene, —CH_3— which is free to rotate so that they **become equivalent.**

(2) Hydrogens on different carbons yield the same absorption signal if they are structurally indistinguishable such as marked *a* or *b.*

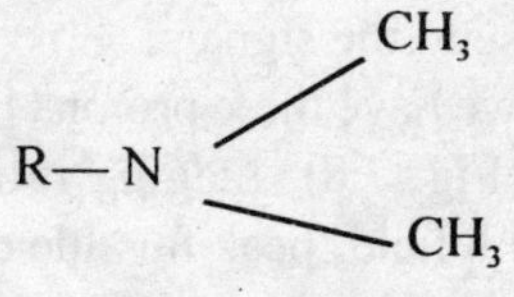

R—N(CH_3)(CH_3)

One Signal

b CH_3, a H, H a, H, H, b CH_3

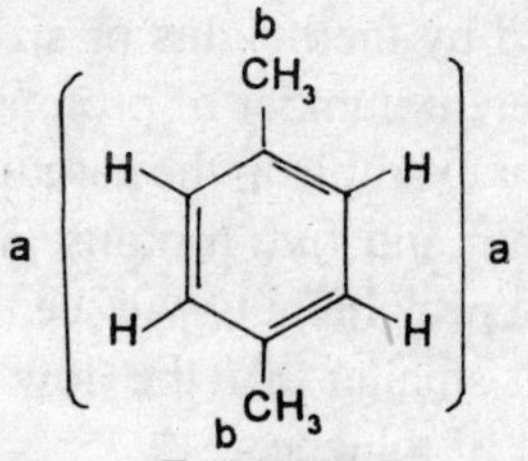

Two Signal

It has also been found that a proton with a particular environment shows almost the same chemical shift. For instance in the case of hydrogens of the primary, secondary and the tertiary types the absorption occurs at about the following value :

RCH_3 τ 9.1	[δ 0.9]
R_2CH_2 τ 8.7	[δ 1.3]
R_3CH τ 8.5	[δ 1.5]

All these protons, in turn, differ a lot from aromatic protons which absorb far down field.

$$Ar—H\ \tau\ 1.6—4\ (\delta\ 6—8.5)$$

Attachment of chlorine to the carbon being the proton produces a down field shift. If the chlorine is attached to the carbon once removed from the carbon bearing the proton, there is again a downward shift but this time much weaker.

CH_3—Cl δ 3.0 CH_3—C—Cl τ 1.5

R—CH_2—Cl δ 3.4 R—CH_2—C—Cl δ 1.7

R_2CH—Cl δ 4.0 R_2CH—C—Cl δ 1.6

This can be explained on the basis of inductive effect. The chlorine causes a greater down field shift. Other halogens show similar effects. The inductive effect (*i.e.*, a downward shift) depends on the halogen atoms. More the number of halogens, more the downward shift.

Sometimes the τ or δ values given for each type of proton in tables show slight variations in different NMR spectrum. This is because of the fact that chemical shift of a proton depends on (i) the over-all structure of the molecule (ii) solvent and (iii) temp. (iv) concentration used while recording the NMR spectrum.

NMR, Peak Area and Proton Counting

If in a NMR spectrum, we observe the intensities of signals, indicated by the heights or sizes of the peaks, they are different for the different number of protons giving rise to the signals.

For example in the case of toluene, we have three protons in the side chain and five protons in the ring (Fig. 7.8). Judging roughly from the peak heights, we see that the (high field) peak for side chain proton is smaller than the (low field) peak for the aromatic protons in the case of toluene.

More exact comparison based on the *areas under the peaks* shows that peaks for side-chain and aromatic protons have sizes in the ratio 3 : 5 for toluene. This illustrates a general quality for all NMR spectra.

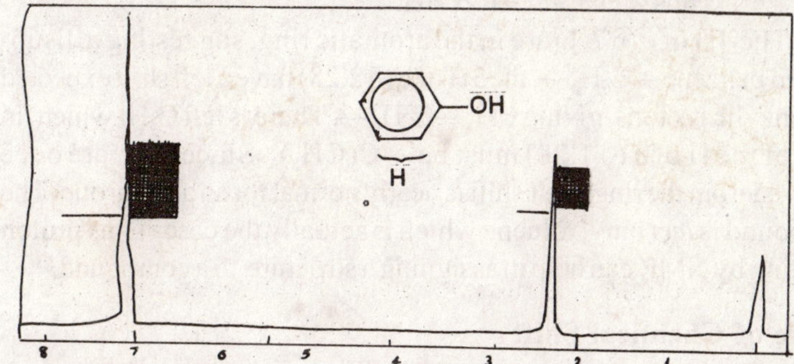

Fig. 7.8: NMR spectrum : Chemical shift in Toluene

The areas under the NMR signals is *directly proportional to the number of protons giving rise to the signal.*

The areas under the NMR signals are measured by an electronic integrator and are usually given on the spectrum chart in the form of a stepped curve : the heights of steps are proportional to peak areas. NMR chart paper is cross-hatched and we can easily estimate step heights by simply counting squares. The number of squares are in the same ratio as the numbers of different kinds of protons. We convert this set of number into a set of smallest whole number. The number of protons giving rise to each is equal to the whole number for that signal or to some multiple of it. Fig. 7.9 explains this point.

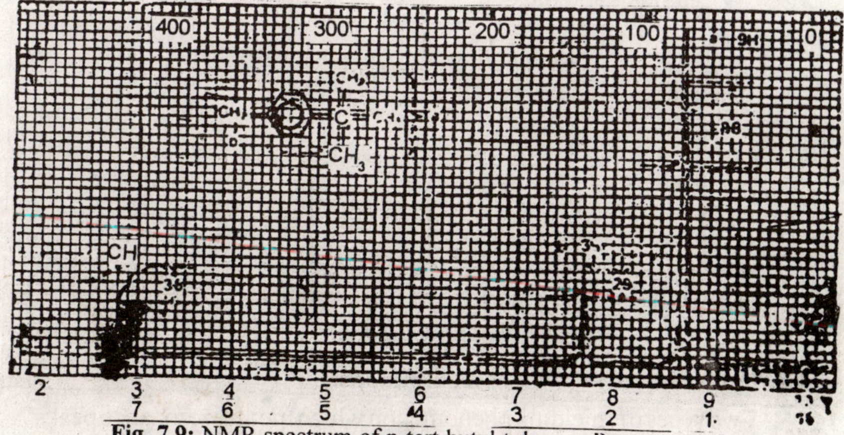

Fig. 7.9: NMR spectrum of p-tert butyl toluence Proton counting

The ratio of the heights $a : b : c$ is

8.8 : 2.9 : 3.8 = 3.0 : 1.0 : 1.3 = 9.0 : 3.0 : 3.9

Thus a, 9H; *b*, 3H; *c*, 4H.

The 4H of *c* (δ 7.1) are in the aromatic ring, suggesting a disubstituted benzene—C_6H_4—, the 3H of b (δ 2.28) have a left shift expected for benzylic protons, giving CH_3—C_6H_4—. There is left C_4H_9 which, in view of the H of *a* (δ 1.28) must be —$C(CH_3)_3$. Since they are once removed from the ring, their shift is nearly normal for an alkyl group. The compound is *tert* butyl toluene which is actually the case. Thus proton counting by NMR can help in assigning a structure to a compound.

Cause of Chemical Shift

The bands in NMR spectrum are observed at different chemical shifts δ or τ values. The cause of this difference is the different magnetic field experienced by different nuclei as a result of secondary magnetic fields associated with the molecules. These secondary fields arise due to applied magnetic field H_0. The secondary field affects as under:

(i) When the secondary field opposes the applied field then the effective field experienced by the nucleus is less than the applied field. In such a case called **positive shielding** the resonance position moves upfield in the NMR spectrum.

(ii) When the secondary field reinforces the applied field then the effective field experience by the nucleus is more than the applied field. In this case called **negative shielding** the resonance position moves down field.

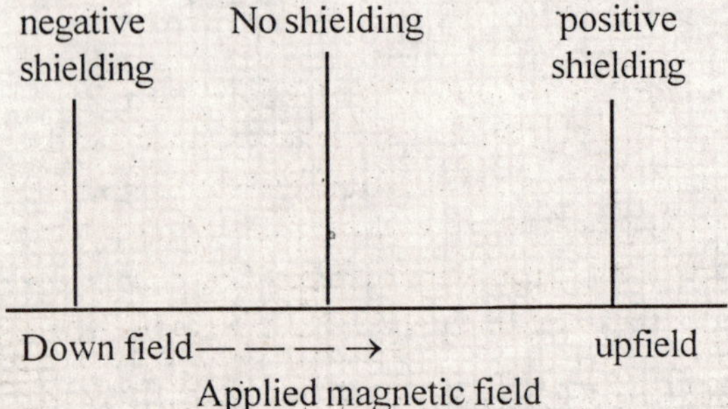

Fig. 7.9 (a) Effect of shielding on resonance peaks

Two types of shielding phenomenon which modify resonance peaks in NMR spectrum are (i) Local shielding and (ii) Low range shielding.

(i) Local Shielding

Sometimes the field experienced by a nucleus may be modified by fields due to induced circulation of electrons localised on that nucleus. This is termed local shielding. Its extent depends upon electron density in the immediate vicinity of protons. Two order of magnitude of δ can be predicted if the electron density is known.

	CH_3—H	CH_2—I	CH_3—Br	CH_3—Cl	CH_4—F
δ (ppm)	0.2	2.2	2.7	3.6	4.3

So we find that is electronegativity increases value of δ also increases.

It is shown in Fig. 7.10.

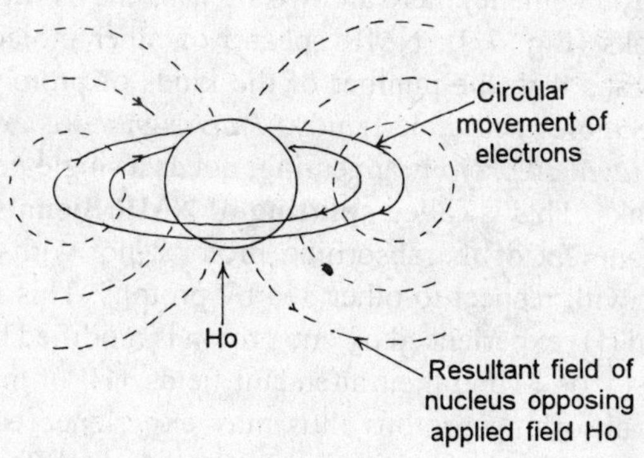

Fig. 7.10: Local shielding

(ii) Low Range Shielding

In aromatic compounds the secondary fields set up the induced circulation of π-electron often influence the fields experienced by nuclei not directly associated with the π-electrons.

Because of this, additional shielding and deshielding effects may be included within the molecule.

The protons on an acetylenic linkage lie in a shielded region but those on a carbonyl group or ethylenic linkage lie in a deshielded region. This explains the observation that acetylic protons are resonating at relatively high field (7—8τ) as compared with protons on a carbonyl group (0—0.7τ) and ethylenic protons (3.5 to 5.5 τ).

The long range shielding effects are more pronounced in unsaturated groups than saturated groups.

Splitting of Signals—Spin-spin Coupling

We have studied earlier that an NMR spectrum indicates a signal for each type of proton in a molecule, *i.e.,* the number of signals is equal to the number of kinds of the protons present in the molecule. This is true only in few cases such as in *p*-xylene. In this case there are two signal consisting of single peak as there are two types of protons in the molecule—(aromatic and methyl).

In actual practice most spectra appear to be much more complicated. For example in the NMR spectrum of $CHBr_2$—CH_2Br (1, 1, 2—Tribromoethane) there are two signals, one of which is split into three peaks (Fig. 7.9). NMR spectra of other molecules also show more peaks than the number of the kinds of protons that are present in the molecules under study. *Thus signal we expect from each set of equivalent protons* appearing, not as a single peak, but as a group of peaks. This is called **splitting of NMR Signals,** splitting reflects environment of the absorbing protons; not with respect to electrons, but with respect to other nearby protons. This is because the actual field (H) experienced by one proton is modified by the spin arrangements (↑ or ↓) and their attendant fields $\pm H^1$ of the adjacent protons. The absorbing protons thus may experience each of the modified fields and its absorption singal get split into different peaks. This is called **spin-spin coupling.**

Thus signal splitting is caused by the magnetic influences of hydrogen's adjacent atoms. For example in 1,1, 2—tribromoethane CH_2Br—$CHBr_2$, Fig. 7.9 signal from two equivalent protons of —

a b

CH_2Br b group is split into two peaks (a doublet) under the magnetic influence of the non-equivalent proton of the neighbouring hydrogen in the $CHBr_2$ group. Similarly signal from hydrogen—$CHBr_2$ group is split into three peaks (a triplet) under the magnetic influences of the two equivalent protons of the neighbouring —NH_2Br group. This is due to **spin-spin coupling** of the absorbing and neighbouring protons.

Spin-spin splitting is observed only when non-equivalent protons is present in the neighbourhood. Thus no signal splitting is observed in the signal from six equivalent protons of ethane.

CH_3—CH_3....No signal splitting

Taking the case of two adjacent carbon atoms, carrying respectively, two secondary protons and a tertiary proton.

t s	(t=tertiary)
—CH—CH_2	(s=secondary)

Considering first the absorption by one of the secondary protons. The effective magnetic field felt by a secondary proton at a particular instant is modified by the spin of the tertiary proton. It may be slightly increased if the tertiary proton happens at that instant aligned with the applied field (↑ ↑) or it may be decreased if the tertiary proton happens to be aligned against the applied field (↑ ↓). Now it can be assumed that at any time probability of alignment with and against the applied field are equal. In other words for half the molecules, the absorption by a secondary proton is shifted slightly downfield and for the half of the molecules the absorption is shifted slightly upfield. *This results in splitting the secondary proton signal into two peaks* : a **doublet** with peak intensities Fig. 7.11 and Fig. 7.12 (a)

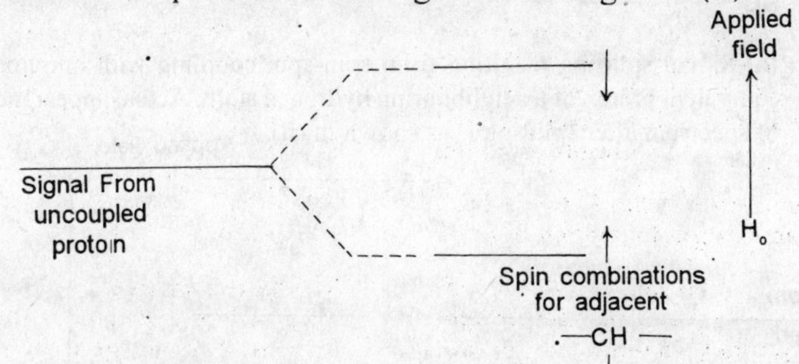

Fig. 7.11: Spin-spin coupling with one proton gives 1:1 doublet

The distance between the two centres of peaks of the 1:1 doublet is termed coupling constant J_{ab} measured in Hertz. The coupling constant is independent of the applied magnetic field and operating frequency of the spectrometer used.

The absorption of the tertiary proton in turn is also affected by the spin of the secondary protons. Now there are two protons on the neighbouring carbon atom, whose alignments in the magnetic field need consideration. There are four equally probable combinations of spring alignments for these protons (↑↓, ↑↑, ↓↑, ↑↓), of which two (underlined) are equivalent. *At any instant, therefore, the tertiary proton experiençes any one of three fields and its signal is split into three equally spaced peaks called* a **triplet.** The relative intensities of this triplet of peaks would be 1 : 2 : 1, the central peak being almost double in intensity due to the combined probability of two equivalent combinations.

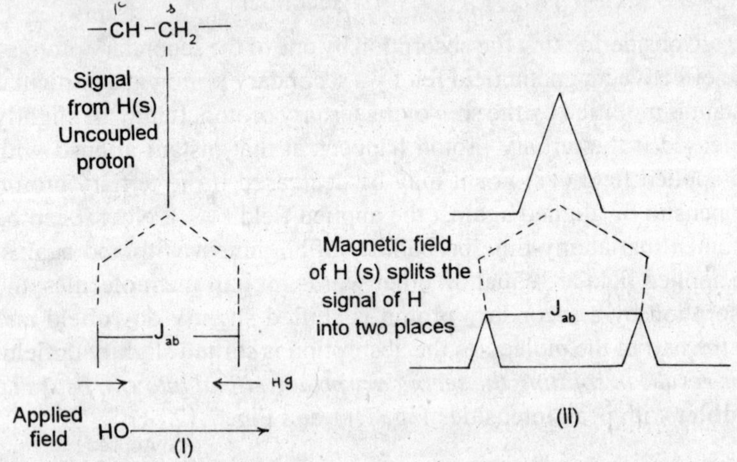

Fig.7.12: (a) Signal splitting resulting from spin-spin coupling with one non-equivalent proton of a neighbouring hydrogen atom. Actual appearance of spectrum after splitting is as shown in (ii).

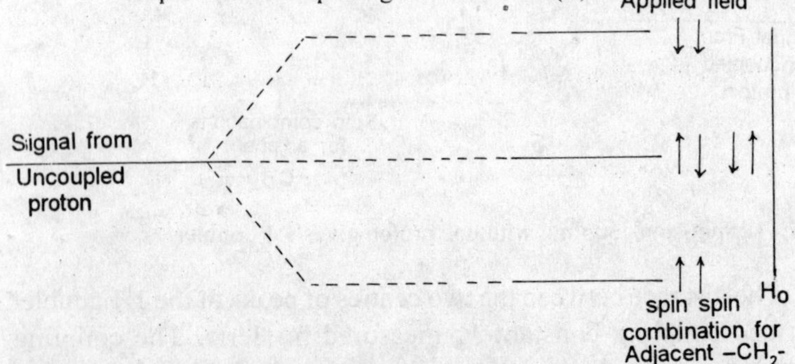

Fig.7.12: (b) Spin-spin coupling with two protons gives a 1 : 2 : 1 triplet

It may be noted that total area under the doublet (from—CH_2— both the peaks) is twice as big as the total area under the triplet (from—CH—) *i.e.,* all the three peaks. This is because the doublet is due to absorption by two secondary protons and the triplet is due to only one tertiary proton.

The actual spectrum of 1, 1, 2—tribromoethane having structure similar to —CH—CH_2—discussed above shows a doublet with equal peak intensities and a triplet with relative peak intenstities 1 : 2 : 1. The area under doublet is twice as big as the area under the triplet.

The distance of separation of peaks (the coupling constant) in the doublet is exactly the same as the distance of separation of peaks in the triplet. Spin-spin coupling is a reciprocal after and the effect of the secondary protons on the tertiary proton must be identical with the effect of tertiary proton on the secondary proton.

Even if they were to appear in a complicated spectrum of many absorption peaks, the identical peak separations would tell us that this doublet and triplet were related that the (two) protons giving doublet and the (one) proton giving the triplet are coupled, and hence are attached to adjacent carbon atoms.

Let us now consider the system having two adjacent carbons atoms, carrying respectively three primary and one tertiary protons.

$$\overset{1t}{—CH}—\overset{p}{CH_2} \qquad \begin{cases} p = \text{primary} \\ t = \text{tertiary.} \end{cases}$$

By arguing in a way similar to that given above this system should give the NMR spectrum comprising a **doublet** (*due to primary protons*) **with 1 : 1 equal peak intensities,** and a **quartet** (*due to tertiary proton*) **with relative intensitise 1 : 3 : 3 : 1.**

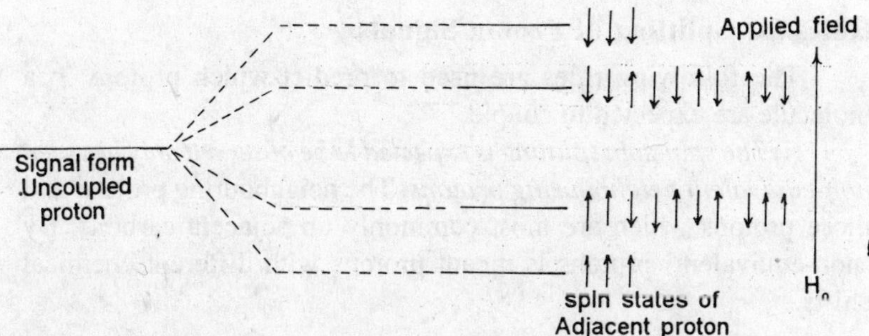

Fig. 7.13: (c) Spin-spin coupling—Coupling with three protons gives a 1 : 3 : 3 : 1 quartet

From Fig. 7.13 (*d*) it is clear that there are 8 possible combinations of spin-states of three protons. Six combination are equivalent. Consequently only four combinations are effective. So the tertiary proton experiences any one of the four combinations giving rise to a signal split into four peaks. The area under the doublet is thrice the area under quarter.

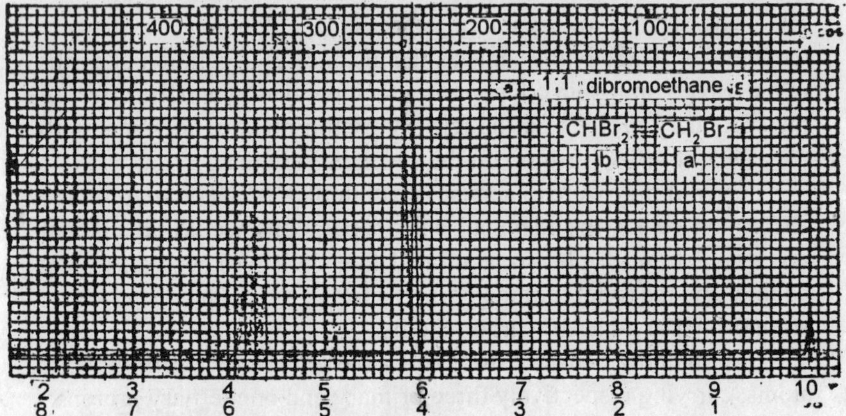

Fig. 7.14: (d) NMR spectra; spitting of signals of 1,1,2-Tribromo ethane

NMR spectrum of 1 ; 1 dibromoethane ($CHBr_2$—CH_2)

containing a structure similar to —C—CH3 confirms the above findings.

Rules for Splitting of Proton Signals

The following rules are used to predict which protons in a molecule are expected to couple.

(i) *The spin-spin splitting is expected to be observed only between non-equivalent neighbouring protons.* The neighbouring protons and those protons which are most commonly on adjacent carbons. By 'non-equivalent' protons is meant protons with different chemical shifts.

For example,

No splitting is observed due to coupling between the protons on C—1 and C—2 of 1, 2 dichloroethane.

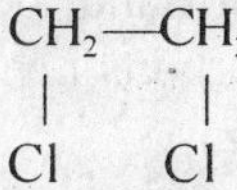

$$\begin{array}{cc} CH_2 - & CH_2 \\ | & | \\ Cl & Cl \end{array} \quad \text{No splitting}$$

1,2-Dichloroethane

because even though on adjacent carbon atoms, they are also equivalent.

(ii) *Splitting of one proton by another proton on the same carbon is quite uncommon because they are generally equivalent.* For example no splitting is observed due to coupling between the protons making up the same—CH_2 group, since they are equivalent.

Note. If protons on the same carbon atoms are non equivalent they may show coupling.

For example no splitting is observed between two vinyl protons of isobutylene.

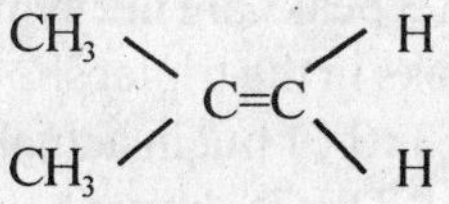

$$(CH_3)_2C{=}CH_2 \quad \text{No splitting}$$

isobutylene

since they are equivalent but there is splitting between the two vinyl protons on the same carbon as in 2-bromopropene since they are non equivalent.

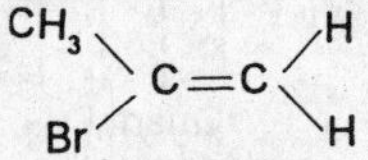

$$CH_3(Br)C{=}CH_2$$

2-bromopropene

(iii) *Mutual splitting of protons separated by more than two atoms (i.e., which are not adjacent) is also very rare.* For example in the spectrum of 1,2—dibromo—2—methyl propane.

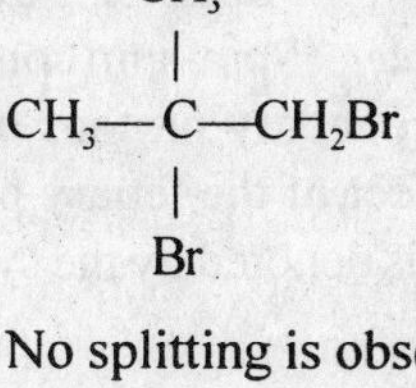

$$CH_3-\underset{Br}{\overset{CH_3}{C}}-CH_2Br \quad \text{No splitting}$$

No splitting is observed between the six methyl protons, on the one hand and the two—CH_2—protons, on the other hand, because these non-equivalent protons are not on adjacent carbons *i.e.,* these protons are separated by more than two carbon atoms.

Note. Sometimes protons farther removed from each other may also be couple, particularly if π bonds intervene.

(iv) *The number of peaks (N) into which a proton signal is split equals one more than the number of vicinal protons (n) i.e., N= n+* 1.
N = 2 (one vicinal H) = **doublet** with peak intensities (1 : 1)
= 3 (two vicinal H) = **triplet** " " " (1 : 2 : 1)
= 4 (three vicinal H) = **quartet** " " " (1 : 3 : 3 : 1)
= 5 (four vicinal H) = **quintet** " " " (1 : 4 : 6 : 4 : 1)

(v) *All the peaks of a given multiplet (i.e., doublet, triplet, quartet etc.) do not have the same intensity.* It has been found that the inner peaks *i.e.,* the peaks near the other coupled multiplets, are relatively larger than he outer pekas. So in case of NMR spectrum of ethyl bromide (CH_3CH_2Br) the triplet and quartet peaks are not symmetrical as expected. The relative intensities of peaks in the triplet should be 1 : 2 : 1 and those of the quartet should be 1 : 3 : 3 : 1 but in actual practice the triplet and quarter are not symmetrical. This is shown below:

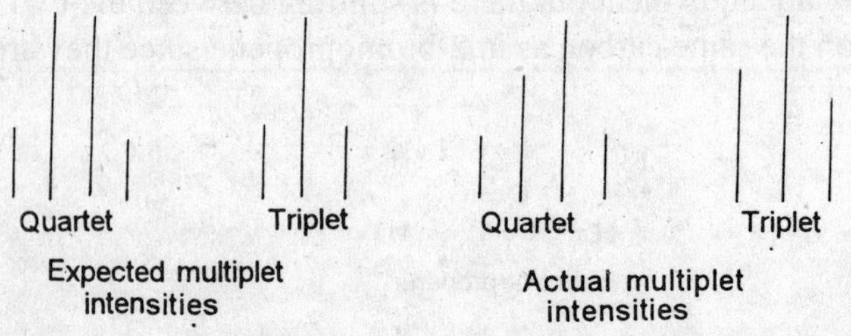

Thus various multiplets do not show quite the symmetry that is attributed to them.

(vi) *The magnitude of the splitting (separation of peaks) or coupling constants* (distance between peaks) *in the doublet is exactly the same as the separation of peaks in the triplet.* (Spin-spin coupling is a reciprocal affair and the effect of the secondary protons on the tertiary proton must be identical with the effect of the tetiary proton on the secondary protons). Coupling constant is denoted by the symbol 'J'.

Examples of Splitting of Signals due to Spin-Spin coupling

(a) In the spectrum of 1, 1, 2-tribromoethane, $CHBr_2$—CH_2Br, we have doublet of —CH_2Br due to $CHBr_2$— and triplet of $CHBr_2$— due to —CH_2Br.

(b) In the spectrum of ethyl bromide, CH_3—CH_2Br, we have a triplet of CH_3—due to —CH_2Br while a quartet of —CH_2Br due to —CH_3.

(c) In the spectrum of 1, 1-dibromoethane CH_3—$CHBr_2$, we have a doublet of CH_3—due to —$CHBr_2$ while a quartet of —$CHBr_2$, due to —CH_3.

(d) In the spectrum of CH_2Cl—CH_2Cl, there is no splitting as both protons on vicinal carbons are equivalent.

(e) In the compound, 1, 2-dibromo-2-methylpropane

$$CH_3-\overset{\overset{\displaystyle CH_3}{|}}{\underset{\underset{\displaystyle Br}{|}}{C}}-CH_2Br$$

there is no splitting, as the protons are not on adjacent carbons.

(f) In the spectrum of iso-butylene, there is no splitting because two vinyl protons are equivalent.

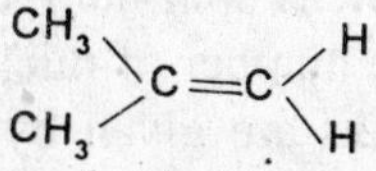

$$(CH_3)_2C=CH_2$$

Iso-butylene

(g) In the spectrum of 2-bromopropene, splitting may be observed due to two vinyl protons on the same carbon if they are non-equivalent

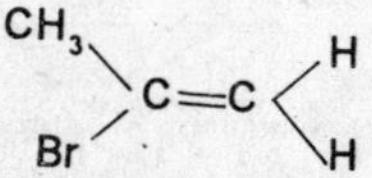

$$CH_3(Br)C=CH_2$$

2-bromorpropene

(h) In the spectrum of acetaldhyde, (CH_3—$\overset{\overset{\displaystyle H}{|}}{C}$=O), we get a double of CH_3—due to—C(H)=O and where a quartet—CHO due to —CH_3.

(i) In the spectrum of toluene, C_6H_5—CH_3, we get four peaks but no splitting because the protons are not on adjacent carbons.

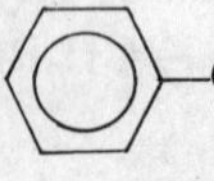

(j) In the spectrum of *p*-nitrotoluene, NO_2—C_6H_4—CH_3 (c, b) there are three types of protons, *a, b, c.* There is no splitting of —CH_3 since it has no proton on the adjacent carbon, *b* and *c* protons give doublets each.

Determination of Number of Peaks and Ratio of Areas

As always given by, the number of peaks into which a proton signal is split is $n+1$ (where n is number of vicinal protons). An alternate method to determine the number of peaks is given by the formula.

$$\text{No. of peaks} = 2n\,I+1$$

Where I is the nucleus spin number $(\frac{1}{2})$ of the nuclei causing the splitting and n is the number of nuclei busing the splitting. The relative areas under peaks are given by the coefficients preceding each term in the expansion formula of $(X+1)^m$ where m is equal to number of peaks minus one.

Example 1

Indicate what NMR spectra would you expect for compounds given ahead.

(i) $CH_3CH_2CH_2CH_2$

(ii) CH_3—C(CH_3)(CH_3)—CH_3

Solution :

$\quad\quad a \quad b \quad b \quad a$

(i) $CH_3CH_2CH_2CH_3$ has two types of non-equivalents proton *a* and *b*. —CH_3 gives a triplet due to —CH_2 while —CH_2 gives a quartet due to —CH_3. The ratio of intensities is 6 : 4 or 3 : 2.

$$\begin{array}{c} CH_2 \\ | \\ CH_3—C—CH_3 \\ | \\ CH_3 \end{array}$$

(ii) CH_3— C—CH_3 has only one type of protons.

These do not split as there is no hydrogen on the adjacent carbon.

Example 2

Predict the fine structure and intensity of peaks for the (*a*) and (*b*) protons in CH_3—CH_2—Cl.
(*a*) (*b*)

Solution :

Ratio of areas for (*a*)

1. No. of peaks of (*a*) = 2n I+1 = $2 \times 2(\frac{1}{2})+1 = 3$

$(n=2) \quad\quad (X+I)^2 = X^2 + 2X + 1$

$= 1 : 2 : 1$

2. No. of peaks for (*b*) = 2n I+1 = $2(3)\times(\frac{1}{2})+1 = 4$

$(n=3) \quad\quad (X+1)^3 = X^3 + 3X^2 + 3X + 1$

$= 1 : 3 : 3 : 1$

UNSOLEVED PROBLEMS

1. Will there be any spin-spin splitting in the spectrum of the molecules?

(i) $ClCH_2$—CH_2Cl

(ii) CH_3—CCl_2—CH_2Cl

[**Ans.** (i) No. All protons are equivalent.]

[(ii) No. The non-equivalent protons are more than three bonds removed from one another].

2. Predict the fine structure and intensity of peaks for (*a*) and

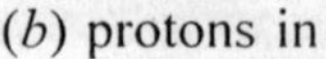

(*b*) protons in

(i) $CHCl_2—CH_2—Cl$ (ii) $Cl—CH_2—O—CH_3$
(*a*) (*b*) (*a*) (*b*)

$$\underset{(a)}{CH_2}—\underset{(b)}{CH_2}—\overset{\overset{\displaystyle O}{\|}}{C}—Cl$$

(iii) $CH_2—CH_2–C—Cl$ (with C=O), (*a*) (*b*)

[**Ans.** *n No. of peaks* *Ratio of Areas*

		n	*No. of peaks*	*Ratio of Areas*
(i)	(*a*)	2	3	1 : 2 : 1
	(*b*)	1	2	1 : 1
(ii)	(*a*) } (*b*) }		No splitting	
(iii)	(*a*)	2	3	1 : 2 : 1
	(*b*)	3	4	1 : 3 : 3 : 1]

Spin-Spin Coupling Notation

An alpabetical notation for referring to types of nuclei engaged in spin-spin coupling has evolved. The following is a guide to this notation.

(1) Equivalent nuclei are referred to by the same letter of the alphabet, with sub-numbers indicating the number of equivalent nuclei.

$$H—\underset{\underset{\displaystyle H}{|}}{\overset{\overset{\displaystyle H}{|}}{C}}—H \text{ is an } A_4 \text{ system}$$

and

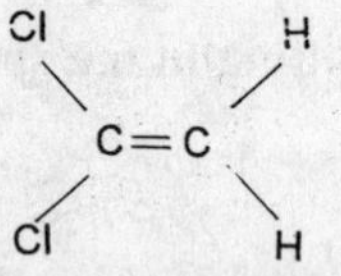

$$\begin{matrix} Cl & & H \\ & C{=}C & \\ Cl & & H \end{matrix} \quad \text{an } A_2 \text{ system}$$

(2) Nuclei with resonance signals separated by only a small chemical shift difference are designated by letters from the same region of the alphabet.

R H

C=C is AB

R′ H

and

CH_3—C—H is A_3B

(3) Where chemical shifts are widely separated, letters from quite different parts of the alphabet are employed.

F

H—C—H is A_2X_2

F

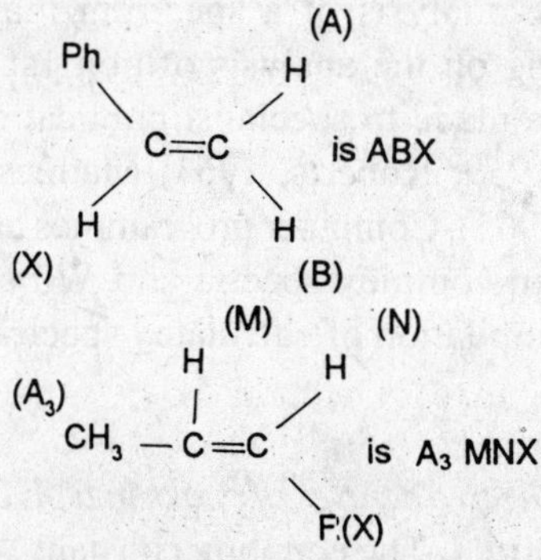

(4) Where more than one coupling constant between equivalent groups is possible, primed letter are resorted to :

H F

C=C is an AA ′XX ′ system, not A_3X_2

H F

Each proton does not couple equally to each fluorine, and *vice-versa,* thus the J_{cis} and J_{trans} values will values will be different. Similarly, in *para*-substituted aromatic compounds, *e.g.,*

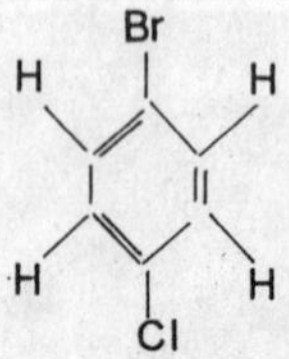

(1-bromo-4-chlorobenzene) is AA'BB' and (4-nitrotoluene) is AA'XX'

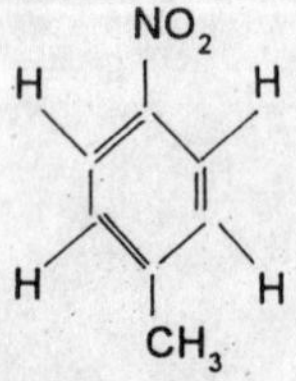

In both cases, *meta* and *para* couplings occur, as well as *ortho* coupling. Coupling to the CH_3 protons has been ignored.

Higher Order spin-spin Coupling

Values for chemical shifts and coupling constants are readily arrived at for spectra or portions of spectra arising from first order coupling, however when first order coupling rules are not obeyed, these values are more difficult to deduce. Higher order spin systems, for example ABX, A_2B_2, etc., may be solved by recognising certain band spacings and substituting these values in established equations. In more complex cases the analysis of a spectrum may get very involved. For further reading on the analysis of non 1st order spin systems reference should be made to specialist publications (Pople, Schneider and Bernstein, 1959; Roberts, 1961; Mathieson, D. W. 1967: Bible, 1956; Becker, 1965). Computer programmes are available for the calculation of certain complex spectra and Wibeg and Nist (1962) have published a compilation of calculated spectra.

Coupling Constant (J)

The distance between the two centres of peaks in a given multipler is known as **coupling constant J.** The coupling constant, 'J' signifies the magnitude of splitting effect.

Its value is independent (unlike chemical shifts) of the external induced magnetic fields. The value of coupling constant remains the same, whatever the applied magnetic field.

The value of 'J', however, depends on the structural relationship between the coupled protons and is expressed in cycles per second (*i.e., cps*) or Hz and is commonly in the range 0—15 *cps.* The size of the coupling constant depends on the number and kind of intervening chemical bonds and the special relations between the protons as summarised below:

(i) *Geminal Protons.* For protons attached to the same carbon atom *i.e.,* **Geminal protons,** J varies with bond angle (dihedral angle). Thus at 125°, J is zero while at 105° the value of J is 20 *cps.*

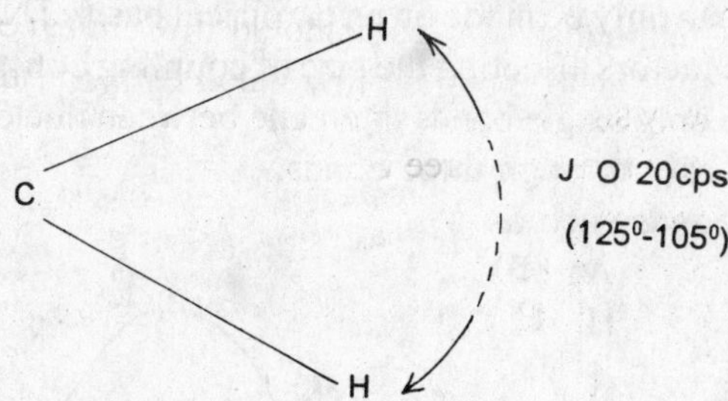

(ii) *Veicinal Protons.* For protons attached to adjacent carbon atom *i.e.,* vicinal protons J varies from 2—18 *cps* depending on positions of protons in space and the structure of the molecule. It means J varies with dihedral angle.

Protons with anti or trans-conformation (in case of freely rotating groups) *J has values varying from 5—14 cps whereas protons with gauche or cis conformations have J=2—6 cps.*

J 5-14 cps

H

Anti or Trans

H H H H

Gauche

(iii) *Vinylic Protons.* In vinylic protons when rotation of bonds is restricted due to double bonds, the *cis protons have J=6 –14 cps while trans protons have J=11 – 18 cps. The coupling constant depends on the structured relationship between coupled protons.* For example in any substituted ethylene or in any pair of geometric isomers J is always larger between trans protons than between cis protons, further more size of J varies in a regular way with electronegativity of the substituents, so that one can often assign configuration without having both isomers in hand.

Factors Influencing the Value of Coupling Constants

It is not fully understood why coupling constants vary considerably with proximate structure, and at present the prediction of approximate J-values can only be made on an empirical basis. The following are some of the factors affecting the size of coupling constants:

(1) When only single bonds intercede between nuclei, coupling usually occurs only through three bonds.

In compounds such as

```
      (A) (B)
       H   H
       |   |
    R—C——C—R'''
       |   |
       R'  R''
```

J_{AB} is in the order of 5—8 Hz.

Sometimes, when molecular geometry is favourable, long rang low order coupling can take place through four or even five bonds.

(2) Couplings through multiple bonds can take place over greater distances, may be because of the mobility of π-electrons. Coupling values of olefinic protons decrease in order :

```
 H
  \    /
   C=C          J_AB Trans     11 — 18 Hz
  /    \
        H

 H      H
  \    /
   C=C          J_AB cis       6—14 Hz
  /    \

        H
  \    /
   C=C          J_AB gem       0. —3.5 Hz
  /    \
        H
```

In unsaturated compounds coupling of the order of 1 Hz through four bonds in often observed.

H CH$_3$
C=C $J_{A,3x} \simeq 1$ Hz

(3) In aromatic systems *ortho* coupling (J≃8Hz) *meta* (J≃2–3 Hz) > *para* (J≃0 –1Hz).

(4) Coupling between geminal protons on saturated carbon in dependent, in part, upon to bond angle θ, J_{AB} decrease from about 20 Hz when θ is 105° to zero when θ is 125°.

(A)
R R
C)0
R H
(B)

It should be emphasised that this is an approximate relationship since other factors influenced J_{AB}. Cookson, Crabb, Frankel and Hudec (1966) have published a comprehensive account of geminal coupling constants and factors influencing them.

(5) The size of the vicinal proton coupling constant is dependent upon the dihedral angle 0. A plot of 0 against J (Fig. 7.15) shows that J_{AB} is maximum (≃10 Hz) when the dihedral angle is 180°, and a

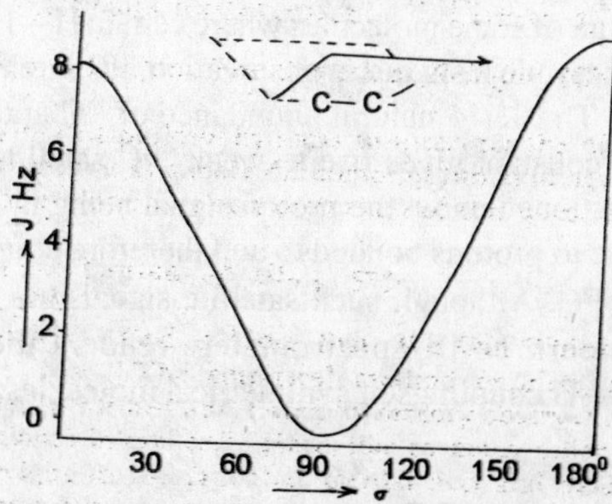

Fig. 7.15: Karplus relationship between the dihedral angle and coupling constant for vicinal protons protons

minimum when it is 90° and about 8 Hz when 0 is 0°. In cyclohexane axial-axial coupling has $J_{axax} \simeq 8 - 12$ Hz, whereas $J_{eqax} \simeq 3 - 4$ Hz. Karplus (1959), who has studied vicinal coupling entensively, warns against attempting to determine dihedral angles with any degree of accuracy from vicinal coupling constants.

(6) The electronegativity of substitutent X in a system :

```
    H   H
    |   |
  — C — C — X
    |   |
```

has little effect on the vicinal coupling constant (< 2 Hz). In the limited number of cases investigated increasing substitutent electronegativity gives a decreases in the J-value.

(7) Coupling values for splitting between protons and other nuclei, for example ^{15}N, ^{31}P, ^{18}C, etc., are often very large; an exception is proton-deuterium coupling, where the J-values are about one-seventh of those between protons. Advantage is taken of this in *spin-decoupling* by deuterium replacement of hydrogen.

Measurement of Coupling Constants

A major problem encountered in comparing NMR data from different sources is the variability of conditions under which spectral parameters are measured. To minimise errors in the measurement of coupling constants the use of scale expanded spectra is recommended.

In the spectrum of some molecules where certain H—H coupling constants are inaccessible from direct observation, ^{13}C satellite signals may be exploited. The 1.1% natural abundance of ^{13}C and the large ^{13}C—^{1}H coupling constant gives rise to weak '^{13}C satellites' almost symmetrically positioned about the proton signal being investigated. The signals are due to protons bonded to and therefore coupled to ^{13}C ($I=\frac{1}{2}$) rather than ^{12}C. Although such satellite signals are weak, the sensitivity of modern NMR spectrometers renders them easily observable and H—H couplings of symmetrical molecules may often be extracted.

Magnetic Equivalence of Protons

Equivalence or non-equivalence of protons is quite important in the study of NMR specta since it determines both the number of signals and their splitting. The equivalence depends on the rate at which the molecule undergoes certain changes. Each of these molecular changes can change both electronic and protonic environment of a given proton, thereby affecting both its chemical shift and its coupling with other protons. The NMR spectrometer has the ability to see the proton in each environment as well as in an average of all of them. In this lies the usefulness of NMR spectroscopy.

To understand this aspect of NMR spectroscopy, the NMR spectrometer is linked to a slow camera. Such a camera photographs the spokes of a wheel in different ways depending upon the speed with which the wheel spins: as sharp, individual spokes if spinning is slow; as blurred spokes if spinning is faster; as a single circular smear if the spinning is faster yet. In the same way if the molecular change is relatively fast, the NMR spectrometer sees a proton in its average environment—a smeared picture; if the molecular process is slow, the spectrometer sees the proton in each of its environment.

Let us now study the molecule of ethyl bromide in its staggered conformation. There are three such conformations—I, II and III, which differ from each other as much as a different individual proton occupies the anti-position with respect to bromine. A particular proton of the methyl group is anti to bromine in one of the conformations and gauche in the other two. In any given conformation, one of the methyl protons is anti and the other two are gauche. The anti proton is in a different environment from the two gauche protons, and for a

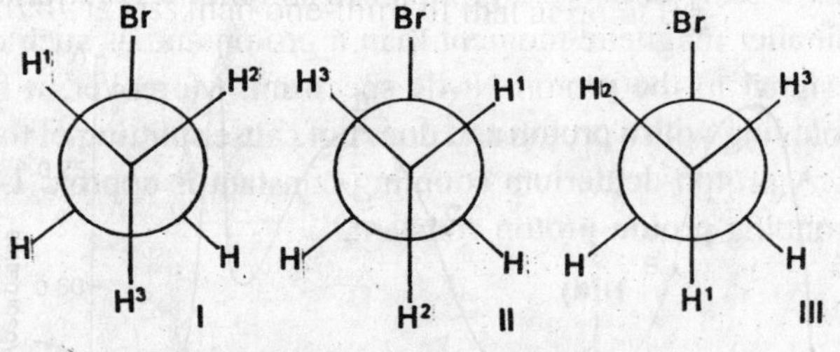

Fig. 7.16: Conformations of Ethyl bromide

moment it is not equivalent to them. Yet the three methyl protons of ethyl bromide give a single signal (a triplet, because of the adjacent methylene group), suggesting the magnetic equivalence of the three methyl protons. This interconverstion of conformations due to rotations about single bond is thus very fast and the NMR spectrometer sees each proton in an average environment. This phenomenon in NMR spectrum analysis is called **Magnetic equivalence of protons.** It should be understood that if this interconversion of the three conformations is slowed down considerably by lowering temperature, the NMR spectrum will reveal each proton in its individual environment.

This is exactly what happens when the NMR spectrum of cyclohexane is taken at ordinary temperature and at temperatures below –70°C. At room temperature it has single sharp peak and at –70°C, it has two peak.

At ordinary temperature, the change of one chair form of cyclohexane into another by rotation about single bonds is very fast (Fig. 7.17) and hence there is only one NMR signal for all the twelve protons, even when six equatorial (*e*) hydrogens are non-equivalent to six axial (*a*) hydrogens.

When the temperature is lowered to –100°C and the rate of rotation around single bonds, is lowered the NMR spectrum records the individual environments of each set of equivalent protons. Thus in the NMR spectrum of cyclohexane at a temperature below –70°C *i.e.,* at –100°C, two clearly separated peaks (one for equatorial protons and the other for the axial protons) are observed.

Deuteration (*Deuterium labelling*)

In this technique a deuterium (^{2}H or D) is introduced into a molecule usually to detect a group or to simplify a spectrum. It has a much smaller magnetic moment than a proton and as such does not give a signal in the proton NMR spectrum. Moreover, it has very weak coupling with a proton and does not cause splitting of the proton signals. A proton-deuterium coupling constant is approx. 1/7 of the corresponding proton-proton constant.

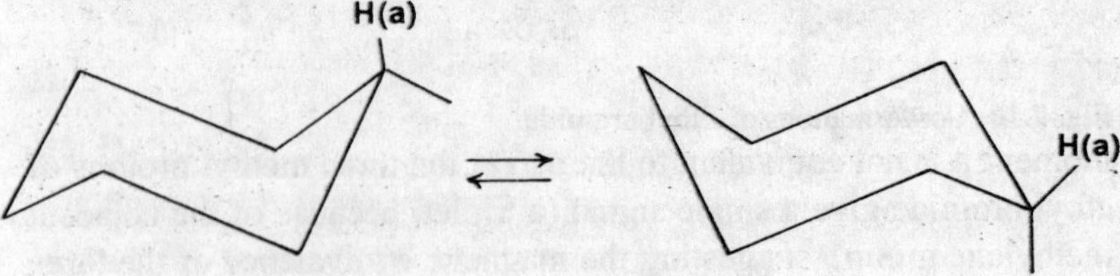

Fig. 7.17: Interconversion of chair forms of cyclohexane.

In the NMR spectrum of ethyl bromide, for example, we have one triplet due to the methyl protons and quartet due to methylene protons.

Let us successively replace the protons by deuterium. We shall observe that this procedure can help us to find out which signal is produced by which proton or protons.

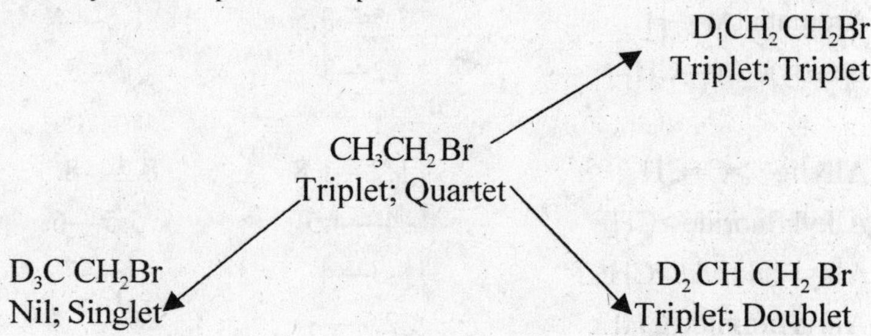

Similarly suppose the protons on the α-carbon atom of a ketone are replaced by deuterium

$$\overset{\gamma}{X-CH_2}-\overset{\beta}{CH_2}-\overset{\alpha}{CH_2}-\overset{\overset{O}{\|}}{C}-$$

The original spectrum would be,
triplet for α protons, a quinter for β and a triplet for γ-protons.

$$\overset{\gamma}{X-CH_2}-\overset{\beta}{CH_2}-\overset{\alpha}{CD}-\overset{\overset{O}{\|}}{C}-$$

In the deuterated compounds there shall be nil for α-proton, triplet for β (broadened) and triplet for γ-proton (unaffected). Evidently deuterium labelling can be used to know which signal is given by which proton or protons.

Table 7.1: Typical Proton Chemical Shifts

Type of Proton	*Chemical Shifts (ppm.)* δ	τ
1	2	3
Primary R_2—CH_3	0.9	9.1
Secondary R_2CH_2	1.3	8.7

1	2	3
Teritiary R_3CH	1.5	8.5
Vinylic >C=CH_2	4.6—5.9	4.1—5.4
Acetelenic—C≡CH	2—3	7—8
Aromatic Ar—H	6.5—8.5	1.5—4
Benzlyic Ar—CH_3	2—3	7—8
Allylic >C=CH_3 (with | above C)	1.7—1.8	8.2—8.3
Alkyl fluoride >CHF	4—4.5	5.5—6
Alkyl chloride >CHCl	3—4	6—7
Alkyl bromide CHLr	2.7—4	6—7.3
Alkyl idoide >CHI	2—4	6—8
Alcohol >CHOH	3.4—4	6—6.6
Ethers R—O—CH<	3.3—4	6—6.7
Aldelydlic —C—H (with ‖ O)	9—10	0—1
Ketonic —CH—C— (O, |, ‖)	2—2.7	7.3—8
Esters —C—O—CH (O, ‖)	2—2.2	7.8—8
Hydroxylic R—OH	1—6	4—9
Phenolic R—OH	4—12	(—2)—6
Amino R—NH_2	1—15	5—9
Carbxylic R—COOH	10—12	(—2)—0

Spin-Spin Decoupling

A spectrum is often rendered very complex and difficult to interpret by spin-spin coupling. Simplification may be achieved in following two ways:

(1) By using an instrument with a more powerful homogeneous magnetic field, for example, a 100 MHz instrument in preference to one operating at 60 MHz. A greater line separation (higher resolution) results, but there is a practical limit to the magnet strength available.

(2) By a spin-spin decoupling technique.

Such techniques not only reduce the amount of the fine structure and clarify the spectrum, but enable interacting protons to be identified.

Isotope exchange. The hydroxyl protons of ethanol, as we have seen are able to exchange with other protons or deuterium (^2H) causing decoupling and simplification of the $—CH_2—$ proton resonance. Dimethylsulphoxide, as a solvent, facilitates the study of coupling through exchangeable protons of this sort by hydrogen bonding to them and slowing the exchange rate.

Isotopic exchange may be used to advantage in the deuteration of compounds with active (acidic) hydrogens bonded to carbon, as for example, in protons *alpha* to ketone or nitrile.

$$R—CH_2—\overset{\overset{\large O}{||}}{C}—R \xrightarrow[OH]{D_2O} R—\underset{\underset{\large D}{|}}{CH}—\overset{\overset{\large O}{||}}{C}—R \xrightarrow[OH]{D_2O} RCD_2—\overset{\overset{\large O}{||}}{C}—R$$

and

$$\underset{(A)}{NC—CH_2—CH_2—OH} \xrightarrow[OH]{D_2O} \underset{(B)}{NC—\underset{\underset{\large D}{|}}{CH}—CH_2—OD} \xrightarrow[OH]{D_2O} \underset{(C)}{NC—CD_2—CH_2—OD}$$

The scheme (A to C) illustrates isotopic exchange in 2-hydroxypropionitrile. In D_2O its spectrum appears as an A_2X_2 pair of triplets, the C-1 and C-2 proton signals being at 2.68δ and 3.0δ respectively (Fig. 7.18). The addition of hydroxide ions to this solution causes the activated C-1 protons to exchange successively (B and C) with deuterium and show the corresponding changes in the PMR spectrum. The rate of exchange is dependent upon the concentration of OH^-. Fig. 7.18 illustrates the change in the spectrum of a solution of 2-hydroxypropionitrile (A) in 1.44N NaOH in D_2O with time.

(a) At zero time only the A_2X_2 pattern of A is present.

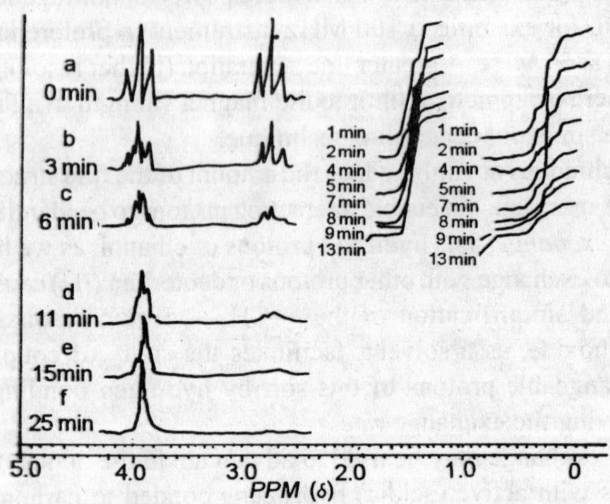

Fig. 7.18: Deuterium exchange of 2-hydroxypropionitrile with D_2O in the presence of sodium hydroxide. Left : Successive spectra. Centre : Integrated intensities of b-proton triplets. Right : Integrated intensities of a-proton triplets

(b) and (c) A doublet due to the C-2 protons split by the single C-1 proton (B) is superimposed upon the original lowfield triplet giving a five line pattern.

(d) Here there are contributions from structures (A), (B) and (C), a triplet, doublet and singlet respectively appear superimposed for the C-2 protons.

(e) Contributions from only (B) and (C) are apparent in the resonance of the C-2 protons.

(f) Finally, after 25 min almost complete exchange has occurred, the C-protons are almost entirely decoupled to yield a singlet and C-protons resonance is lost. Integration surves for C-2 and C-1 protons respectively are shown to the right of the spectra.

The synthesis of deuterated compounds from appropriately deuterated starting materials also helps in the elucidation of molecular fine structure. An example is the investigation of the conformational preference of the nonbonding electron pair in piperidines by the synthesis, and PMR examination of the α-protons, of 3,3,5,5-tetradeuteropiperidines (D).

Nuclear Magnetic Double Resonance (NMDR)

During a normal NMR experiment the sample is subjected to a single radiofrequency field. If a second (or sometimes more) radiofrequency field is applied in a specific manner, then a series of techniques collectively known as **nuclear magnetic double resonance** results. By careful application of these technqiues, particularly in complex spectra with overlapping resonance bands, much valuable information can be obtained.

Spin-Spin decoupling by NMDR. If during an NMR experiment on a simple AX system (Fig. 7.19), the X nucleus is irradiated with a second radiofrequency signal of frequency (v_x) equal to the processional frequency of X, then the X-nuclei will undergo rapid changes of spin state. Nucleus A is aware of only one equivalent (averaged) spind state of X and therefore its resonance signal appears, not as a doublet, but as a sharp singlet. Irradiation of X effectively decouples the signal from A (Fig. 7.19 *b*). Similarly if A is irradiated with a second r/f signal equal to its own resonance frequency (v_A), then the signal from X will collapse to a singlet (Fig. 7.19).

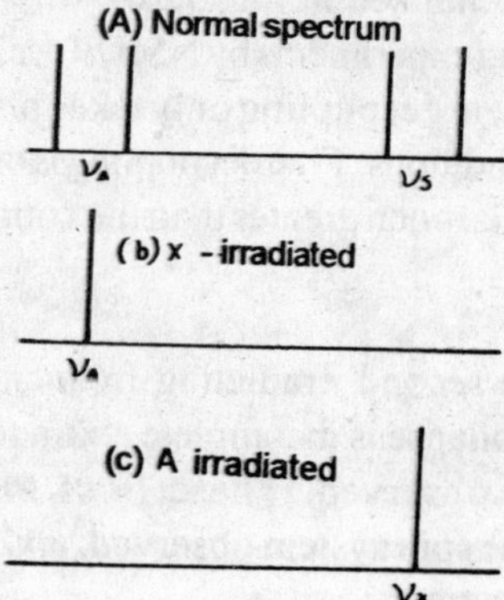

Fig. 7.19: Double irradiation

Double resonance experiments can be carried out either by: (1) keeping both the applied frequency (v_1) and the irradiation frequency (v_2) constant, and varying the magnetic field (*field sweep*), or (2) by having a constant magnetic field (H_0) and irradiation frequency (v_2), and sweeping with the applied frequency (v_1) (*frequency sweep*). The irradiating frequency (v_2) may be changed after each scan, permitting an extensive investigation of spin-spin coupling. **Nuclear triple resonance,** where two irradiation frequencies are applied simultaneously and the signal due to a third nucleus examined, is valuable in the investigation of multiple and long-range couple.

The irradiation frequency (v_2), in order to decouple the nuclei completely, is a powerful r/f signal; v_1 is seak in comparison.

Heteronuclear decoupling is the decoupling of different isotopes where the difference in resonance frequencies is large, *e.g.*, ^{1}H and ^{19}F. The experiment is expressed as

$$^1H[^{19}F]$$

Nucleus being studied   Nucleus irradiatied

Homonuclear decoupling involves nuclei of the same isotope, where the difference in resonance frequencies is small. It is expressed by ^{1}H[^{1}H]. Similarly, the irradiation of the X nucleus of an ABX system is expressed as AB [X].

For effective double irradiation a good chemical shift difference between the interacting nuclei is required.

Spin decoupling techniques, as exemplified by ethyl alcohol spectra, are summarised in Fig. 7.20.

Decoupling experiments by NMDR are, generally, easy to conduct, however complete decoupling only takes place under rather restricted experimental conditions. True decoupling is only seen when the chemical shift separation is much greater than the coupling constant.

Spring Tickling

When the second irradiating frequency (v_2) is relatively weak and spectrum collapse is incomplete, extra resonance lines or 'residual splittings' are observed. These lines may be related to energy transitions of the spin system observed, and may aid the interpretation of complex spectra.

Nuclear Overhauser Effect (NOE)

Unlike spin decoupling and spin tickling which are based upon changes in energy levels, the nuclear Overhauser effect results from population changes in the energy levels which cause corresponding intensity changes in the spectra. The NOE experiment consists of saturating one signal and observing the intenstities of other signals. The resonance bands because of protons spatially close to the group being saturated show an increase in intensity. Such intensity changes are best measured by integration and not by peak height. The theory of the NOE is dependent upon relaxation mechanisms, and the relaxation of any nucleus is affected by all surrounding magnetic nuclei. Because the contribution of any nucleus to the relaxation of a second nucleus is dependent upon the mean square value of the magnetic field produced by the former, so the magnitude of the nuclear Overhauser effect will relate to the internuclear distance.

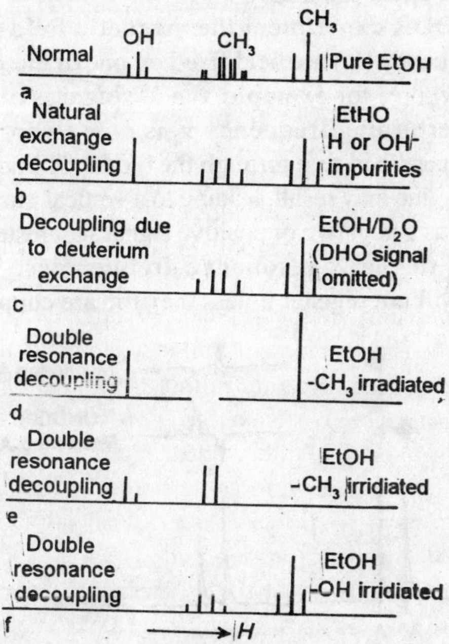

Fig. 7.20: Decoupling—illustrated by ethanol spectra

In the spectrum of dimethyl formamide it is not possible from shift or coupling parameters to assign the two methyl signals. Saturation fo the low field methyl signal causes a 17% increase in the intensity of the CHO signal, whereas saturation fo the high field methyl signal gives no change in the formyl proton band. Clearly the low field line is *cis* to the formyl hydrogen.

Interactions giving rise to nuclear Overhauser effects may be intermolecular or intermolecular. In dilute solutions in operatic solvents only the lafter will be observed. It is important from an experimental standpoint in all NOE experiments to de-gas samples. Oxygen, being paramagnetic, will affect and often dominate the relaxation process and its presence will invalidate results.

INDOR Spectroscopy

Probably the most powerful double resonance technique for the observation of the multiplicity of signals hidden by band overlap is **Internuclear Double Resonance** (INDOR). Here signals are only observed from coupled protons.

In the INDOR experiment the magnetic field (H_0) and the observation frequency (v_1) are held fixed on one of the A signal lines in a spectrum where, for example, the X singal is obscured. The second weak perturbing frequency v_2 is now swept through the spectrum. As v_2 passes in turn through the hidden X lines a change in intensity of the A line may result, leading to a vertical movement of the recorder pen. Thus a negative or positive signal will result at positions corresponding to the X resonance frequencies. The signals overlapping X will not register unless they too are coupled to proton A (Fig. 7.21).

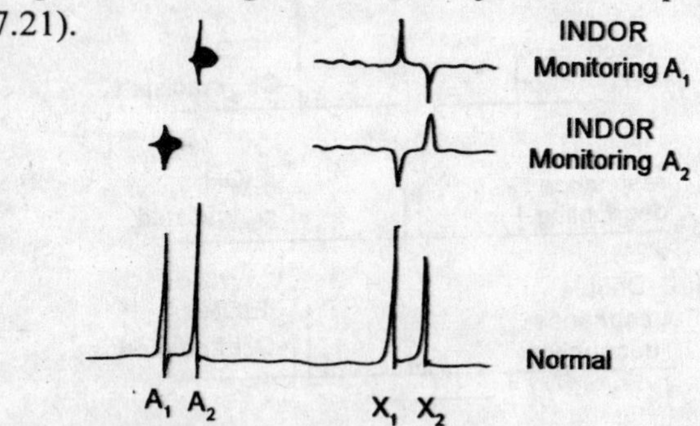

Fig. 7.21: 90 MHz spectra of C_6H_5, CHBr CHBr.NO_2, 300 Hz, scan, demonstrating the INDOR effect.

APPLICATIONS

Structure Elucidation

The versatility of nuclear magnetic resonance spectroscopy as an analytical technique is illustrated here by a selection of its many applications. PMR is particularly useful in view of the abundance of ^1H in organic compounds, its strong resonance signal and convenient spin value of $\frac{1}{2}$. It has been applied extensively in structural elucidation.

The unequivocal proof of an organic structure by PMR spectroscopy is not often possible, but in conjunction with ultraviolet, infrared and mass spectrometry, it is an extremely valuable tool. Theoretically, all the resonance lines of a spectrum can be assigned to some molecular function, but in practice, for complex spectra, this proves difficult and often unnecessary.

An examination of a PMR spectrum must entail careful consideration of the following features:

(1) Resonance line positions, or *chemical shift* values, indicate the electronic envrionemnt of groups of equivalent protons. Correlation charts should be consulted to establish the probable nature of each group. Resonance line overlap occurs frequently, particularly at high field, and spectrum simplification by spin-spin decoupling may be necessary. Examination of a spectrum in the presence of D_2O or CD_3OD will establish the identify of bands due to —OH or NH, bases exhibit chracteristic shift in CH_3COOD solution.

(2) *Line intensity,* as determined by the electron integrator, is proportional to the number of protons responsible for the signal.

(3) *Spin-spin coupling* patterns give the number of protons in interacting equivalent groups. Coupling patterns may be clarified by employing a **lanthanide shift reagent** or by INDOR spectroscopy.

Some main applications of NMR spectroscopy are given below.

(1) Indentification of the Organic Compounds

NMR spectrum of a compound like I.R. spectrum serves as a finger print of that compound. Hence if spectrum of the given sample is identical with the spectrum of the known compound, the sample must be identical with the given compound. Thus NMR spectrscopy helps in the identification of the organic compounds.

(2) Determination of Molecular Structure —Structural Diagnosis

A large number of principles are known which decide the structure of an unknown compound from is NMR—spectrum. Some of these are outlined as follows:

(i) The number main NMR singals should be equal to the number of equivalent protons in the unknown compound.

(ii) The chemical shift indicates what type of hydrogen atoms are present *e.g.*, methylene, methyl, olefins, ethers etc.

(iii) The spin-spin splitting or multiplicity reveals about the possible arrangement of groups in the molecule.

(iv) From the area of peaks, one can draw the conclusion about number of hydrogen nuclei present in each group. For example, the relative area of methyl (CH_3) and methylene (CH_2) peaks in $CH_3CH_2CH_3$ would be 6 : 2. In butane it would be 6 : 4. Complete analysis of a complicated NMR spectrum is quite difficult but often it is not necessary. There may be two or three possible structures for a compound on the basis of other evidence available and NMR spectroscopy may just clinch the structure with the help of isotopic substitution and double resonance—an electric technique. It is now possible to analyse completely even the most complicated NMR spectra.

(3) Analytical Applications

General Consideration

The procedures are limited to structure determinations involving nuclei which possess either an odd mass number such as ${}^{1}_{1}H$, ${}^{13}_{6}C$, ${}^{15}_{7}N$, etc, or an even mass number with an odd atomic number such as ${}^{3}_{1}H$, ${}^{10}_{5}B$, etc. It cannot be applied to nuclei such as ${}^{12}_{6}C$, ${}^{15}_{8}O$, or ${}^{32}_{16}S$, which posses both even mass and even atomic numbers. *The latter group possesses no magnetic properties.*

The nmr spectroscopy techniques are applied primarily to liquid samples. The absorption bands for solids are generally too wide and instruments are, at the present, not sensitive enough for gasesous samples. Since NMR technology is rapidly developing, this limitation might be removed in the future.

Recent developments of far reaching importance in inorganic chemistry have made possible, the observation of high resolution **NMR** spectras on solids. One of these high resolution techniques known as **CPMAS-NMR,** involves high speed sample spinning at '**magic angle**' with respect to the field axis; the technique is a combination of magic angle spinning (**MAS**) with cross polarisation (**CP**). Compounds containing ^{13}C, ^{31}P, and ^{29}Si and many other nuclei have been studied in solid state using these techniques. An example is the use of ^{29}**Si MAS-NMR** spectra to infer the position of Si atoms in natured and synthetic aluminosilicates. It is also used to study molecular compounds in the solid state. For example, the ^{13}**C–CPMAS** spectrum of Fe_2 (C_8H_8) $(CO)_5$, at –160°C indicates that all C atoms in C_8 ring are equivalent on time scale of the experiment. The interpretation of this observation is that the molecule is fluxional in the solid state.

Liquid sample sizes range from 0.1 to 0.4 ml. If the sample itself is a liquid it may be determined directly; if not, it may be dissolved in a suitable solvent. The minimum concentration which may be detected in a liquid is approximately 1%. The choice of a liquid solvent is made on the basis of its purity, chemical stability, and volatility. Most important, it must have an NMR absorption spectrum which interferes with neither that of the sample nor the internal standard. When a limits sample size is available it may frequently be recovered after NMR procedures by various physical methods. The procedure is non-destructive.

While solids cannot be determined directly with normal NMR, they can be dissolved in a solvent, or a wide-line NMR procedure may be used for the study of solids. Chemical shifts from functional groups cannot be observed; however, the total amount of a particular nuclear species may be determined. It is used in the study of fats, oils, polymers, and moisture contents. Gas samples may be liquefied by cooling or a suitable derivative may be synthesized and studied.

4. Rates of Reactions

Certain reactions are very fast for measuring them by ordinary devices. They can be studied with the help of NMR spectroscopy. Such reactions involve a fast exchange of protons between the two reacting molecules such as between water and acetic acid.

$$\underset{(a)}{CH_3COOH} + \underset{(b)}{HOH} = \underset{(b)}{CH_3COOH} + \underset{(a)}{HOOH}$$

5. Qualitative Analysis

Were it not for the chemical shift and the spin-spin splitting phenomena, NMR would offer little information which is not available by other less expensive methods. The NMR is primarily applied in organic chemistry and is used in the elucidation of the structure of and identification of organic molecules.

As we have seen, the chemical shift yields information as to the types of hydrongens which are present, whereas the spin-spin splitting yields information concerning the nature of the neighbouring groups. These two features, combined with the fact that the relative areas of absorption peaks are in direct proportion to the number of equivalent hydrogens in each group, make NMR a powerful structure indentification tool.

The actual identification may be made by comparison against a laboratory file, a commercially available spectra file, or tables which relate the chemical shift of coupling constant with structure.

From the NMR spectrum of a compound it is possible to detect the presence or absence of a functional group. For example appearance of an NMR signal at 9.5 shows the presence of a —CHO group in the compound.

6. Quantitative Analysis

The fact that the areas of nmr absorption peaks are in direct proportion to the types of hydrogens which are causing the respective peaks is the basis for the NMR quantitative analysis of organic liquids. These areas may be rapidly determined with the integration recorders. In order for quantitative procedures to be applied, the components must be known. In addition, each component of the mixture must provide at least one absorption band which is isolated from the absorption bands of any other components. Impure samples may be determined by the addition of a known pure compound as an internal standard.

Quantitative applications have been used to elucidate chain length or the percentage of hydrogen in a pure compounds. The degree of saturation in a compound of known chain length and known percentage composition may be calculated.

In a chemical reaction where both the reactants and products are known, the rate of the reaction may be determined. For example, the rate constants of hydrogen exchange in chlorophyll have been reported.

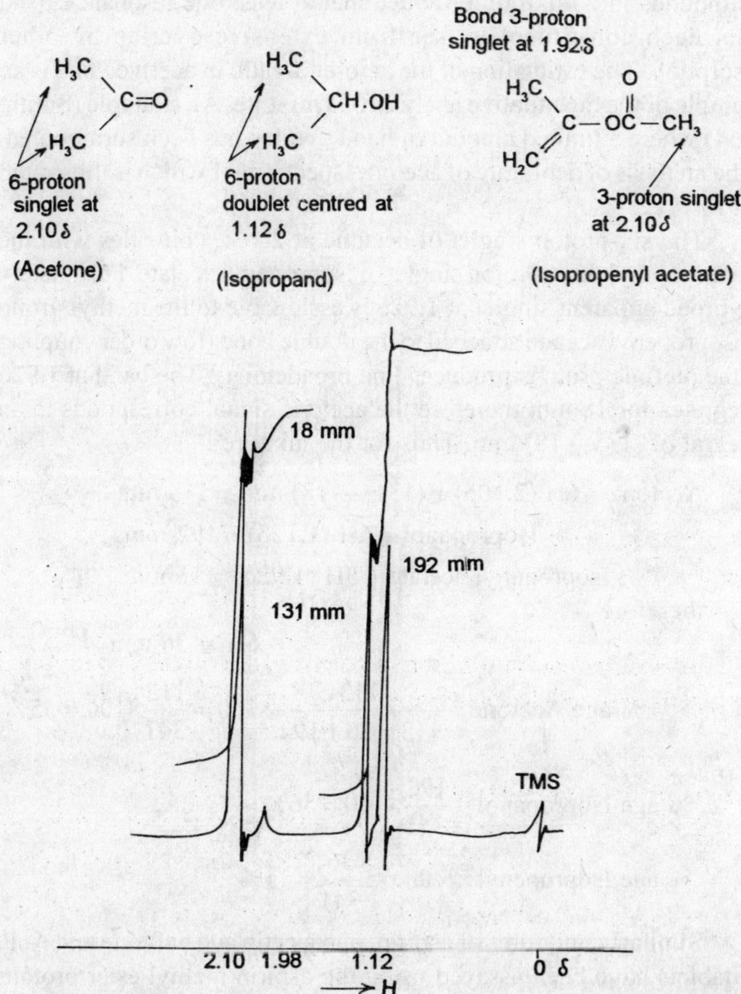

Fig. 7.22: NMR spectrum of a mixture of isopropenyl acetate, isopropanol and acetone

Attention has been focused largely on nuclear magnetic resonance spectroscopy as an aid to structure elucidation and to the study of molecular dynamic processes. Far less heed has been paid to its applications to quantitative analysis, though this is now beginning to receive attention. Automatic integration of resonance bands affords an easy and rapid quantitative means of determining the ratio of compounds in a mixture, provided that at least one resonance band from each constituent is free from extensive overlap by other absorption. The estimation of the keto-enol ratio in acetlycetone is an example of the quantitative analysis of a mixture. An example (Smith, 1964) where a limited amount of band overlap has been surmounted, is the analysis of a mixture of acetone, spectrum of which is illustrated in Fig. 7.22.

The six-proton singlet of acetone at 2.10δ, coincides with the acetate methyl three proton singlet of isopropenyl acetate. Fortunately the broad apparent singlet at 1.92δ is assignable to the methyl group of isopropenyl acetate attached to the double bond (low order coupling to the olefinic protons produces line broadening). The band at 1.92δ integrates for 18 mm therefore the acetone signal corresponds to an integral of (131—18) mm. Thus for the mixture :

$$\text{Acetone} \equiv 6\text{H } (2.10\delta) \equiv (131 - 18) \text{ mm} \equiv 113 \text{ mm}$$

$$\text{Isopropanol} \equiv 6\text{H } (1.12\delta) \equiv 192 \text{ mm}$$

$$\text{Isopropenyl acetate} \equiv 3\text{H } (1.92\delta) \equiv 18 \text{ mm}$$

therefore

$$6\text{H} \equiv 36 \text{ mm}$$

$$\%\text{ age Acetone} = \frac{113}{113 + 36 + 192} \times 100 = \frac{113}{341} \times 100 = 35\%$$

$$\%\text{ age Isopropanol} = \frac{192}{341} \times 100 = 56\%$$

$$\%\text{ age Isopropenyl acetate} = \frac{36}{341} = 11\%$$

Similarly, mixtures of aspirin, phenacetin and caffeine and A, P, C, tablets have been assayed using the aspirin methyl ester protons (2.3δ), the protons of the phenacetin acetyl group (2.1δ) and for caffeine the 1 and 3 N—CH_3 protons (3.4δ and 3.6δ). The procedure takes 15—20 minutes and is reasonably accurate (Hollis, 1963).

The absolute concentration of a constituent in a pharmaceutical raw material or formulation may be obtained by adding a reference compound to the solution for NMR examination. Weighed amounts of both sample and reference are dissolved in an appropriate solvent and the spectrum and integral are obtained.

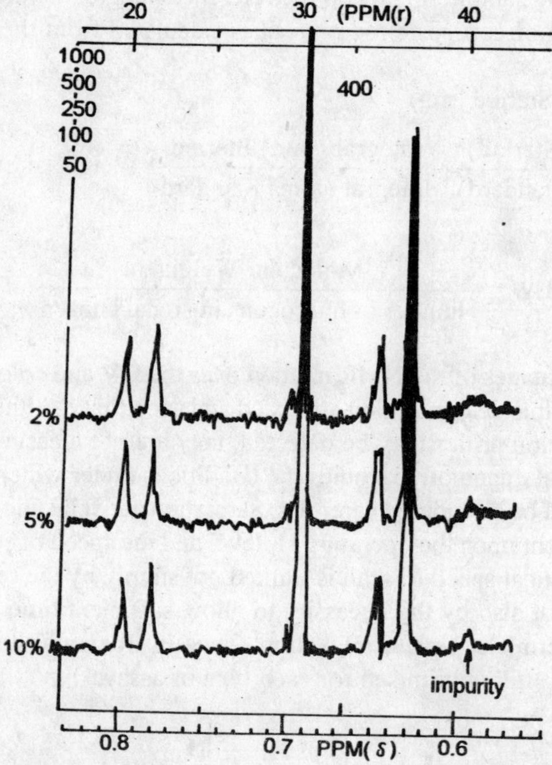

Fig. 7.23: Partial NMR spectrum of steroidal 1,4-dien-3-one in dimethyl sulphoxide containing fumaric acid and 2.5, or 10% of added steroidal 4.en-3-one.

Corticosteroids both in bulk and in formulations have been assayed in this manner (Avdovich 1970). **Prednisolone** and Triamcinolone are 1, 4-dien-3-ones and exhibit a characteristic olefinic region spectrum (Fig. 7.23). The C-1 proton is seen as part of an AB quartet at 7.9δ coupled to the C-2 proton at 6.4δ which in turn is partly obscured by the C-4 (1H) resonance at 6.3δ. To the 1, 4-dien-3-one sample in dimethylsulphoxide solution is added a known weight of a pure stable reference standard, fumaric acid, the olefinic resonance line of which is seen at 7.1δ. At 5.9δ a peak due to the 4-position proton of a 4-ene-3-one steriod is observed, but does not interfere with the assay.

Several integrals are run for each determination and the integrals from the 7.9δ sample singal and the reference integral are measured. The amount of corticosteirod present is calculated from the sample expression :

Amount of Steriod (mg)

$$= \frac{\text{EW (Steriod)}}{\text{EW (Standard)}} \times \frac{\text{Integral (mm) Steriod}}{\text{Integral (mm) Standard}} \times \text{Weight of Standard}$$

where

$$\text{EW} = \frac{\text{Molecular Weight}}{\text{Number of hydrogens in signal chosen}}$$

Advantages of the NMR method over the UV and colorimetric methods in this determination are speed and specificity. Although the 4-en-3-one impurities may be detected, they are not measured.

During quantitative studies the conditions under which spectra are run must be considered carefully. Saturation and relaxation effects are dependent upon the operating H_1 level and the speed of scan. The fastest practical speed of scan is limited not simply by the resolution required, but also by the necessity to allow sufficient time between scans to permit complete relaxation. Thus in trumental conditions must be carefully optimised for each type of assay.

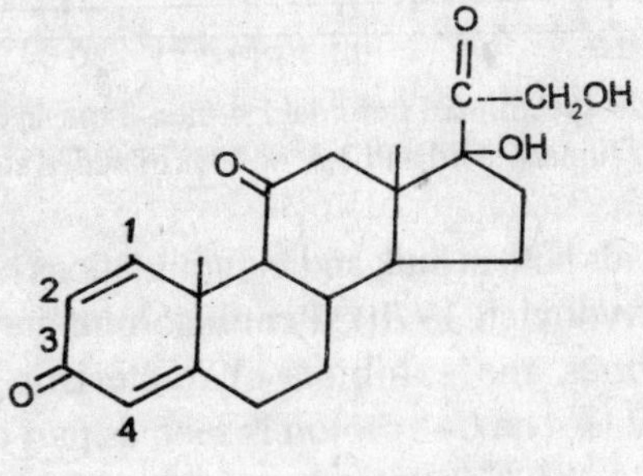

Other spectrum factors which must be considered in quantitative studies are spinning side-band and ^{13}C satellites. These are part of the main proton signal and must be treated consistently in each assay.

7. Surfactant Chain-Length Determination

To establish the ethylene ether chain length of a non-ionic surface-active agent by chemical methods is a lengthy procedure. Nuclear magnetic resonance provides a rapid, accurate and non-destructive method of analysis.

A spectrum and at least five repeat integrals are obtained for a 10% solution of the surfactant in carbon tetrachloride. The ratio of the ethylene ether proton integral to the cetyle terminal chain of the surfactant is determined. From the knowledge that the cetyl end group integral is equivalent to 28 H, the value of *n* for the $—O—(CH_2—CH_2—O)_n$ chain is obtained from a simple proportion calculation. Table 7.2 illustrates how the **NMR** results compare with the manufacturers specifications.

Table 7.2

Detergent	*Chain Length (n)* *Specification*	*NMR Value*
Texofor A 10	10	10
Cetomacrogol 100	22-25	23
Texofor A 45	~ 47	47
Texofor A 60	~ 60	62

8. Hydrogen Analysis

The percentage of hydrogen in an unknown sample may be determined easily and rapidly from the total integral of its PMR spectrum. Results as accurate as those obtained by conventional microanalytical means are claimed. The hydrogen percentage (H%) is given by the expression :

$$H\% = \frac{\text{total integral area}}{kCV} \qquad ....(1)$$

where

k = proportionality constant for the sample tube
C = weight of sample per unit volume
V = effective volume of solution

Two sample tubes of identical diameter are required and both are charged with a known volume (0.532 ml) of a solution *n*-octane in carbon tetrachloride (0.2040 g/ml). A volume calibration mark is made on each tube and one tube is sealed as a permanent reference standard. Total integral values for each sample should be the same if the tube diameters are identical. The unknown compound (10-150 mg) is weighed directly into the matched sample tube and made up to the calibrated volume with carbon tetrachloride. The percentage hydrogen is calculated from the total integral areas, using, eq. (1). Samples, which may be recovered, should be as large as possible in order to minimise errors introduced by the integration of background noise.

9. Iodine Value

Triglycerides have in their PMR spectra four characteristic sets of signals from the resonance of olefinic protons, the four C-1 and one C-3 glyceride methylene protons, methylene protons directly linked to a double bond, and the remaining protons on saturated carbons. The integration curve of the combined C-1 and C-3 glyceride methylene protons, occurring in isolation around 4δ, can be measured accurately, and with these as an internal calibration, the olefinic protons (the degree of unsaturation) and the total number of protons (a measure of the average molecular weight) may be integration (Johnson and Shoolery, 1962) and the molecular weight, agree well, in general, with those estimated by Wijs method (Table 7.3). An exception is Tung oil which, because it possesses conjugated unsaturation, a function of a high oleostearic ester content, gives a low iodine value by Wijs method. The value calculated from PMR data corresponds well with that determined by vapour phase chromatography.

Table 7.3: Iodine Values for a Selection of Natural Oils

Oil	*NMR method*	*Wijs method*
1	*2*	*3*
Coconut	10.5 ± 1.3	8.0 — 8.7
Olive	80.8 ± 0.9	83.0 — 85.3

1	*2*	*3*
Peanut	94.5 ± 0.6	95.0 — 97.2
Soybean	127.1 ± 1.6	125.0 — 126.1
Sunflower seed	135.0 ± 0.9	136.0 — 137.7
Safflower seed	142.2 ± 1.0	140.0 — 143.5
Whale	150.2 ± 1.0	149 — 151.6
Linseed	176.2 ± 1.2	179.0 — 181.0
Tung	225.2 ± 1.2	146.0 — 163.5

10. Moisture Analysis

Water adsorbed on biological materials, such as food products, appears in an NMR spectrum as a relatively sharp band superimposed on the very broad proton absorption of the solid. The moisture content of a powdered solid may be determined from the adsorption curve (Pople, Schneider and Bernstein, 1959), which is rather broader than that of pure liquid water, or more usually from the derivative curve. Comparison is made with a calibration plot of signal amplitude against percentage moisture for samples of known moisture content. The technique is non-destructive and is independent of the particle size of the powder.

11. Hydrogen Bonding

It is possible to make a study of hydrogen bonding in organic compounds and metal chelats using NMR spectroscopy. Hydrogen bonding if present shifts the signals low fields.

If up field shift occurs it may be due to breaking of intermolecular hydrogen bonding.

Schneider, Bernstein and Pople have found that hydrogen bonding arises due to

(i) The electrostatic effect of donor atom

(ii) The diamagnetic anisotropy of the donor atom or group and

(iii) Changes in long range shielding of the acceptor group.

Of the three factors given above factor (i) and factor (ii) are found to predominate in strong and weak hydrogen bonds respectively. Factor (iii) is used to explain gas-to-solution shifts for acetylene and hydrogen bond.

Table 7.4 gives the 5 values (low field shifts due to hydrogen bonding) for different types of acidic protons.

Table 7.4: δ Values of Acidic Protons

Compounds	*δ(ppm)*	*Compounds*	*δ(ppm)*
ArSH	3-4	ROH	0.5-4.5
ArOH	4.5-6.5	RSH	1-2
$ArNH_2$	3-6	R COOH	9-13
ArNHR	3-6	$R\ CONH_2$	5-13
RNH_3	1-5	R CONHR	5-12
RNHR	1-5	$>C=N-OH$	9-12

12. Structure Determination

(a) Complexes of WF_6 : The general formula of complexes of tungsten hexaflurodie is (WF_6L) where L is the ligand. In these complexes, it was assumed by earlier chemists that the tungsten atom might be seven coordinate. When the low-resolution ^{19}F NMR spectrum of these complexes is recorde, it shows three lines of relative intensity 4:1:1, which does not appear to be consistent with a seven coordinate structure. Therefore, it is assumed that the probable structure is $(WF_5L)^+F^-$ in which the tungsten atom is coordinated to five fluorine atoms and one ligand in an octahedral manner. The four coplanar fluorine atoms result in the largest peak, the axial fluorine to one of the other peaks and the fluoride anion forms the remaining peak. The assignment of these three peaks has been confirmed by the high-resolution NMR spectrum of these complexes.

(b) *Structure of* SOF_4 : When the ^{19}F spectrum of SOF_4 is recorded, it shows only one signal in the low field. This reveals that all the four fluorine atoms in the SOF_4 molecular are equivalent. Therefore, the NMR spectrum supports the following structure of SOF_4.

F O F
S
F F

(c) *Structure of ClF_3* : When the ^{19}F spectrum of ClF_3 is recorded, it shows two types of flourine signals. This reveals that there are two types of fluorine atoms in ClF_3. Thus, the NMR spectroscopy supports the following structure :

$$\begin{array}{c} F_B \\ | \\ F_A—Cl—F_A \end{array}$$

(d) *Structure of HF_2^-* : If the ^{19}F magnetic resonance spectrum of HF_2 is recorded, it shows only one signal, supporting a linear structure, F^- —H^+—F^- for the anion.

(e) *Linkage Isomerism* : If a ligand can be bonded to a metal in several alternative ways, NMR spectroscopy can be used in some cases to show which particular linkage isomer is present. An interesting example is the elucidation of the structure of the unusual chelate K[Pt(*acac*)2Cl]. According to Figgis and co-workers, the structure of this complex I is shown Fig. 7.24.

When a solution of K[Pt(*acac*)$_2$Cl] is acidified, a yellow light-sensitive complex is obtained which is readily soluble in organic solvents. when the NMR spectrum of the yellow complex is recorded, it shows three methyl signals at 7.37, 7.94, and 8.08τ in a ratio of 1 : 1 : 2 and a peak at —2.47τ. The latter peak (–2.4τ) is due to a hydrogen-bonded proton. When this data provided by NMR are interpreted alongwith IR information, one may assign the formula II (shown in fig. (7.25) to the yellow complex.

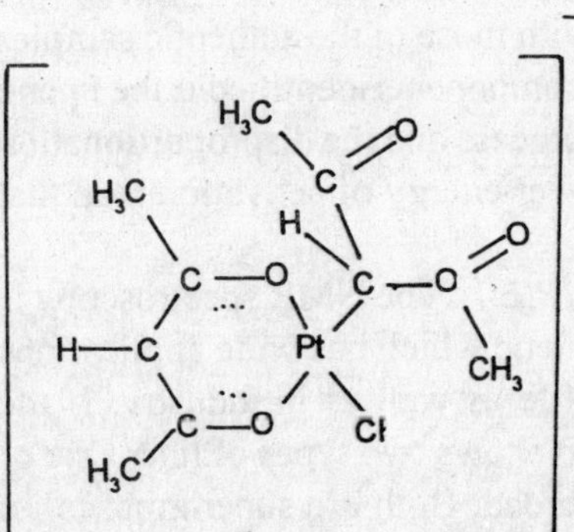

—1

I

Fig.7.24

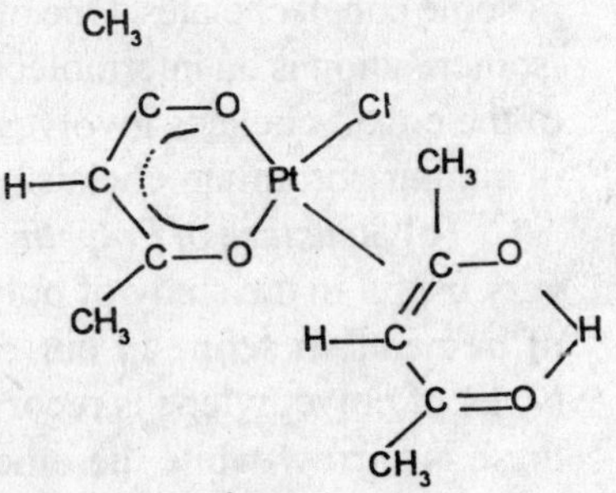

II

Fig. 7.25

Another interesting application of NMA spectroscopy is the elucidation of structure of bis-(dipivaloylmethanido) mercury (II) complex. Its NMR spectrum in $CDCl_3$ shows a resonance peak at 5.08τ. This peak is 1/18 the size of the methyl peak. This is due to the γ-proton, and its position is about 0.5 ppm higher than that in metal chelate. These data together with UV and IR can be interpreted if the non-chelated structure is assigned to the mercury complex. This structure is shown in Fig. 7.26.

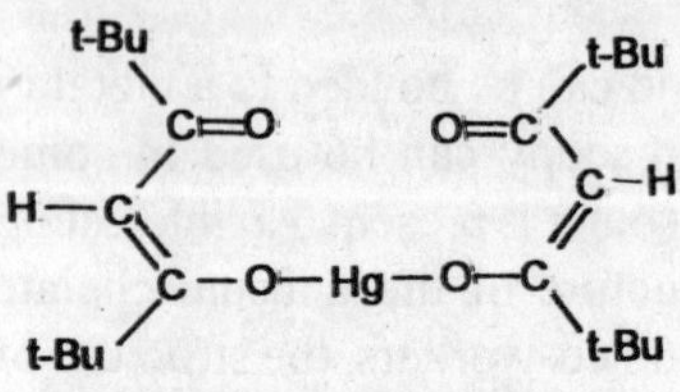

Fig. 7.26

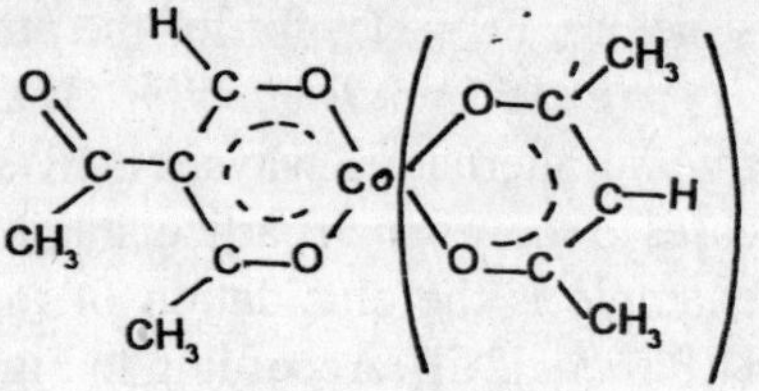

Fig.7.27

(f) *Ligand Isomerism* : NMR spectroscopy has been used by Collman and Jui-Yuan Sun to study the lignad isomerisation and disproportionation of some unsymmetrically substituted metal acetylacetonates. When formyl cobalt acetylacetonate is heated in boiling toluene at 111° for 48 hours, the chelate undergoes ligand isomerization of yield a mixture of six components in the resulting mixture which has been confirmed by thin layer chromatography. Identification of the various components in the mixture is done by comparison of their NMR spectra with those of the authentic samples of some cobalt chelates. One of the components identified is the ligand isomerization is an intermolecular process and the disproportionation of the cobalt chelates involves a lower energy of activation than that of similar chromium chelats.

(g) *Structure of Polyethylene (P.E.)* : The NMR spectroscopy is very useful in the study of polyethylene which has wide applications of tremendous scope in our daily life as well as in industry. If the NMR of polyethylene is recorded, it shows two types of line. One of these is narrow while the other is broad. Both are superimposed on each other. The narrow component is ascribed to the amorphous phase whereas the broad component to the crystalline phase. This result is in agreement with the result obtained by X-ray analysis.

13. Keto-enol Tautomerism

The NMR spectroscopy is very successful in studying keto-enol tautomerism. In the case of acetyleacetone, the following equilibrium exists :

$$CH_3 - CO - CH_2 - CO - CH_3 \rightleftharpoons CH_3 - \underset{\underset{OH}{|}}{C} = OH - CO - CH_3$$

(Keto) (Enol)

The NMR spectrum of acetylacetone is recorded in Fig. 7.28 and then the peaks are assigned to —OH, =CH— and —CH_3 groups of the enol form, and to the —CH_2 and —CH_3 groups of the keto. The assignment of bands is based upon the fact that the ratio of the integrated areas for the methyl and methylene protons in the keto form should be 3:1 and that for the methyl and methine protons in the enol form should be 6 : 1.

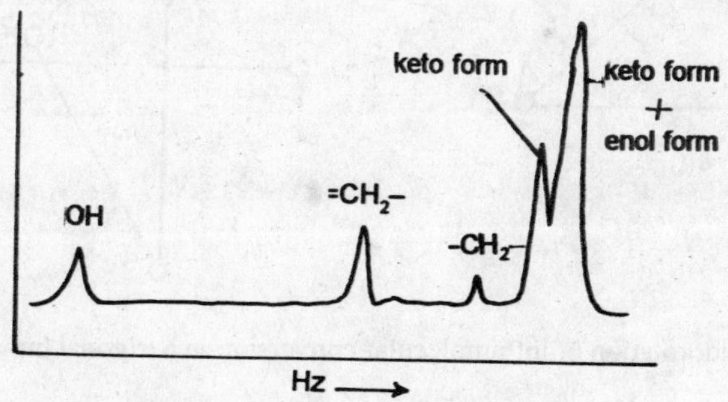

Fig. 7.28: NMR spectrum of liquid acetyle acetone

From the NMR spectrum, the areas of —CH_2— and = CH— peaks are carefully measured and it has been found that the acetylacetone at 25°C contains 18.6 ± 0.6% keto form, this value is more accurate as compared with the value 24% obtained by bromine method.

14. Intermolecular Conversion

From the electron diffraction study of PF_5, it is clear that the compound is a trigonal bipyramid and contains two non-equivalent kinds of fluorine atoms. This clearly shows that the NMR spectrum of PF_5 should give different NMR signals. When the NMR spectrum (^{19}F) of PF_5 is recorded, it shows a 1 : 1 doublet whose separation does not depend upon field strength. This reveals that five fluorines are equivalent and their signal is splitted to a doublet by coupling with the ^{31}P nucleus (spin $\frac{1}{2}$). This rather anomalous result has been explained by suggesting that in PF_5 there occurs intermolecular rearrangements which are rapid enough to prevent the appearance of separate signals for atoms lying at structurally non-equivalent sites in a molecule. Berry called this type of behaviour as "pseudorotation" and gave a schematic representation for this (Fig. 7.29). The pseudorotation is a vibrational process and is not involving any bond breaking process. Thus, this could explain the equivalence of the five fluorines as well as the spin-spin coupling between the ^{31}P and ^{19}F in PF_5.

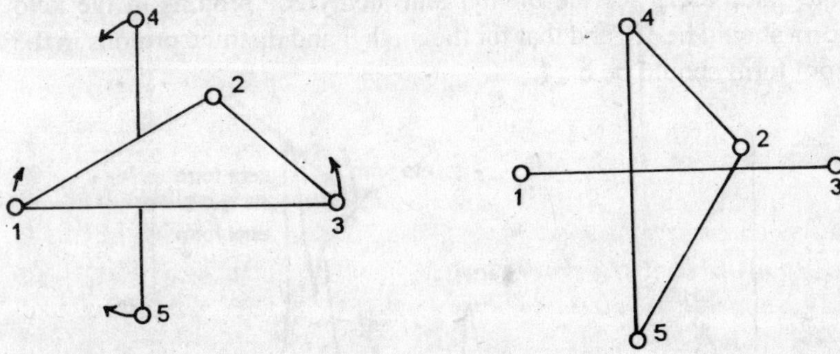

Fig. 7.29 : Pseudorotation or intramolecular conversion in a trigonal bipyamidal molecule.

15. Intermolecular Exchange Reactions

In discussing the high resolution spectrum of ethyl alcohol it should be kept in mind that spin-spin interaction between the hydroxyl group proton and the—CH_2 group is to be taken into consideration. Such a spectrum arises due to the presence of acid or base, because of the following reactions :

$$C_2H_5\text{-}OH + H_3O^+ \underset{}{\overset{k_1}{\rightleftharpoons}} C_2H_5OH_2^+ + H_2O$$

$$C_2H_5OH' + C_2H_5OH_2^+ \overset{k_2}{\rightleftharpoons} C_2H_5O\,HH'^+ + C_2H_5OH$$

$$C_2H_5OH + OH^- \overset{k_3}{\rightleftharpoons} C_2H_5O^- + H_2O$$

$$C_2H_5OH + C_2H_5O'^- \overset{k_4}{\rightleftharpoons} C_2H_5O^- + C_2H_5OH$$

If a nucleus undergoes exchanges between two chemically and hence magnetically different sites, the resonance of the nucleus in these two sites will be broadened if the frequency of exchange is of comparable magnitude to the difference in chemical shifts. From the Heisenberg's uncertainty principle it follows that :

$$\Delta E \Delta t \sim \frac{h}{2\pi}$$

$$\Delta E = h\Delta v$$

$$\Delta v = \frac{1}{\Delta T}$$

Thus the uncertainty in frequency is approximately equal to the reciprocal of the uncertainty in time for which the nucleus has a given energy, which is actually mean lifetime τ, *i.e.*, $\Delta T = \tau$.

The broadening of the hydroxyl lines in ethanol proceeds roughly as shown below:

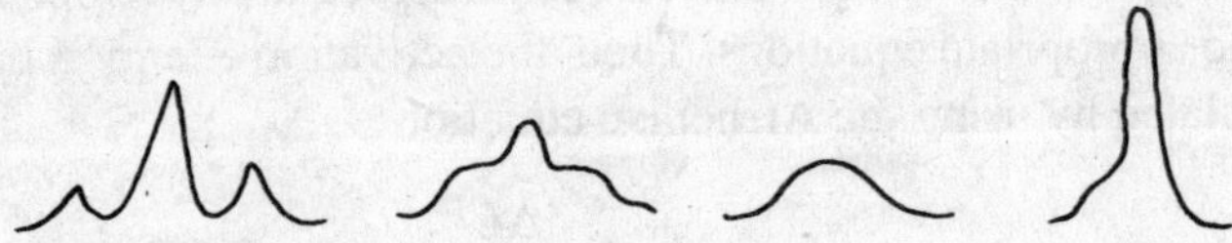

Fig. 7.30 :

As the rate of exchanges increases the—OH triplet because of the spin-spin splitting by the—CH_2 group begins to broaden and finally coalesces into a single sharp line. The –CH_2 lines are similarly affected by the exchange process. The evaluation of rate constants from line broadening measurements is fairly easy for molecules having simple N.M.R. Spectra. For ethyl alcohol, we get

$$k_1 = 2.8 \times 10^6 \ l \text{ mole}^{-1} \text{ sec}^{-1},$$

$$k_2 = 1.1 \times 10^6 \ l \text{ mole}^{-1} \text{ sec}^{-1},$$

$$k_3 = 2.8 \times 10^6 \ l \text{ mole}^{-1} \text{ sec}^{-1},$$

and $$k_4 = 1.4 \times 10^6 \ l \text{ mole}^{-1} \text{ sec}^{-1}.$$

16. Determination of Activation Energy

This application can be illustrated by taking a well studied example, *i.e.*, N. N'-dimethyl-acetamide (DMA)

$$\begin{array}{c} CH_3 \diagdown \diagup CH_3 \\ C-N \\ O \diagup\!\!\!\diagup \diagdown CH_3 \end{array}$$

Because of the large separations, there is no spin-spin coupling between protons on different methyl groups. When the NMR spectrum of DMA is recorded at—30°C, the spectrum shows a single peak and a doublet, indicating that the two methyl groups attached to the nitrogen are not equivalent, Magnetic equivalence of these methyl groups can be realised by increasing the rate of rotation, *i.e.,* by heating the compound. Thus, as the temperature of DMA is increased, the peaks of the doublet begin to broaden and ultimately convert into a single line at about 90°C. The rate of rotation may be given by

$$k = 1/2\tau$$

where k is the rate of rotation and τ is the mean lifetime of the particular configuration which is related to the linewidth and line separation of the methyl peaks. The value of τ can be measured at any temperature using the appropriate equations. Then, the activation energy Δ E may be calculated by using the Arrhenisu equation

$$Ink = In\,A - \frac{\Delta E}{\mathrm{RT}}$$

where A is a constant which is characteristic of the reaction. For DMA, Δ E is 12 cal/mole.

Study of Isotopes other than Protons

In addition to proton a few nuclei which have magnetic moments have been studied by NMR spectroscopy. Some of these are ^{13}C, ^{18}O, ^{19}F, ^{31}P, ^{79}Br etc.

Fluorine (^{19}F)

Its spin quantum number is $\frac{1}{2}$ and it has a magnetic moment of 2.6285 nuclear magnetons. So its frequency is slightly lower than the proton. (It is 56.5 MHz where as that of proton is 60 MHz at 1400 gauss). It is therefore possible to use proton NMR spectrometer with slight modification for study of ^{19}F. Experiments have revealed that fluorine absorption is also sensitive to environment and solvent also plays an important role in determine the peak position.

Phosphorus (^{31}P)

It has a spin number $\frac{1}{2}$ and gives sharp NMR peaks with chemical shift extending over 700 ppm. Its resonance frequency at 1400 gauss is found to be 24.3 MHz.

Carbon 13 NMR (CMR) Spectrosocopy

During the last ten years natural abundance ^{13}C-NMR spectroscopy has progressed from infancy to maturity (Levy and Nelson, 1972; Grutzner, 1972). Instrumentation has developed to such an extent that the technique is rivalling ^{1}H-NMR in versatility.

Almost all the foregoing discussion for ^{1}H-NMR may be applied to CMR, but there are important difference, Table 7.5.

Table 7.5: Nuclear Properties of ^{1}H and ^{13}C

	^{1}H	^{13}C
Nuclear Spin (I)	$\frac{1}{2}$	$\frac{1}{2}$
Resonance Frequency at 23.5 Kgauss	100 MH_2	25.2 MH_2
Natural abundance	99.9%	1.1%
Sensitivity for an equal number of nuclei	1.00%	0.016
Shift Range	20 ppm	600 ppm

Sensitivity

A natural abundance of 1.1% for ^{13}C is clearly a factor that renders CMR less sensitive than PMR. However, because of its low abundance, the probability of a ^{13}C atom residing adjacent to another ^{13}C atom is low. Complications arising from ^{13}C—^{13}C coupling are, therefore, negligible.

A second factor affecting sensitivity is the low magnetogyric ratio (g) of ^{13}C, about $\frac{1}{4}$ that of ^{1}H. Since sensitivity is proportional to γ^3, then ^{13}C affords about 1/60 the sensitivity of ^{1}H in an NMR experiment. Effectively, ^{1}H is about 6000 times more sensitive than ^{13}C in NMR terms, *i.e.*, $\simeq$ 60 (γ factor) $\times$ 100 (abundance factor).

The low sensitivity of CMR is overcome by the use of large samples, up to 2 ml in 15 mm tubes, and by enhancement and decoupling techniques in conjunction with highly stable spectrometers operating at high fields.

Both CAT and *Fourier Transform* have been described and are used to obtain ^{13}C spectra. The latter is particularly useful because of the speed of spectrum accumulation.

Coupling of ^{13}C to ^{1}H not only complicates spectra from the standpoint of interpretation but it give, instead of a single sharp resonance signal, a multiplicity of bands often of low intensity. Heteronuclear decoupling of each ^{1}H individually from each ^{13}C signal would be a lengthy and tedious process. The problem is overcome by *proton noise decoupling.* During a noise decoupling experiment, all the protons in the sample are decoupled simultaneously and each ^{13}C resonance band is seen as a single line. Proton noise decoupling also disturbs the ^{13}C energy level populations and results in a nuclear Overhauser effect affording an enhancement of each ^{13}C signal intensity by up to a factor of three.

By employing a combination of the above techniques, a ^{13}C natural abundance spectrum may be obtained from as little as 10 mg of sample. Much smaller amounts of compound are needed if the sample is ^{13}C-enriched.

Chemical Shift

Chemical shift is usually the most important spectral parameter in CMR spectroscopy. Whereas ^{1}H resonance occur over a relatively narrow range (10 – 10 ppm), ^{13}C shifts cover a range of about 600 ppm, with most resonances falling within a 200 ppm range.

The signal spread and the application of proton noise decoupling results in spectra with each carbon atom in a molecule being displayed as a single discrete sharp band. As in PMR, ^{13}C shifts are measured relative to a TMS internal standard. In general, ^{13}C shifts fall into well defined ranges according to the electronic and magnetic environment of the carbon (Fig. 7.31).

Lanthanide shift reagents are occasionally employed in CMR spectroscopy, but because of the low incidence of resonance band overlap, they are much less useful than in ^{1}H investigations.

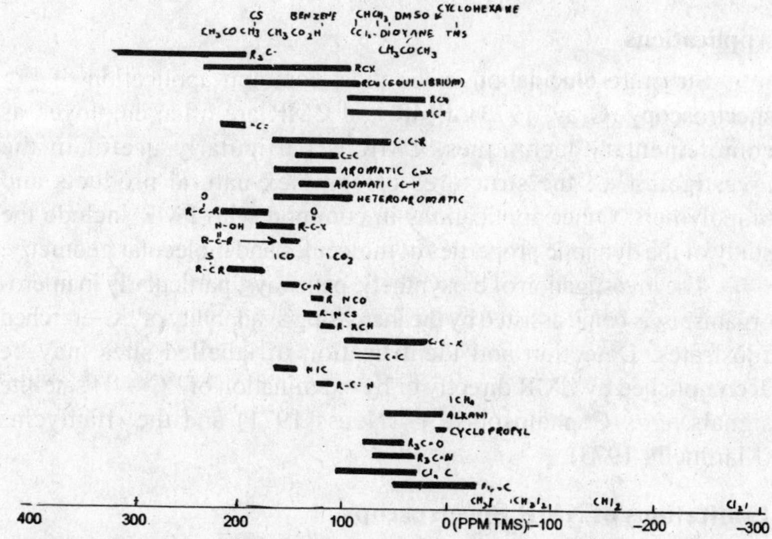

Fig. 7.31: General ^{13}C chemical shift Chart

Spin-Spin Coupling

For the reasons outlined, above $^{13}C—^{13}C$ and $^{13}C—^{1}H$ couplings are not usually seen in CMR spectra. The former may be examined in ^{13}C-enriched samples and the latter can sometimes be extracted from ^{13}C satellites in ^{1}H spectra ($J_{13O-1H} \simeq 150$ Hz). In the absence of proton noise decoupling, ^{13}C spectra are very complex. However, selective decoupling techniques may help to determine specific $^{13}C—^{1}H$ coupling constants.

Integration

At present in normal CMR spectra, there is an correlation between integrated peak areas and the number of ^{13}C nuclei in each signal. Integrals, are therefore, not generally measured. A Fourier Transform derived spectrum is based upon the spin-lattice relaxation of each ^{13}C atom and the intensity of each band is not dependent solely upon the number of ^{13}C atoms responsible for the signal. Similarly, because of the variation in relaxation times, proton noise decoupling results in variable nuclear Overhauser enhancement.

Modification of the Fourier Transform technique is being explored as a means of quantitating ^{13}C signals.

Applications

Structure elucidation is the most common application of $^{13}_{6}C$ spectroscopy (Gray, 1973). PMR and CMR are often employed as complementary techniques. CMR is particularly useful in the investigation of the structures of complex natural products and biopolymers. Other applications in common with PMR include the study of the dynamic properties of molecules and molecular geometry.

The investigation of biosynthetic pathways, particularly in micro-organisms, is being assisted by the increasing availability of ^{13}C-enirched substrates. Detection and identification of labelled sites may be accomplished by CMR directly or by examination of $^{13}C—^{1}H$ satellite signals, *e.g.,* Cephalosporin C (Neuss 1971) and the rifamycins (Martinelli, 1973).

Limitations of NMR Spectroscopy

The various limitations of NMR spectrosocpy are listed below:

(i) One of the serious problems with NMR is its lack of sensitivity. The minimum sample size is about 0.1 ml having minimum concentration of about 1%.

(ii) In some compounds, two different types of hydrogen atoms resonate at similar resonance frequencies. This results in an overlap of spectra and makes spectra difficult to interpret.

(iii) In most of the cases only liquids can be studied by NMR.

(iv) While characterising the organic compounds, no information about molecular weight is given but the relative number of different protons are only known.

In spite of the above mentioned limitations, NMR is one of the most useful analytical tools yet developed for the elucidation of many organic structural problems.

NMR Spectrum—Analysis

The position of proton resonance signals has been shown to be dependent on shielding and deshielding arising from inductive effects, hydrogen bonding, etc. However, proton resonance in certain regions of the NMR spectrum is indicative of certain types of protons. For example :

Aliphatic—C—H protons which do not have an adjacent atoms except hydrogen and sp^3 hybridized carbons, absorb in the region

(8.2—10τ) (0—1.2δ).

Aliphatic—C—H protons which have adjacent halogen, O or N atoms (Cl—C—H, O—C—H, N—C—H) or sp^2 or *sp* hybridized carbons (>C=C—C, —H, —C ≡C—C—H) absorb in the region 5.0—8.5τ (1.55—5δ).

Acetylenic—C≡C—H protons absorb in the region 7.0—8.2τ (1.8—0.0τ).

Olefinic—C=C—H protons absorb in the region 2.5—5.5τ (4.5—7.5δ).

Aromatic—(Benzenoid and heterocyclic) protons absorb in the region 0.5—4.0τ (1.6—9.5δ)

Adehydic —C(=O)—H protons abosrb in the region 0.0—1.0τ (9–10δ). Even though there is a considerable overlap, yet the generalizations made are very useful.

Some important information of the NMR analysis for a few compounds is given below :—

Toluene

C_6H_5—CH_3
b *a*

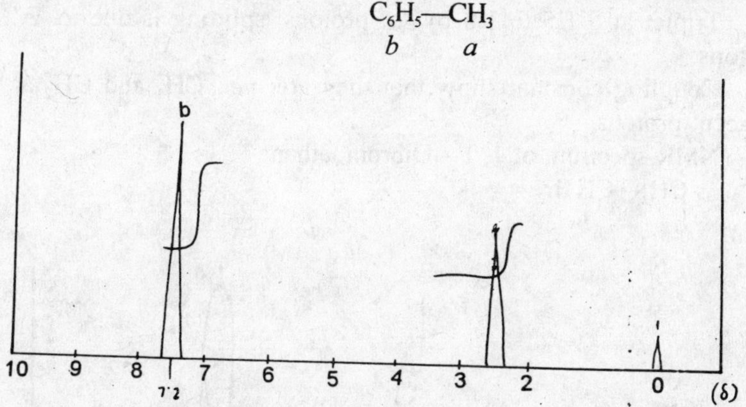

NMR spectrum of Toluene

Peak area counting show that the ratio of *b* : *a* is 5.3
i.e., 5 protons are of same type.
and 3 protons are of same type.

Singlet at δ 2.3 indicates 3 aliphatic hydrogen, *i.e.*, CH_3 group.

Singlet at 7.2 indicates 5 aromatic protons.

NMR spectrum of Ethyl bromide

CH_3—CH_2—Br
a *b*

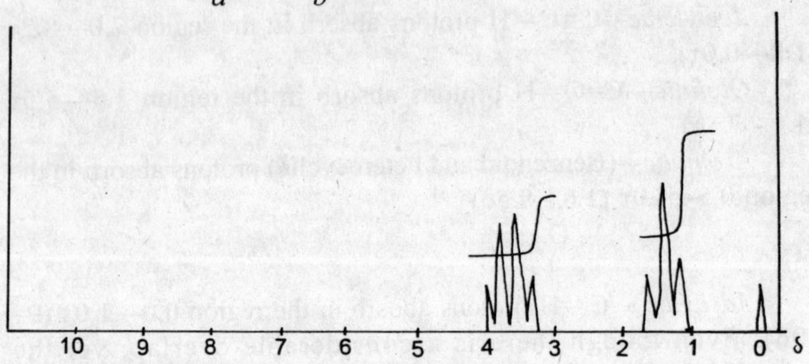

Peak area counting show the ratio of protons, *i.e.*, *a* : *b* is 3 :2

Quartet at δ 3.2 to 3.6 is formed by *'b'* protons due to splitting by *'a'* protons.

Triplet at δ 1.5 to 1.8 by *'a'* protons, splitting is due to *'b'* protons.

Coupling constant show that they are, *i.e.*, CH_3 and CH_2 at adjacent protons.

NMR spectrum of 1, 1—Dibromoethane

CH_2—CH Br_2
a *b*

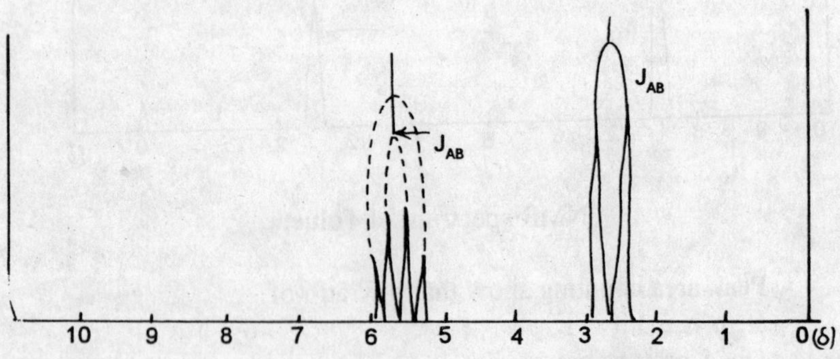

Integration show the ratio of protons *a* : *b* is 3 : 1.

A doublet at δ 2.4 to 2.6 (of CH_3 group) splitting caused by proton *'b'*.

quarter at δ 5.7 to 6 indicates—CH splitting by—CH_3.

Coupling constant *Jab* shows that —CH_3 and CH are coupled *i.e.,* they are at adjacent position.

NMR spectrum of Ethyl alcohol

c	*b*	*a*
CH_3—	CH_2—	OH
1.17	3.62	5,37 δ

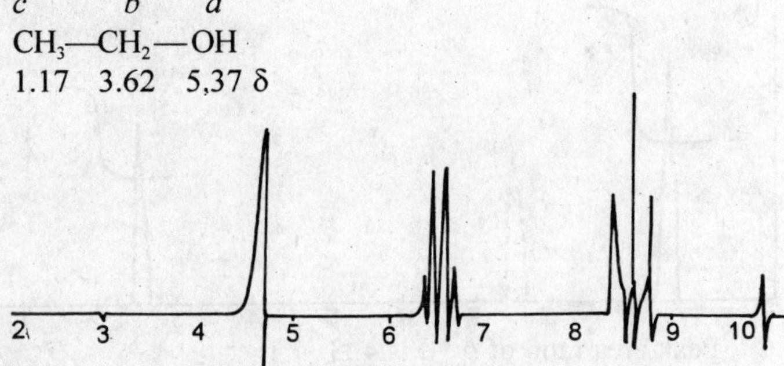

Peak area counting indicates proton ratio of *a* : *b* : *c* is 1 : 2 : 3.

a triplet at 1.17 (of CH_3) is due to *'b'* protons

a quartet at δ 3.62 (of CH_2) due to *'c'* protons

Singlet at δ 5.37 is of *'a'* protons.

Coupling constant J 7.1 *cps* indicates that

—CH_3 and —CH_3 are at adjacent position.

NMR spectrum of acetaldehyde

Peak area *b* : *a*

1 : 3

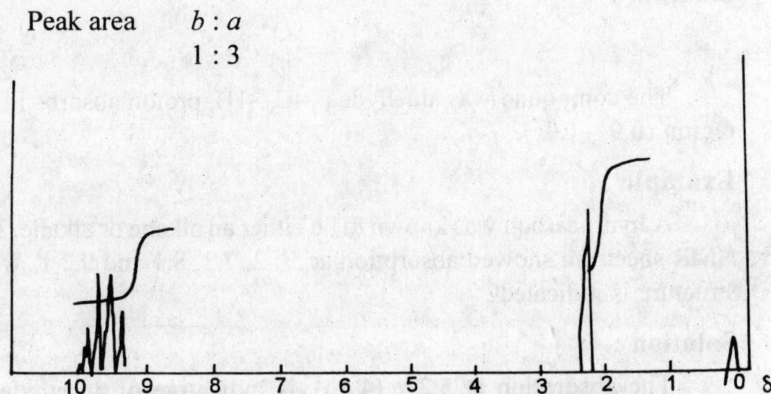

Singlet at ∂ 2.15 indicates —CH_3 group.
quartet at δ = 9.68 is of *'b'* protons
Splitting is by *'a'* protons.

$$\overset{b}{NO_3}\text{—}C_6H_4\text{—}\overset{a}{CH_3}$$

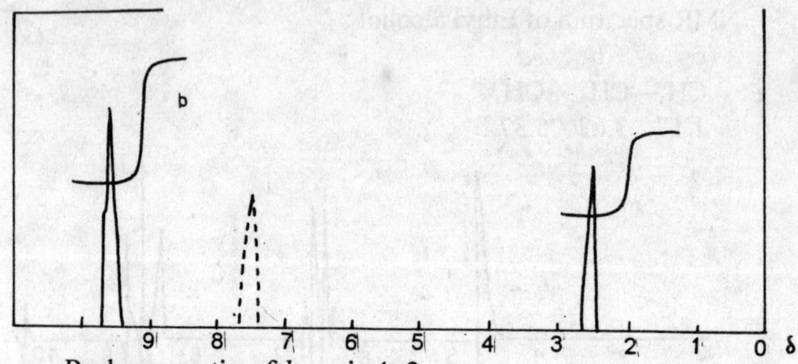

Peak area ratio of *b* : *a* is 4 :3
Singlet at 2.3 δ of —CH_3 group.
In toluene the aromatic protons are at δ = 7.2.

Due to —NO_2 group the curve shifts to lower field by δ = 2.4 and is obtained at δ = 9.6.

Example

A compound was believed to be either an alipahtic aldehyde or a ketone. The NMR spectrum showed and absorption at 0.10 τ (9.9δ). What conclusion could be drawn?

Solution :

The compound was aldehyde ($\text{—}\overset{\overset{\displaystyle O}{\|}}{C}\text{—}H$), proton absorbs in the region (0.0—1.0τ).

Example

A hydrocarbon was known to be either an alkene or alkane. The NMR spectrum showed absorption at 5 : 2, 7.3, 8.1 and 9.2 τ. What structure is indicated?

Solution :

The absorption at 5.2 τ (4.8δ), is indicative of an ethylenic proton. The hydrocarbon must be an alkene.

SOLVED PROBLEMS

1. In what region of the spectrum would the labelled protons of the following compounds be expected to absorb?

(i) $\underset{(a)}{CH_3} - \overset{O}{\overset{\|}{C}} - \underset{(b)}{CH_2} - \underset{(c)}{CH_3}$ (ii) $\underset{(d)}{CH_3} - \underset{(c)}{CH_2} - \underset{(b)}{CH_2} - \overset{O}{\overset{\|}{C}} - \underset{(a)}{H}$

(iii) $\underset{(a)}{CH_3} - \underset{(b)}{CH} = \underset{(c)}{CH} - \overset{O}{\overset{\|}{C}} - \underset{(d)}{H}$ (iv) $\underset{(a)}{CH_3} - C \equiv C - \underset{(b)}{H}$

Ans. (i) (a) and (b) in 5—8.5τ and (c) in 8.2—10τ region.

(ii) (a) in 0.0—1.0 τ, (b) in 5—8.5τ, (c) and (d) in 8.2—1.0τ region.

(iii) (a) 5—8.5τ, (b) and (c) in 2.5—5.5τ and (d) in 0.0—1.0τ region.

(iv) 5—8.5τ, and (b) in 7.0—8.2τ region.

2. In what regions of the spectrum would the labelled protons of the following compounds be expected to absorb?

(i) $Cl—\underset{(a)}{CH_2}—\underset{(b)}{CH_3}—OH$ (ii) $C_6H_5CH(CH_3)_2$ *(a) (b) (c)*

Ans. (i) (a) and (b) in the 5—8.5τ region.

(ii) (a) in 0.5—4.0τ, (b) in 8.2—10.0τ region.

Example 3

Sketch out integrated spectra for ethyl alcohol and acetaldehyde. You are provided with a table of chemical shifts for various protons.

Solution :

Use table of chemical shifts.

Sketch is given in (a) and (b) below :

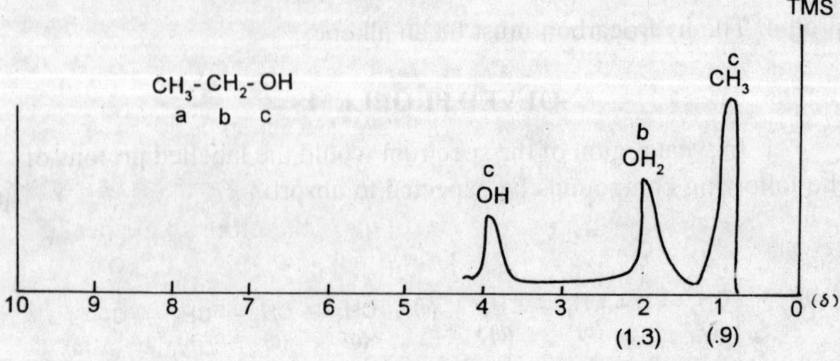

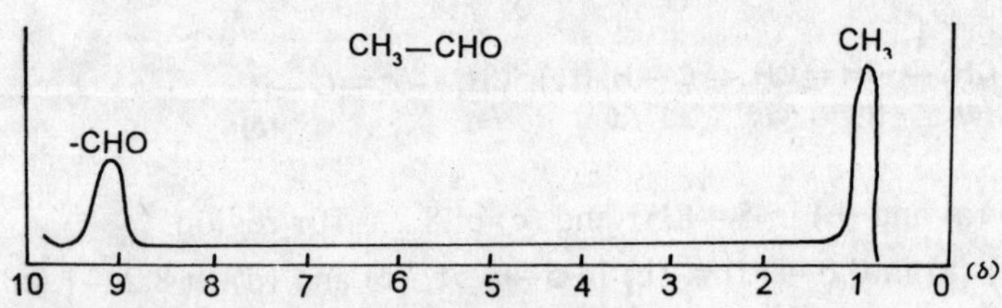

Example 4

Give the relative positions of the signals and their multiplicity in the compound given below :

(i) $CH_3 - CH_2 - CHO$

(ii) $CH_3 - \overset{\overset{\displaystyle O}{\|}}{C} - O - H$

(iii) $C_6H_5-CH_3$ (benzene ring bearing CH_3)

(iv) $CH_3 - CH_2I$

Solution :

(i) $\underset{a}{CH_3}-\underset{b}{CH_2}-\underset{c}{\overset{\overset{\displaystyle H}{|}}{C}}=O$

Sequence *a,* then *b* then *c* ——→ down-field.

(a) protons give triplet, and (b) protons produce quintet and (c) protons give triplet.

(ii) $CH_3—\overset{\overset{O}{||}}{C}—O—H$
(*a*) (*b*)

Sequence *a, b* ——→ down field.

For (*a*) and (*b*) protons no splitting, therefore single peaks.

H H CH_3 *b* H H H *a*

Sequence *b* to *a* ——→ downfield.

No splitting between *a* and *b* protons.

(iv) $CH_3—CH_2—I$
a *b*

Sequence *a* to *b* downfield.

(*a*) protons give triplet and (*b*) protons give quartet.

Example 5

The NMR spectrum of a molecule exhibits a multiplet complex splitting pattern having a singlet, a heptet and a doublet. The peak area ratios are 4 : 3 : 1 : 6. Give the possible number of hydrogens in the molecule? If the compound contains only C and H and its molecular weight is 133 ± 2, give the formula of the compound?

Solution :

The total number of protons is a multiple of the sum of the simplest ratio of the peak areas.

The number of hydrogens could be a multiple of

4+3+1+6 = 14, *i.e.,* 14, 28 or 42......

The number of carbon atoms per molecule is calculated using the formula,

No. of C atoms per molecule

$$= \frac{\text{M. wt.} - \text{Mass of H in molecule} - \text{Mass of other atoms}}{12}$$

$\therefore$ No. of C atoms per molecule $= \dfrac{133 \pm 2 - 14}{12} = 10$

Formula is=$C_{10}H_{14}$

Higher multiples of 14 for the number of hydrogens are not possible.

e.g., for 28 the value is $\dfrac{133 \pm 2 - 28}{12} = 9$

And hence formula would be C_9H_{28} which does not exist.

Example 6

What type of NMR spectrum is expected of 1, 2, 2 trichloropropane $CH_3CCl_2CH_2Cl$.

Solution :

The structural formula of the compound is

```
      H   Cl   H
      |    |   |
  Cl—C—C—C—H
      |    |   |
      H   Cl   H
     (a)      (b)
```

In the above structure, two kinds of protons (*a*) and (*b*) are present on 1 and 3 carbons. since these are not adjacent no splitting of NMR signals is expected by either of them.

Hence the chemical shift expected shall be

```
                     H
                     |
(i) Singlet due to —C—H at τ 7.80 (δ 2.20) 3H
                     |
                     H
```

and

```
                       H
                       |
(ii) Singlet due to H—C— at τ 5.98 (δ 4.20), 2H.
                       |
                       Cl
```

Example 7

A hydrocarbon which has a molecular weight of 140 ± 3 and it has two sets of NMR peaks with proton ratios of 3 : 7. Assign the possible formulae of the compound? Further the NMR, IR and UV spectra for the compound indicated the presence of an aromatic ring. What is the formula of the compound?

Solution :

The number of hydrogens = 3 + 7 = 10 or 20......

$$\text{Number of carbon atoms} = \frac{140 \pm 3 - 10}{12} = 1$$

$$\text{or} \qquad = \frac{140 \pm 3 - 20}{12} = 10$$

$$\text{or} \qquad = \frac{140 \pm 3 - 30}{12} = 9$$

∴ Possible formula is $C_{11}H_{10}$ or $C_{10}H_{20}$ because C_9H_{30} does not exist.

A compound with C_nH_{2n}, *i.e.,* $C_{10}H_{20}$ cannot be aromatic since the simplest aromatic system has the general formula C_nH_{2n-6}.

Therefore $C_{11}H_{10}$ could be aromatic.

(It would contain other units of unsaturation in addition to one aromatic ring).

Example 8

In the given organic compound the three kinds of Proton appear at 50, 150, and 200 Hz using 60 MHZ NMR spectrometer. What will be their relative positions 90 MHZ spectrometer?

Solution :

(i) $\frac{50}{60} \times 90 = 75$ Hz

(ii) $\frac{150}{60} \times 90 = 225$ Hz

(iii) $\frac{200}{60} \times 90 = 300$ Hz

Example 9

Integrated NMR spectra for Organic compound having molecular formula :

C_7H_8 is shown below. Predict the structural formula for this compound.

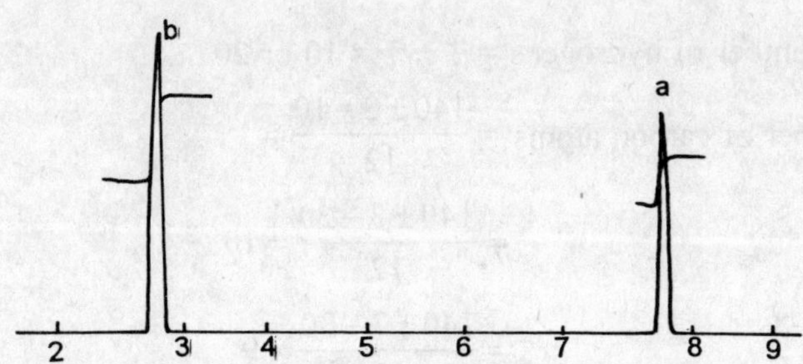

Solution :

(i) C_7H_8. Two signals = 2 types of protons
Peak intensity = $a : b$ or 3 : 5
'b' Protons are at low field, *i.e.,* in the aromatic region
'a' Protons are at high field. Possibility is *b* *a*

C_6H_5—CH_8

Example 10

NMR spectrum of a molecule with a mol. wt. of 90 ± 2 gives two singlets with a ratio of 1 : 3. The molecule has only C, H and O. Assign the possible formulae assuming one oxygen atom per molecule?

Solution :

The number of hydrogens = 1+3 = 4, 8, 16.....
Number of carbon atoms

$$\text{(i)} = \frac{90 \pm 2 - 4 - 16}{12} = 6; \ (C_5H_4O)$$

(ii) $= \dfrac{90 \pm 2 - 8 - 16}{12} = \dfrac{66 + 2}{12}$ (not a whole number)

(iii) $= \dfrac{90 \pm 2 - 16 - 16}{12} = 5$ ($C_2H_{16}O$) does not exist.

This formula is C_6H_4O.

Example 11

What is the likely number of proton signals and their spin-spin splitting for the following compounds :

(i) CH_3CH_2OH (ii) $CH_3—CH_3$

(iii) $CH_3—CH_2—\overset{\overset{\displaystyle O}{\|}}{C}—CH_2$

Solution :

(i) CH_3CH_2OH (ii) $CH_3—CH_3$ (iii) $CH_2—CH_2—\overset{\overset{\displaystyle O}{\|}}{C}—CH_3$

a b c No. of signals *a b c*

No. of signals = 3 = one No. of signals = Three

For (*a*) protons=triplet No. splitting as For (*a*) signals = Triplet

Example 12

Give structural formulae for organic compounds whose NMR spectra are represented below. The molecular formulae : $C_2H_3Br_2$, $C_2H_4Br_2$, C_2H_5Br and $C_6H_{14}O$ are given in the sequence (i) to (iv).

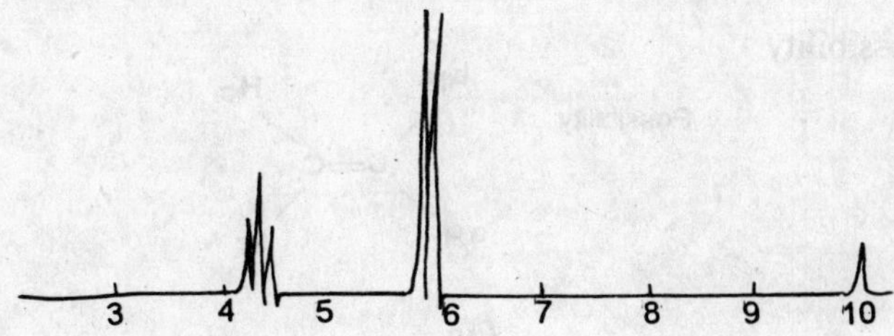

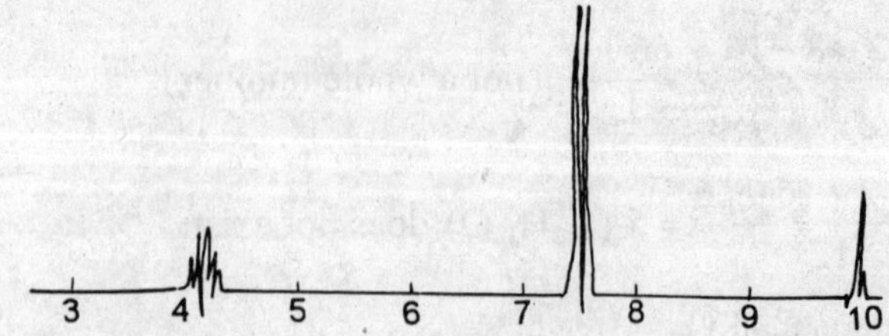

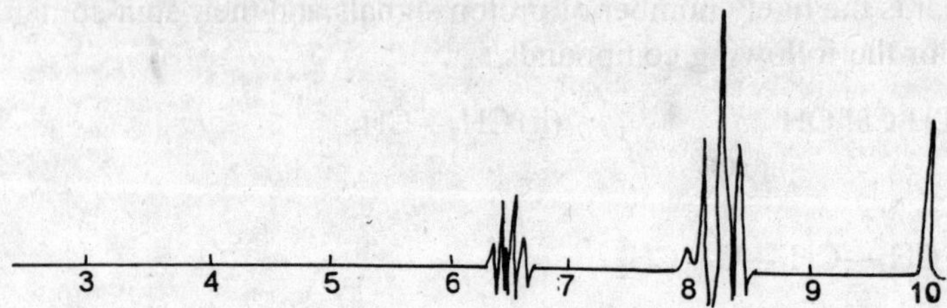

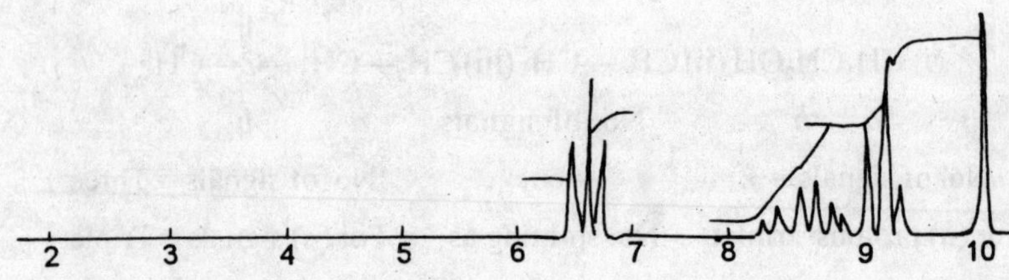

Solution:

(i) C_2H_2Br Three signals. Three types of protons.

Triplet due to $—CH_2$

Also the compound is unsaturated

Possibility

Possibility

^bH H^c

C=C

^aH Br

Fig.

8

NQR Spectroscopy

Introduction

NQR (nuclear quadrupole) is a property associated with all nuclei with spin quantum number $1 \geq 1$. Such a nuclei possesses an electric quadrupole moment eQ in addition to magentic moment. eQ is a measure of deviation of positive charge in a nucleus from its spherical symmetry. Transitions in these level are observed in the radio frequency region of electromagnetic spectrum. This technique can be used fro the study of isotopes of many elements such as chlorine, bromine, iodine, nitrogen, arsenic, antimony, tin etc. An arrangement of charges having a quadrupole moment is found in the classical linear quadrupole + – – +. Since it possesses no charge, no force or torque is experienced by them in a homogeneous field which tend to align it in a preferred orientation. But in contrast to the classical quadrupole, the nuclear quadrupole has an intrinsic spin angular momentum which couples with the torque and produces a precession and its frequency can be found out by quadrupole moment (eQ) and the gradient of electric field (q).

Theory of NQR

Those nuclei which possess, integral and half integral nuclear spins such as 1, $\frac{3}{2}, 2, \frac{5}{2}$....etc. are found to possess nuclear electric quadrupole moments which interact with the electric field gradients which are created at the nuclei due to the symmetric distribution of charge arising from the extranuclear electrons or the bonding electrons in the molecules of which the nuclei form a part. The NQR energy levels for both the integral and half integral spins (*i.e.*, I $\geq$1) are given by

$$E_m = \frac{eQq}{1I(21-1)}[3m_i^2 - I(I+1)]$$

$$= A[3m_i^2 - I(I+1)]$$

where $$A = \frac{eQq}{4I(21-1)}$$

and $m_i = -I, -I+1, \ldots, +I.$

Here Q (or *e*Q) is the nuclear quadrupole moment, *q* is the electric field gradient (EFG) and *e*Q.*q* is the product of these quantities and is known as nuclear quandrupole coupling constant. I is the nuclear spin, *m*, is the component of I along the axis of quantization which is taken to be *z*-axis. The electric field gradient which is a tensor quantity has nine components. These nine components are defined as under:

$$q_{xx} = -\frac{\partial F_x}{\partial x} = \frac{\partial^2 V}{\partial x^2}$$

$$q_{xy} = -\frac{\partial E_x}{\partial y} = \frac{\partial^2 V}{\partial x \partial y}$$

....

....

$$q_{2x} = -\frac{\partial F_x}{\partial z} = \frac{\partial^2 V}{\partial z^2}$$

It is assumed that nucleus is situated at the origin of the co-ordinate system *x, y, z* and V is the time averaged electrostatic potential that is produced at the nucleus by all charges outside it.

The electric field (E) of all the charges is given by

$$\vec{E} = \vec{E}_{xi} + \vec{E}_{yj} + \vec{E}_{zk}$$

where *i, j, k* are the unit vectors and they specify the directions of *x, y, z* axis respectively. There relationship between the components of electric fields and electrostatic potential is given by the following equations:

$$E_x = -\frac{\partial V}{\partial x},\ E_y = \frac{\partial V}{\partial y},\ E_z = -\frac{\partial V}{\partial z}$$

Thus $$\vec{E} = -\left[\frac{\partial V}{\partial x}\vec{i} + \frac{\partial V}{\partial y}\vec{j} + \frac{\partial V}{\partial z}\vec{k}\right]$$

The field gradient components can be written as 3 × 3 matrix

$$\begin{Bmatrix} q_{xx} & q_{xy} & q_{xz} \\ q_{yx} & q_{yy} & q_{yz} \\ q_{zx} & q_{yz} & q_{zz} \end{Bmatrix}$$

They are symmetric matrix (tensor) with $q_ij = qj_i$, thus

$$q_{xy} = q_{yx},\ q_{zx} = q_{xz},\ q_{yz} = q_{xy} \qquad \text{.... (1)}$$

So the electric field gradient (EFG) tensor can be represented as

$$\begin{Bmatrix} q_{xx} & q_{xy} & q_{xz} \\ q_{xy} & q_{yy} & q_{yz} \\ q_{xz} & q_{yz} & q_{zz} \end{Bmatrix} \qquad \text{.... (2)}$$

The diagonal component of this tensor will obey the Laplace equation.

$$q_{xx} + q_{yy} + q_{zz} = 0 \qquad \text{.... (3)}$$

i.e., $$\frac{\partial^2 V}{\partial x^2} + \frac{\partial^2 V}{\partial y^2} + \frac{\partial^2 V}{\partial z^2} = \nabla^2 V = 0$$

The Laplacian operator ∇^2 is given by

$$\nabla^2 = \frac{\partial^2}{\partial x^2} + \frac{\partial^2}{\partial y^2} + \frac{\partial^2}{\partial z^2}$$

The relations as given by equations (1) and (3) reduce from nine to five, the number of independent tensor components. Generally such coordinate systems are chosen in which the cross components q_{ij} are zero. Such a system of coordinates is generally referred to as principal axis system. The electric field gradient (EFG) in such a system are designated as

$$q_{xx},\ q_{yy},\ q_{zz}.$$

Such a system can be defined as

$$|q_{zz}| \geq |q_{yy}| \geq q_{xx} \qquad \text{..... (4)}$$

Here the modulus sign || is used to indicate absolute value of the components

From equations (3) we know that the sum of the components of EFG is zero so we can easily conclude that only two parameters are needed to completely define the EFG tensor

The asymmetry parameter can be defined as under

$$\eta = \frac{q_{xx} - q_{yy}}{q_{zz}} \quad \text{.... (5)}$$

From equations (3), (4) and (5) it can be easily seen that the asymmetry parameter lies between 0 and 1.

When $\eta = 0$, it is axial symmetry around z-axis.

In case of axial symmetry around z-axis, we have

$$q_{xx} = q_{yy} = \tfrac{1}{2} q_{zz} \quad \text{..... (6)}$$

If $q_{xx} = q_{yy} = q_{zz}$, then the electric field gradient is referred to a spherical. In such a case no interaction of quadrupole moment takes place with electronic charge distribution.

If $q_{xx} \neq q_{yy} \neq q_{zz}$, then the electric field gradient is called as non-symmetric.

Instrumentation

In case of crystalline solids it is possible to observe quadrupole splittings in the NMR spectra if the coupling constant is not too large, *i.e.*, $eQ.q_{zz} \sim 1$ MH_2 approximately. The much larger splittings can be observed in dilute gases using microwave spectroscopy. If the value of coupling constant lies between 3 MHz and 500 MHz then the most accurate measurements can be carried out by direct NQR spectroscopy as given below:

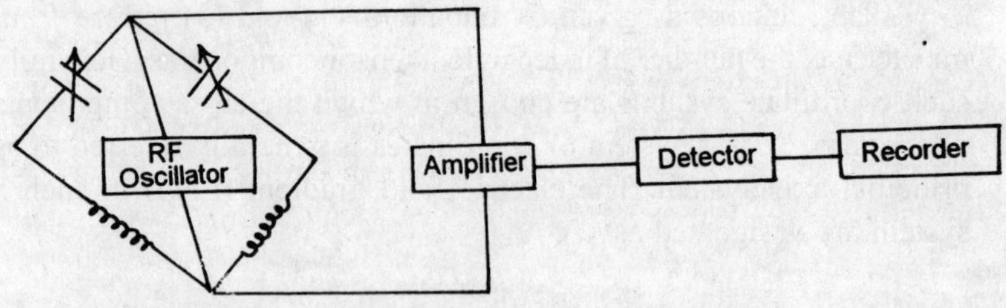

Fig. 8.1: Depicts a NQR spectrometer.

Power is supplied to a previously balanced bridge circuit from a *r—f* oscillator. At resonance, sample absorbs energy in the inductive coil which causes an imbalance in the circuit. The resultant output voltage is then amplified and recorded. The spectra so obtained are known as **pure quardrupole resonance spectra.**

NQR spectra can be obtained only for solids.

In case of liquids and gases, at normal pressure collision between molecules changes the axis of rotation continuously and the EFG at the nucleus will be averaged to zero.

Due to the limitations given above the field of application of NQR is limited in comparison to the field of application of NMR.

Sample Requirements

The sample should satisfy the following requirements so as to produce observable resonance.

(i) The sample should be a crystalline solid. In most of the cases it is desirable as well as necessary to cool the sample to 77 K.

(ii) Sample size is found to depend on the strength of resonance. In case of *p*-dichlorobenzene only 50 mg of sample produces observable ^{35}Cl resonance line but in case of hexamethylene tetra amine we need a few grams of sample to produce a detectable ^{14}N resonance.

(iii) The sample coil should be completely surrounded with the sample while searching for a new resonance.

Applications of NQR

NATURE OF CHEMICAL BOND

1. Nature of Chemical Bond

It is possible to determine the nature of chemical bond by making use of NQR spectroscopy. For this a comparison is done in atomic and molecular state for the same nucleus. This comparison provides necessary information about the the extent of hybridisation and the % age of ionic character of the bond. Since a large number of unknown quantities are involved in these so NQR data can not provide exact information about the type of bond present and the same data could be interpreted differently by different scientists. For example, taking the case of H_2S about which we know that the electronic configuration of S is $3s^2\ 3p^4$ and HSH bond angle is 90°. It was generally believed that pure *p*-orbitals of sulphur are participating in bond formation but by NQR measurements on H_2S molecule it was found that it has a large value for asymmetry parameter ($\eta = 0.60$) for sulphur. In any asymmetric molecule, the EFG along the three axis may be given as —

$$eQq_{xx} = -\left(\frac{N_y + N_z}{2} - N_x\right) eQq \text{ atom}$$

$$eQq_{yy} = -\left(\frac{N_x + N_z}{2} - N_y\right) eQq \text{ atom}$$

$$eQq_{zz} = -\left(\frac{N_x + N_y}{2} - N_z\right) eQq \text{ atom}$$

Where N_x, N_y and N_z give the effective electron population in the p_x, p_y, p_z orbital of the atom respectively.

$$\eta = \frac{q_{xx} - q_{yy}}{q_{zz}}$$

$$= \frac{3(N_y - N_x)}{N_x + N_y - 2N_z}$$

$$= -0.60$$

If a lone pair is present in p_z orbital then $N_z = 2$ and $1.33\ N_y = N_x + 1$ but N_x can not be equal to N_y, d hence there can not be any pure *p*-bonding. To explain the large value of asymmetry parameter we have to consider about 15% *d*-character.

2. Study of Chloromethanes and Chloracetyl Chlorides

By the comparison of coupling constants for ^{35}Cl in various chloromethanes it was found that if we increase the number of chlorine atoms, and increase in coupling constant by ~3.5 MHz per chlorine atoms takes place (Table 8.1).

This extraordinary chemical shift can be explained on the basis of increase of electronegativity of the entity to which chlorine atom is bonded, through an inductive effect.

Table 8.1: Coupling Constants in Chloromethanes and Chloromethyl Chlorides

[*eq*Q/*h* Mhz]

Molecule	*R = H*	*R = COCl*
RCH_2Cl	68.0	60 .8
$RCHCl_2$	72.0	65.0
$RCCl_3$	76.6	67.4
CCl_4	81.2	—

In case of chloroacetyl chlorides, an increase of coupling constants takes place with the effective electronegativity of the group bonded to the carbonyl chloride. Generally the chlorines which are bound to non-carbonyl carbons show coupling constants which are approximately 10 MHz higher in comparison to those of carbonyl chlorines.

3. Structure of Group III Halides

A number of group III halides (MX_3) have been studies by making use of NQR technique. While studying the spectrum of these halides three resonance lines are observed. Out of these two are closely spaced and are much above the third. It means the three halogen atoms are not chemically equivalent. From X-ray data of these halides it has been concluded that these halides exist as dimers and they contain two types of halogen which correspond to the bridge and end positions.

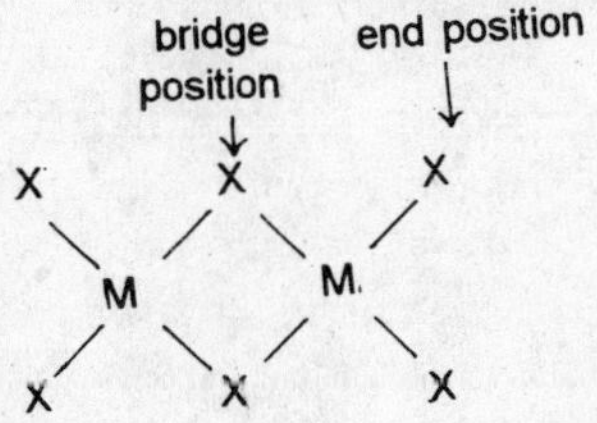

The small difference of one set of resonance frequencies can now be explained on the basis of a slight difference in the field around the chemically equivalent set of atoms.

4. Study of Charge Transfer Compounds

It was believed that charge transfer compounds are formed when CBr_4 and *p*-xylene, CCl_4 and *p*-xylene or bromine and benzene are brought together. Frozen solutions of these mixtures were studied by Hooper and hefound that the resonance frequencies of the halogens are not much different in such compounds than those in the pure molecular state. It was therefore concluded that these charge transfer compounds are not formed in ground state.

5. The simplest application of nuclear quadrupole resonance spectroscopy is for the detections of the same isotope in different environments. Thus it has been shown that all the bromine atoms in $K_2[SeBr_6]$ are equivalent above – 33°C, but that between this temperature and – 52°C two types of crystallographically non-equivalent bromine atoms are present. Below – 52°C there are three types of bromine atoms present. This suggests a progressive drop in lattice symmetry (*e.g.,*cubic → tetragonal → orthorhombic). Phase transitions in several other complex halides of formula type M^I_2 [M^{IV} × 6] [x = Cl, Br, I] have been found.

Limitations of NQR Spectroscopy

(i) We can obtain NQR spectra only for solids so it has only a limited application.

(ii) Moderately large sample requirement for NQR spectroscopy also limits the applications of this technique.

9

Raman Spectroscopy

Introduction

When a beam of light is passed through a transparent substance, a fraction of radiant energy is scattered at all angles from the direction of the beam by the molecules or aggregates of molecules present. The fraction of the beam that is scattered is undetectable to the eye when the particles are of molecular dimensions. This type of scattering is called *Rayleigh scattering.*

If monochromatic radiation, or radiation of a very narrow frequency band is used the scattered energy will consist almost entirely of radiation of the incident frequency (Rayleigh scattering) and in addition certain discrete frequencies above and below that of the incident beam will be scattered. This scattering of certain frequencies above and below that of incident beam is called *Raman scattering.* Such a study was carried out by sir C.V. Raman in 1928. This study also confirmed the predictions made by Smekel in 1923 that if a substance in the gaseous, liquid or solid state is irradiated with monochromatic light, the scattered light should contain radiation with different frequencies than the frequency of incident light. The line whose wavelengths have been modified in Raman effect are known as *Raman lines.* The lines whose wavelength is greater as compared to that of the wave length of incident light are called *stocks's lines* and lines having wavelength shorter than the wavelengths of incident light are called *Anti-stokes lines.*

Stokes radiations is generally more intense than Anti-stokes. A sensitive apparatus is needed for its study. The Raman shift, Δv is given by the relation

$$\Delta v = v_i - v_z$$

Where v_i is the frequency of incident radiation and v_z the radiation scattered by the given molecular species.

Raman shift is independent of the frequency of incident light but it is considered as a characteristic of the substance that causes Raman effect, Δv is positive for stokes lines and Δv is negative for Anti lines.

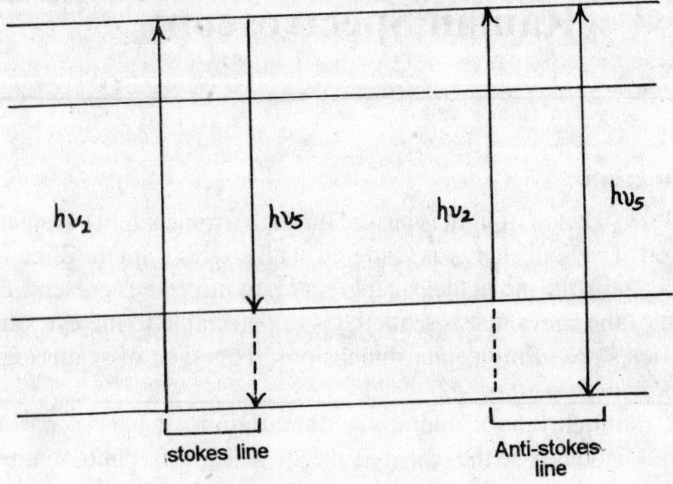

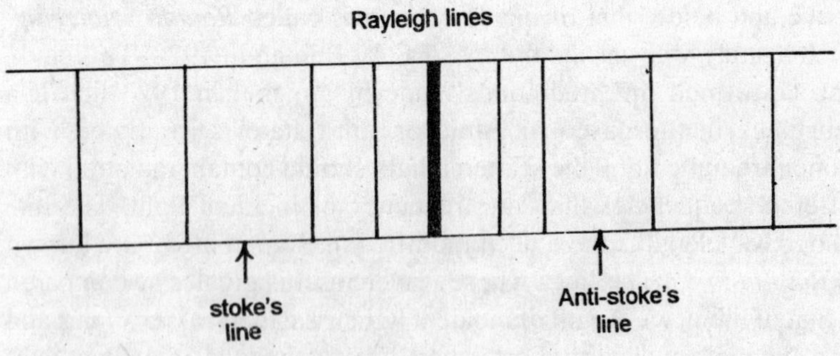

Fig.

Characteristic Properties of Raman Lines

Raman lines exhibit the following characteristics:

(i) The intensity of stoke's lines is always greater in comparison to that of corresponding Anti-stoke's lines.

(ii) Raman shift (ΔV) is generally observed in Infrared (far and near) region of the spectrum.

(iii) The Raman lines are symmetrically displaced about the parent lines. With increase in temperature there occurs a decrease in separation of individual Raman lines from the parent lines.

(iv) The frequency difference between the modified and parent line represents the frequency of absorption band of the material under investigation.

Vibrational Modes Associated with the Raman Effect

The transition responsible for Raman effect are similar to those as for infrared absorption. *The transitions between the ground state of molecule and its first vibrational state are involved in both phenomenon.* Therefore, for a given molecule the energy shifts observed in a Raman experiment should be identical to the energies of infrared absorption bands, if the virabtional modes involved are active towards both infrared absorption and Raman scattering Fig. 9.1 illustrates the similarity of two types of spectra.

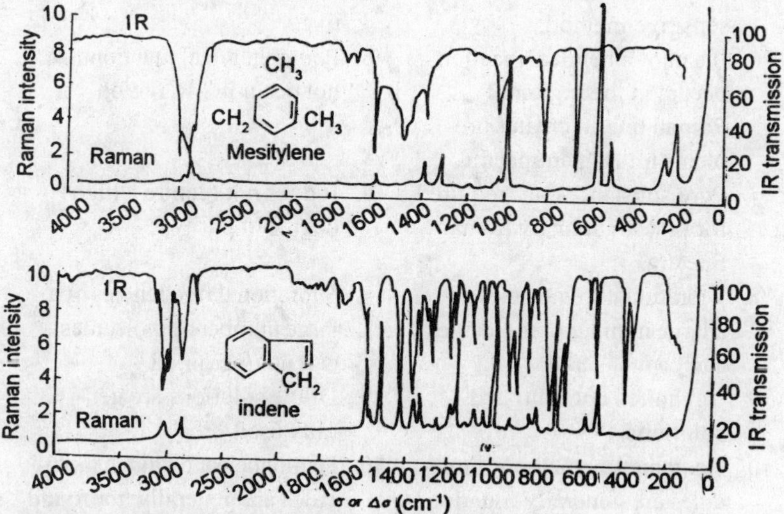

Fig. 9.1: Comparison of Raman and infrared spectra

Also note that the relative size of the corresponding peaks is quite different. Moreover certain peaks which are present in more spectrum are absent in the other.

The main points of difference between Raman Spectrum and Infrared Spectrum are given in Table 9.1.

Table 9.1: Comparison between Spectra and Infrared Spectra

Raman Spectra	*Infrared Spectra*
1. It arises because of scattering of light by the vibrating molecules.	1. It arises because of absorbtion of light by vibrating molecules.
2. Raman spectra depends on the *polarisability* of the bond.	2. I.R. Spectra is given by only such molecules which have a resultant dipole moment.
3. Raman spectra can be recorded only in one exposure.	3. The recording of I.R. spectra requires at least two runs in order to cover the whole of I.R. region.
4. It can be taken with water as solvent.	4. For I.R. spectra water cannot be used as solvent.
5. It is accurate but not a sensitive method.	5. It is accurate as well as sensitive.
6. If a photochemical reaction occurs in the region of Raman lines it creates problems for Raman spectra.	6. Photochemical reactions do not occur in I.R. region.
7. Pure substances are required for investigation by Raman Spectra.	7. It does not require a high degree of purity.
8. Vibrational frequencies of a large number of molecules can be measured.	8. Vibrational frequencies of a large number of molecules can not measured.
9. It requires concentrated solutions.	9. Dilute solutions are genrally used.
10. Hamonuclear diatomic molecules are generally found to be active.	10. Hamonuclear diatomic molecules are generally not found to be active.

Classical Theory of Raman Effect : Molecular Polarizability

The classical theory, though not wholly adequate, is able to explain the existence of the Raman effect and leads to an understanding of a concept basic to this from of spectroscopy—*the polarisability of a molecule.* Polarisability decreases with increasing electron density, increasing bond strength and decreasing bond length.

When a molecule is placed in static electric field a distortion takes place in it because of the attraction of positively charged nuclei towards negative pole of the field and of electron towards positive pole. This separation of charge centres causes an *induced electric dipole moment* in the molecule and the molecule becomes polarised. The magnitude of induced dipole, μ, depends both on magnitude of the applied field E and on the case with which the molecule can be distorted. Thus,

$$\mu = \alpha E$$

where α is the *polarisability* of the molecule.

In case of H_2 molecule (a diatomic) the polarisability is *anisotropic, i.e.,* the electron that form the bond are more easily displaced by an electric field along bond axis than along the one across this direction. The polarisability in various direction can be conveniently represented by polarisability ellipsoid Fig. 9.2.

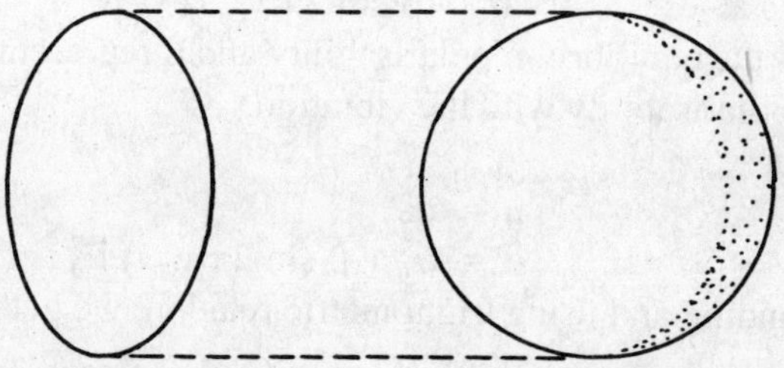

Fig. 9.2: The hydrogen molecule and its polarizability ellipsoid seen from direction at right angles.

The ellipsoid is a three-dimensional surface whose distance from the electrical centre of the molecule (in the case C.G. also) is proportional to $1/\sqrt{\alpha_i}$, where α_i is polarisability along the line joining point *i* on ellipsoid with electric centre. In case of a diatomic molecule the polarisability in the same for all directions at right angles to the bond axis, the ellipsoid has circular cross section in this direction, so it is shaped like a tangerine. The shape of ellipsoid of all diatomic molecules like O_2, HCl and of polyatomic linear molecule are similar to this except that the relatives sizes of their major and minor axis are different.

When a diatomic molecule or a linear polyatomic molecule is subjected to a beam of radiation of frequency ν, the electric field experienced by each molecule varies according to the following equation

$$E = E_0 \sin 2\pi \nu t.$$

and so the induced dipole undergoes oscillations of frequency ν.

$$\mu = \alpha E$$
$$= \alpha . Eo \sin 2\pi \nu t.$$

Such an oscillating dipole gives out radiations of its own oscillation frequency and so it explains the Rayleigh's scattering. We have not taken into consideration the vibration and rotation of molecule while deriving the above equation. These also affect the polarisability of the molecule.

(a) Effect of Vibration

In such a case the oscillating dipole will be have super imposed upon it vibrational oscillation. The change in polarisability due to a frequency of ν_{vib} can expressed as

$$a = \alpha_0 + \beta \sin 2\pi \nu_{vib} . t.$$

where α_0 is the equilibrium polarisability and β represents the rate of change of polarisability with the vibration.

Since,

$$\mu = \alpha E$$
$$\therefore \quad \mu = (\alpha_0 + \beta \sin 2\pi \nu_{vib} t) E_0 \sin 2\pi \nu t.$$

Expanding and using trignometric relation, we get

$$\mu = \alpha_0 E_0 \sin 2\pi \nu t + \tfrac{1}{2} \beta E_0 [\cos 2\pi (\nu - \nu_{vib}) t - \cos 2\pi (\nu - \nu_{vib}) t]$$

$$[\therefore \text{Sin A Sin B} = \tfrac{1}{2} [\cos (A - B) - \cos (A + B)]$$

so the oscillating dipole has frequency components $\nu + \nu_{vib}$ as well as exciting frequency ν. So the Raman shift will be equal to ν_{vib}.

$$\text{Raman shift} = (\nu + \nu_{vib}) - \nu$$
$$= \nu_{vib}.$$

(b) Effect of Rotation

Again we consider a diatomic molecule, when such a molecule rotates its orientation varies with respect to the electric field of rotation. When such a molecule is not optically isotropic, the polarisation will vary with time. In such a case the variation of α can be expressed by the equation

$$\alpha = \alpha_0 + \beta' \sin 2\pi\, (2vr)\, t$$

where v_r is the frequency of rotation.

This equation is identical with the equation obtained while considering the effect of vibration. However in this equation we have used ($2v_r$) and not v_r, this is so because a rotation through an angle π brings the diatomic molecule in such a position in which the polarisation becomes the same as initially.

Substituting the above equation in equation for dipole, *i.e.*,

$$\mu = \alpha E = \alpha E_0 \sin 2\pi\, vt$$

we get,

$$\mu = \alpha_0 E_0 \sin 2\pi\, vt + \tfrac{1}{2}\, \beta'\, E_0\, [\cos 2\pi\, (v - 2\, v_r)$$
$$t - \cos 2\pi\, (v + 2\, v_r)\, t]$$

Thus the frequency of Raman lines will be ($v + 2v_r$) and ($v - 2v_r$) and Raman shift will be $2v_r$. Thus Raman shift in this case will be twice the frequency of rotation of molecule.

It should be clearly understood that if the vibration does not change polarisability of a molecule then β=0 and in such a case dipole oscillates only at the frequency of the incident radiation. Similarly if the rotation does not change polarisability of a molecule then β'=0 and again the diple oscillates only at the frequency of the incident radiation. Thus we get the following general rule.

For a molecule to be Raman active the molecular rotation or vibration must produce some change in a component of molecular polarisability. A change in molecular polarisability is reflected by a change in either the magnitude or the direction of the polarizability ellipsoid.

Quantum Theory of Raman Effect

It is easy to understand Raman effect on the basis of quantum theory of radiations. This theory considers Raman effect to be the outcome of the collisions between the light photons and molecules of the substance. It imagines perfectly elastic collisions taking place between the light photons with energy hv and molecules of mass m in energy state E_p moving with a velocity v. After collision if E_q is the energy state of the molecule and v' is its velocity, the we have

$$E_p + \frac{1}{2}mv^2 + hv = Eq + \frac{1}{2}mv'^2 + hv'$$

where v, v' the velocities of photon before and after collision respectively.

It can be easily proved that the velocity of molecule practically remain unchanged.

so we have

$$E_p + hv = E_q + hv'$$

or
$$v' = v + \frac{E_p - E_q}{h}$$

or
$$v' = v + \Delta v \qquad .....(1)$$

Following three possibilities arise from Equation (1).

(a) If $E_p = E_q$, then $\Delta v = 0$ because $E_p - E_q$ is zero and so $\frac{E_p - E_q}{h} = 0$. Thus in such a collision the molecule simply deflects the photon so it is analogous to Rayleigh scattering.

(b) If $E_p > E_q$, then $v' > v$. This corresponds to anti-stoke's lines. In this case molecule transfers some of its energy to the photon.

(c) If $E_p < E_q$, then $v' < v$. This corresponds to stoke's lines. In this case molecule absorbs some energy from the photon.

Pure Rotational Raman Spectra

Linear Molecules

In case of linear molecules the energy of rotational energy levels is given by

$$E_J = BJ(J+1) - DJ^2(J+1)^2 \text{cm}^{-1} \qquad (J=0,1,2....)$$

In case of Raman spectroscopy, we generally use the simplified form of this equation, *i.e.*,

$$E_J = BJ(J+1) \text{ cm}^{-1} \qquad (J=0,1,2....)$$

Normal selection rules are followed for any transitions between these levels, *i.e.*,

$$\Delta J = 0 \quad \text{or} \quad \pm 2 \text{ only}$$

(contrast it with selection rule of microwave, $\Delta J = \pm 1$)

If $\Delta J = 0$, the frequency of Raman line will be the same as that of incident radiation, so it corresponds to Rayleight scattering.

If ΔJ= ±2, it corresponds either to Stoke's lines (ΔJ= +2) or anti-Stoke's lines (ΔJ–2).

The value of rotational Raman shift in the two cases will be

$$\Delta\nu = \pm\, B(4J + 6)\ cm^{-1}$$

The (+) plus sign if for stoke's lines and (–) minus sign is for anti-stoke's lines.

These lines could be labelled as S **branch lines.** These S branch lines appear to the low wave number side of the exciting line (stoke's lines) if the molecules gains rotational energy from photon and these S branch lines appear to the high wave number side of (anti-stoke's lines) in case the molecule loses energy to the photon.

These transitions are shown in Fig. 9.3.

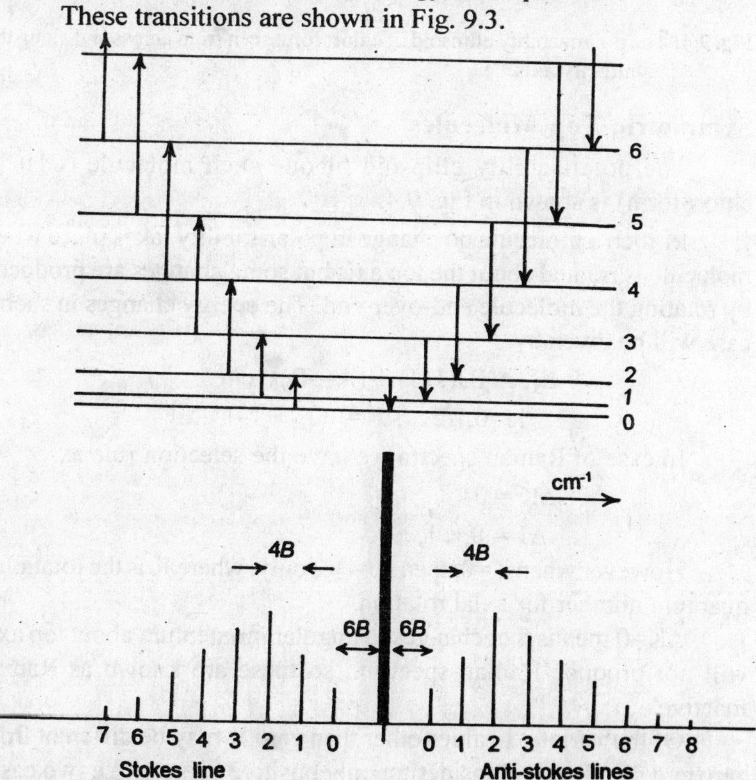

Fig. 9.3: The rotational energy levels of a diatomic molecule and the rotational Raman spectrum arising from transitions between them. Special lines are numbered according to their lower J values.

It has been made clear in this diagram that the separation of first line from the exciting line is 6B cm^{-1} but the separation between successive lines is 4B cm^{-1} only.

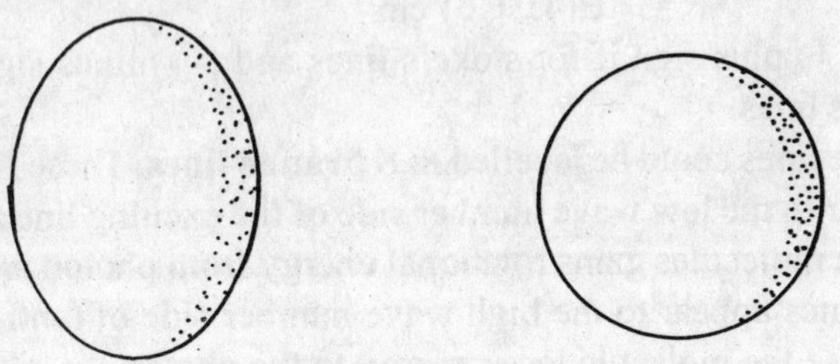

Fig. 9.4: The polarizability ellipsoid of chloroform seen from across and along the symmetry axis.

Symmetric Top Molecules

The polarisability ellipsoid of one such molecule ($CHCl_3$, chloroform) is shown in Fig. 9.4.

In such a molecule no change in polarisability takes place when molecule is rotated about the top axis but some changes are produced by rotating the molecule end-over-end. The energy changes in such a case will be given by

$$E_{J,K} = BJ(J+1) + (A—B)k^2 \text{ cm}^{-1}$$

$$[J=0,1,2...., k = \pm J, \pm(J-1),....]$$

In case of Raman spectra we have the selection rule as

$$\Delta K = 0$$

$$\Delta J = 0, \pm 1, \pm 2$$

However when K=0, then $\Delta J = \pm 2$ only, where K is the rotational quantum number for axial rotation.

$\Delta K=0$ means that changes in angular momentum about top axis will not produce Raman spectrum so these are known as Raman inactive.

Of the various J values other than zero K may be different from zero and so $\Delta J = \pm 1$. Considering only positive value of ΔJ, two cases arise

(a) $\Delta J = +1$ (R-branch lines)

$$\Delta_{VR} = 2B\,(J+1) \text{ cm}^{-1} \qquad (J = 1, 2, 3....)$$

(b) ΔJ= +2 (S-branch lines)

$\Delta \nu R = B\,(4J+6)\ cm^{-1}$ (J= 0, 1, 2....)

Thus is Raman spectrum we shall have two series of lines. Corresponding to **R-branch** and **S-branch.** These are shown in Fig. 9.5.

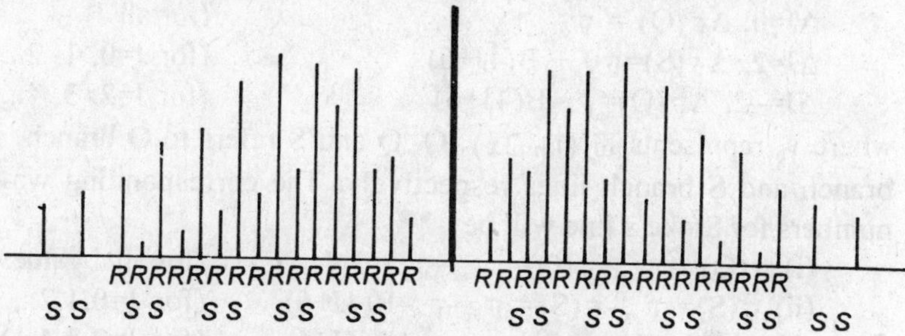

Fig. 9.5: Rotational Raman spectrum of a symmetric top molecule. The R and S branch lines are shown separately in (*a*) and (*b*) respectively, with the total spectrum in (*c*).

Spherical Top Molecules : Asymmetric Top Molecules

Methane (CH_4), Silane (SiH_4) etc. are spherical top molecules. These molecules possess tetrahedral symmetry and the polarisability ellipsoid of such molecules have spherical surface. Thus the rotation of such a ellipsoid produces no change in polarisability. Hence pure rotations of such molecules remain inactive in Raman spectra.

Generally all rotations of a symmetric top molecules are found to be Raman active. However these give rise to complicated Raman spectra and such a spectra can be interpreted by considering the molecule as intermediate between the oblate and prolate symmetric top molecules.

Vibrational-Rotational Raman Spectra

On theoretical considerations it can easily be predicted that vibrational and rotational transitions may take place simultaneously in a Raman transition. Using the selection rule for such simultaneous transitions, we have

$$\Delta J = 0, \pm 2 \qquad \text{and } \Delta v + \pm 1$$

In case of a diatomic molecule, the vibrational rotational energy levels are given by

$$E_{vr} = hc[\omega_e\,(V+\tfrac{1}{2}) - \omega_e x(V+\tfrac{1}{2})^2] + Bh(J(J+1)$$

where V= 0, 1, 2....

and J= 0, 1, 2....

or the frequency $\bar{v}$ is

$$v = \omega_e(V+\tfrac{1}{2}) - \overline{\omega}_e x(V+\tfrac{1}{2})^2 + BJ(J+1)$$

Since J= 0, ± 2, we have

$\Delta J=0$, $\Delta\bar{v}(Q) = \bar{v}_0$ (for all J)

$\Delta J=2$, $\Delta\bar{v}(S)=\bar{v}_0 + B(4J+6)$ (for J=0, 1, 2....)

$\Delta J=-2$, $\Delta\bar{v}(O)=\bar{v}_0 - B(4J+6)$ (for J=2, 3, 4....)

where v_0 represents $\overline{\omega}_e(1-2x)$. O, Q and S refers to O branch, Q branch and S branch lines respectively. The corresponding wave numbers for Stoke's line will be

(i) $\bar{v}(Q) = \bar{v}_{ex} - \bar{v}(Q) = \bar{v}_{ex} - \bar{v}_0$ (for all J values)

(ii) $\bar{v}(S) = \bar{v}_{ex} - \bar{v}(S) = \bar{v}_{ex} - \bar{v}_0 - B(4J+6)$ (for J=0,1,2,...)

(iii) $\bar{v}(O) = \bar{v}_{ex} - \bar{v}(O) = \bar{v}_{ex} - \bar{v}_0 + B(4J+6)$ (for J=2,3,4...)

Fig. 9.6 depicts various rotation-vibration transitions in Raman spectrum. In the Raman spectrum, **Q branch** will be present but is not found in IR spectrum. *Both S and R branches are some what similar to R and P branches found in IR spectra.*

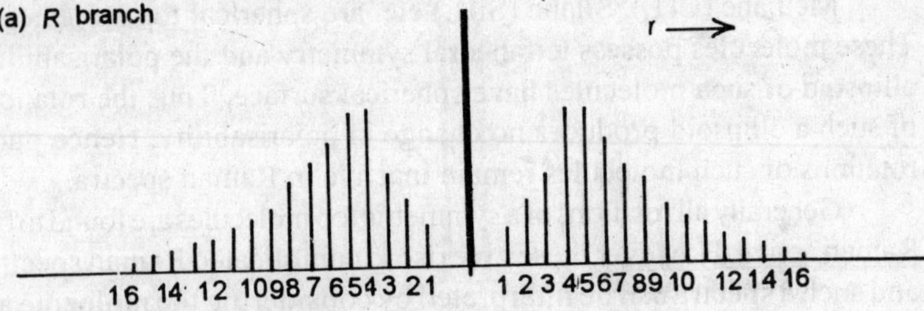

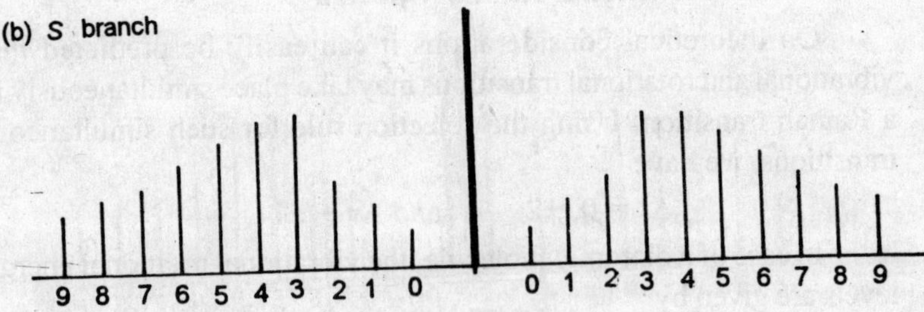

Fig. 9.6:

Techniques and Instrumentation

Raman spectroscopy is essentially *emission spectroscopy*. Instrumentation for modern Raman spectrosocpy consist of three components, namely, an intense source, a sample illumination system and a suitable spectrophotometer. The distinguishing characteristic of Raman work, of course, is the exciting source.

Fig.9.7 shows the basic instrumentation required for obtaining a Raman spectrum.

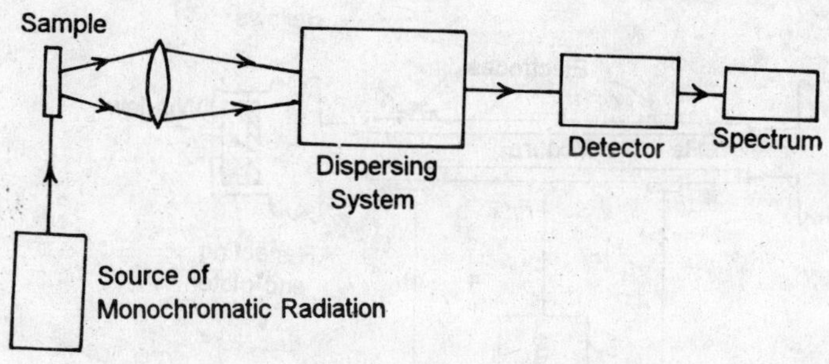

Fig. 9.7: Schematic representation of the Raman apparatus

In consists of a **Raman tube** and the sample (liquid) under investigation is placed in it. This is a glass tube about 1-2 cm in diameter and 10-25 cm in length. One of the ends of this tube is drawn like a horn and blackened outside. It provides a suitable background. The other end of the tube is closed with an optically plane glass plate. The scattered light comes out through the window. A water jacket surrounds the Raman tube. Cold water is circulated through water jacket to avoid overheating of the sample present in sample tube. There is a helium discharge tube (source) and it is the source of light. The light emerging from filter is a monochromatic light. Because of difficulties experienced in obtaining monochromatic light from helium discharge tube it is preferred to use mercury arc. The mercury arc is placed quite close to the sample tube containing the sample and a semi-cylindrical conductor is used for enhancing plane window are directed to the slit of the spectrograph by a lens placed in front of the window. Finally the Raman lines are obtained on the photographic plate.

1. Source

Laser is almost ideal as a Raman source. It gives a very narrow, highly monochromatic beam of radiation, that may be focussed very finely into a small sample and which packs a relatively large power—from several milliwatts to a few watts, depending on the type of laser—into its small frequency spread.

The most widely used Raman source is probably helium/neon laser which operates continuously at a power of 50 mW.

The helium-neon laser is shown in Fig. 9.8.

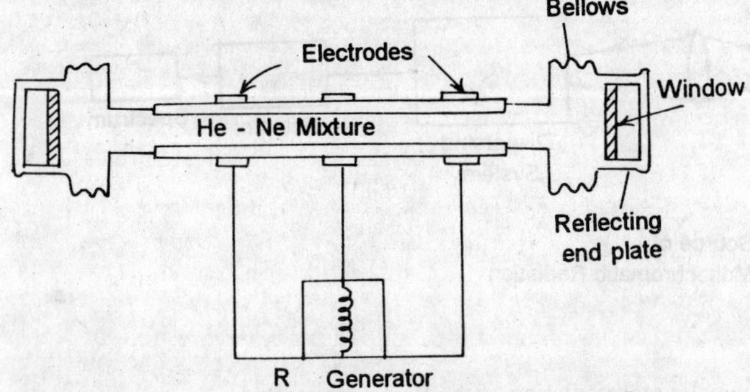

Fig.9.8: Simplified schematic diagram of a helium-neon laser

The tube is filled with a 7:1 mixture of helium and neon gas to get optimum output of 6328 A° line. In the beginning 5—10 *kv* d.c. is needed to start the laser however afterwards the beam can be maintained on lower operating voltage.

The Raman effect arises when a beam of intense monochromatic light passes through a sample that contains molecules that can undergo a change in molecular polarizability as they vibrate. It is strictly a quantum effect. The electric field produced by the polarised molecule oscillates at the same frequency as the passing electromagnetic wave, so that the molecule acts as a source sending out radiation of that frequency in all directions. As shown in Fig. 9.9, the incident radiation does not raise the molecule to any particular quantised level, rather the molecule can be considered as in virtual excited state. As the electromagnetic wave passes, the polarised molecular ceases to oscillate and returns to its original ground level in very short time (approx 10^{-12}s).

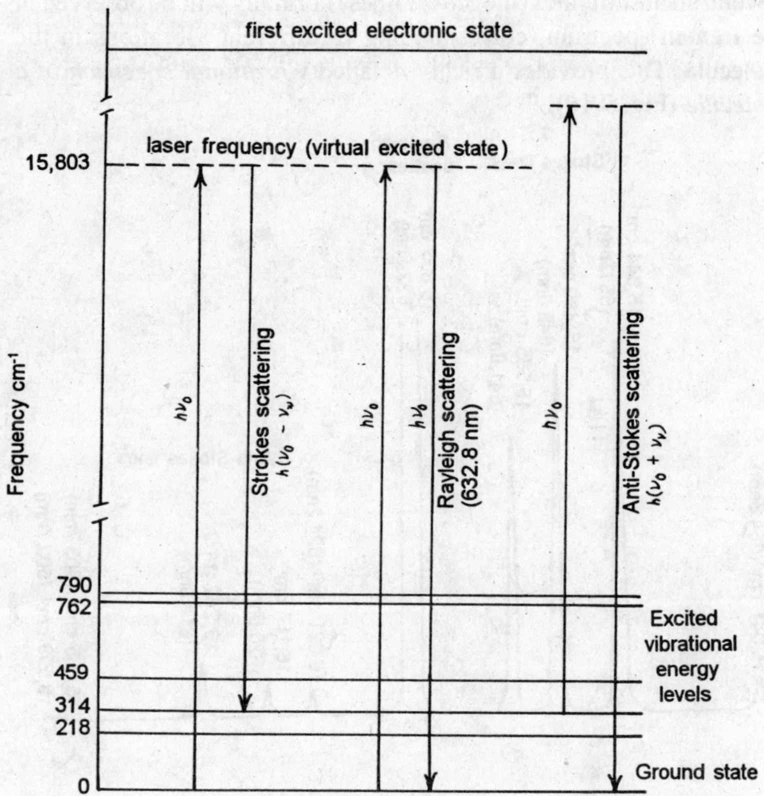

Fig. 9.9: Energy interchange involved in Rayleigh and Raman scattering : the molecule involved in CCl_4 and the source is a He-Ne laser.

A small proportion of the excited molecules (10^{-6} or less) may undergo a change in polarizability during one of the normal vibrational modes. This provides the basis for the Raman effect. Usually incident radiation, v_0, is absorbed by a molecule in the lowest vibrational state. If the molecule remits by returning not to the original vibrational state, but to an excited vibrational level, v_v, of the ground electronic state, the emitted radiation is of lower energy, or lower frequency (v_0–v_v) than the incident radiation. The difference in frequency is equal to a natural vibration frequency of the molecule's ground electronic state.

Several such shift lines (the *Stokes* lines) normally will be observed in the Raman spectrum, corresponding to different vibrations in the molecule. This provides a richly detailed *vibrational spectrum of a molecule* (Fig. 9.10).

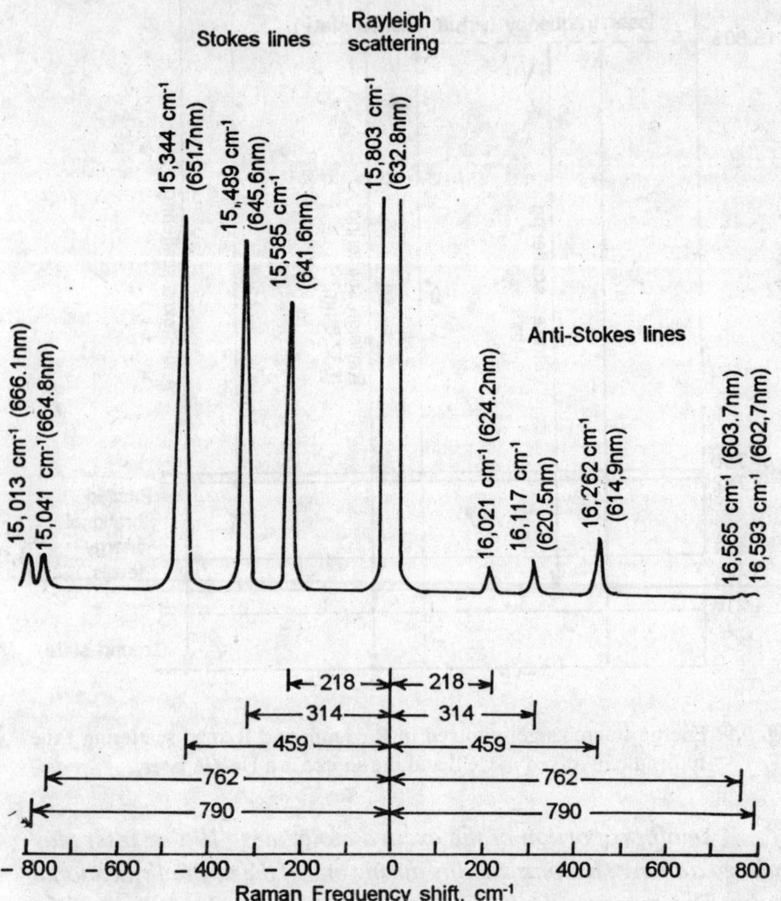

Fig. 9.10: Raman spectrum of CCl_4 obtained with a He-Ne laser

A few of the molecules initially will absorb radiation while they are in an excited vibrational state and will decay to a lower energy level, so that their Raman scattered light will have a higher frequency than the incident radiation. These are called *anti-Stokes* lines. Thus the spectrum of the scattered light consists of a relatively strong component with frequency unshifted (Rayleigh scattering), corresponding to photons scattered without energy exchange, and the two components of the Raman spectrum; the *Stokes lines* and the *anti-Stokes lines. Normally, for chemical analysis, only the Stokes lines are considered.* These are more intense because under usual circumstances most molecules are initially in the lowest vibrational level.

In the usual Raman method the excitation frequency of Raman sources are selected to lie below most S—S* electronic transitions and above most fundamental vibrational frequencies.

An argon-ion laser with lines at 488 and 514.5 nm are used specially when higher intensity is required. Since the intensity of Raman scattering varies as the fourth power of the frequency of the exciting source, the argon line at 488 nm provides Raman lines which are approximately three times as intense as those excited by the helium-neon source given the same input power.

The mercury arc is the most useful source. It is used after obtaining a single wave length using suitable filters. The line corresponding of 4358 A° is the most commonly used radiation.

Other lamp is also in use. It is made up of four low pressure mercury discharge tubes which are completely enclosed in a hollow jacket. The jacket is whitened from inside and acts a reflector and increases the intensity.

Toronto lamp devised by Welsh and Craford is made up of a tube water cooled low pressure mercury lamp. The sample is placed in the central axis of the spiral (Fig. 9.11).

The use of laser source has certain advantages. Some of these are as under :

(a) The multiple lined mercury lamp has been replaced by a single intense frequency source.

(b) The line width of a laser line is smaller as compared to that of mercury excited line. It is thus possible to get better resolution.

(c) Because the laser light is highly coherent in character so it is easier to focus the radiation.

(d) It makes available a large number of exciting frequencies and so it is possible to investigate the coloured samples without causing any electronic transitions. This is particularly useful in study of inorganic salts which are generally coloured.

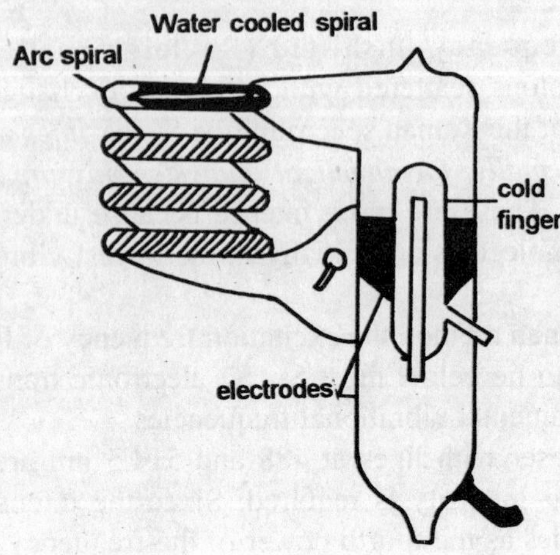

Fig. 9.11: Toronto Lamp

2. Filters

Filters are needed to obtain monochromatic light. If non-monochromatic light is used overlapping of Raman shifts occurs and the interpretation of Raman spectrum becomes difficult. Filters made of nickel oxide or quartz are generally used. Rarely some suitable coloured solution as that of potassium ferricyanide in water or of iodine in carbon tetrachloride is used as monochromator.

3. Sample Container

Various types of sample containers are in use for Raman spectrso-copy. The shape and size of container depends upon the intensity of radiation source, nature of sample and amount of sample available. For gases bigger containers are needed as compared to those for liquids.

Fig. 9.10. Depicts a tube for sample.

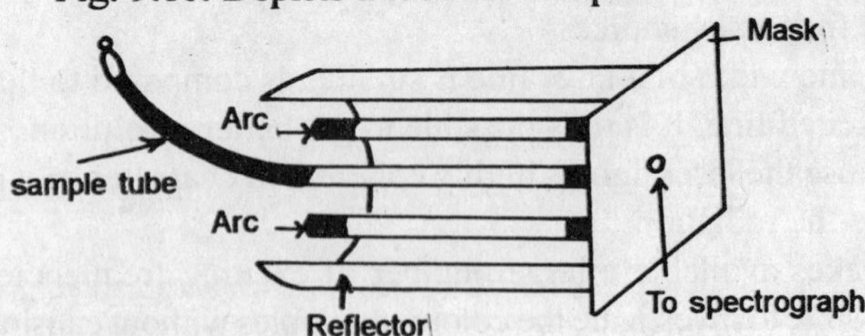

Fig. 9.12: Sample holder tube

Solids are dissolved in a suitable solvent to obtain a Raman spectrography. Any solvent that is suitable for u.v. spectra can be used for Raman spectra as well. In Raman spectroscopy water is considered a good solvent for the study of inorganic salts.

A four traversal tube (Fig. 9.13) is used as a sample container if only a small amount of sample is available.

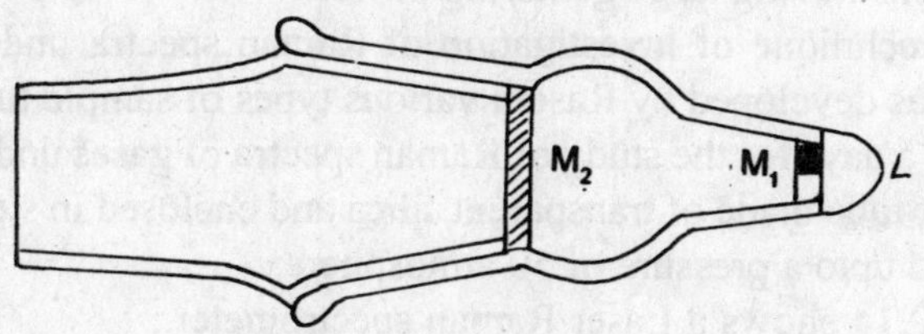

Fig. 9.13: Four Traversal Tube

A Four Traversal tube consists of two concave mirrors (M_1 and M_2). The focal lengths of both these mirrors is equal to half the distance between them. M_1 is only half mirrored and light can pass through the remaining half. A virtual image of M_2 at infinity is produced by the lens $\angle$. When light is allowed to pass through clear half of mirror M_1 and it passes on the M_2. This light then undergoes a series of two reflections *i.e.,* four traversals and emerges out through the same aperture.

Since the intensity of Raman lines in case of gases is very small so highly sensitive detectors and multiple traversal tube are used to obtain their Raman spectra.

4. Spectrograph

The characteristics that a spectrograph for recorded Raman spectra should possess are :

(i) It must possess a large gathering power.

(ii) It should have a prism with high resolving power.

(iii) It should make use of only a short focus camera.

(iv) It should have either photographic or automatic recording devices.

For the study of Raman effect for solids and gases, the experimental arrangement is suitably modified. For study of solids like gypsum or quartz, there is no need to use a sample container.

Instead light from arc is focussed directly, on the material, with the help of a lens. It is possible to study Raman effect for powdered solids by reflecting the light from crystal surface.

Since the intensity of light scattered by gases is very small so they are studied by intense illumination under high pressure and using spectrographs having large gathering power.

The technique of investigation of Raman spectra under high pressure was developed by Rasetti various types of sample tubes are in use now a days for the study of Raman spectra of gases under high pressure. A tube made of transparent silica and enclosed in steel tube can be used upto a pressure of 50 atmosphere.

Fig. 9.14 shows a Laser Raman spectrometer.

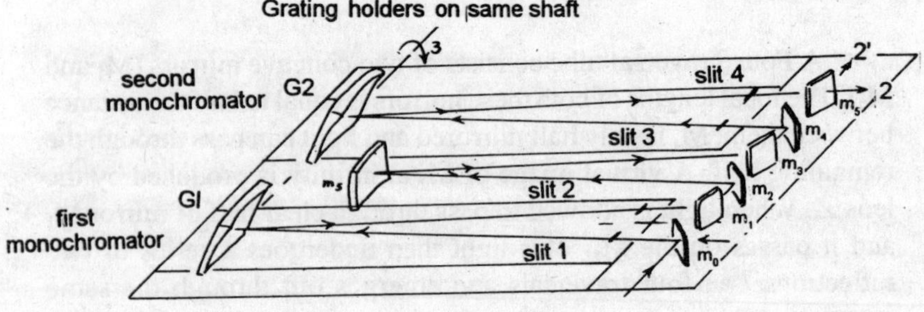

Fig. 9.14: Laser Raman spectrophotometer

The Johin—Yvon instrument (Fig. 9.14) combines the thorough-put of a double system with the rejection levels of a triple system. The *f*/8 monochromator is designed for stray light reduction by using two concave, aberration-corrected, holographic gratings with 2000 grooves/mm. Both gratings are on the same shaft; thus tracking error is zero. The scattered radiation from the sample is gathered by a high aperture objective lens which focuses the light onto the straight horizontal entrance slit F1 of the first monochromator. The concave grating G1 diffracts the beam and focuses the selected wavelength onto the exit slit F2. The monochromatic light exiting from the first monchromator through F2 is now imaged onto the entrance slit F3 of the second monochromator by mirrors M1 through M5. These mirrors are not in the monchromator cavities and work in monochromatic light; therefore, they cannot contribute to scattered stray light. Light entering F3 is then diffracted and focused onto the exit slit F4 of the second monchromator by the concave grating G2. All four slits are individually controlled by the stepping motor; this permits complete versatility when using a compouter in order to obtain either constant slit width spectra or constant bandpass spectra.

Fig. 9.15 shows the optical Schematic of a Laser Raman spectrophotometer.

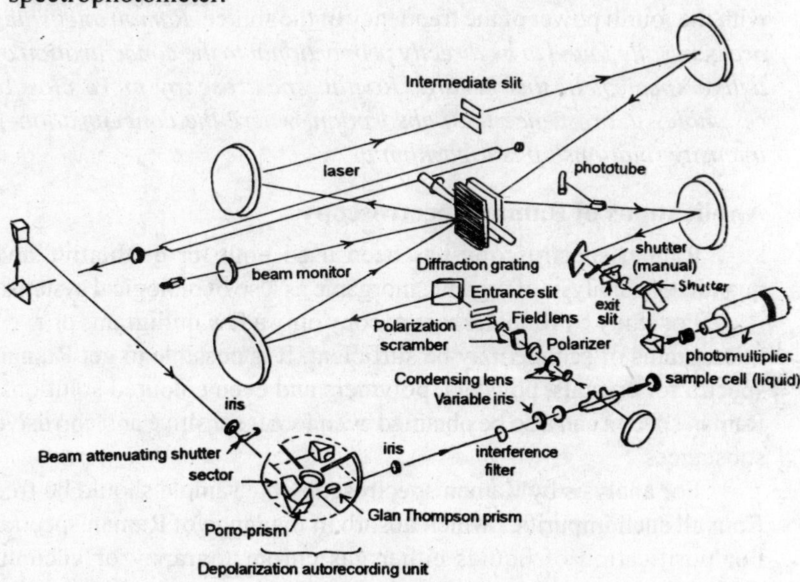

Fig. 9.14: Optical schematic of a Laser Raman spectrophotometer

The primary function of a Raman spectrometer is clear rejection of the intense Rayleigh scattering and detection of the weak Raman-shifted components. A computer adds significantly to the power and versatility of the spectrometer. A intense monochromatic light source, sensitive detection, and high light gathering power, coupled with freedom from extraneous stray light, must be built into a Raman spectrometer. A double grating monchromator keeps stray light from the unshifted laser wavelength to a minimum. Visual type optics are used throughout, and the entire spectral region is covered by a single-type grating.

Intensity of Raman Peaks

The important factors on which the intensity of a Raman Peak depend are :

(a) Polarizability of the molecule.

(b) Intensity of the source.

(c) Concentration of the active group and many other factors.

If no absorption occurs, the power of Raman emission increases with the fourth power of the frequency of the source. *Raman intensities are generally found to be directly proportional to the concentration of active species. In this regard, Raman spectroscopy more closely resembles fluorescence than absorption, where the concentration—intensity relationship is logarithmic.*

Applications of Raman Spectroscopy

Raman spectroscopy has been used both for qualitative and quantitative analysis of organic, inorganic as also of biological systems.

For study by Raman spectroscopy only a few milligrams or even micrograms of sample may be sufficient. It is possible to get Raman spectra for crystals, powders, polymers and even coloured solutions. Raman spectra can also be obtained even for air sensitive and corrosive substances.

For analysis by Raman spectroscopy the sample should be free from all such impurities which absorb in the range of Raman spectra. For purification of liquids either gas chromatography or vacuum distillation is carried out. For analysis of solids we can either dissolve them in a suitable solvent or can use their melt or powder them or even single crystals can be used as such. If a solvent is to be used it is purified by distillation in a grease-free apparatus or by filtration through alumina.

Applications in Inorganic Chemistry

Raman spectroscopy is considered superior to IR spectroscopy for study of inorganic systems because aqueous solutions can be used. In addition to this the metal-ligand bonds are in the range of 100—700 cm^{-1} which is a difficult region for study of IR spectra.

(i) Structure of CO_2

If we considered CO_2 as a symmetrical linear molecule O—C—

O, we expect it to show two fundamental lines (v_2 and v_3) in infrared and one (v_1) in Raman spectra. This has been experimentally verified and the following values have been assigned.

v_1=1340 cm^{-1}, v_2=667 cm^{-1}, V_3=2349 cm^{-1}.

(ii) Structure of N_2O

If it is also considered symmetrical and linear N—O—N then we expect v_1 to be in active in IR and active in Raman spectra but if it is asymmetrical, v_1 should become active in the infrared as well as in Raman spectra. Since this has been observed as additional fundamental in both IR and Raman spectra so we conclude that N_2O is asymmetrical and its structure is therefore N—N—O type.

(iii) Structure of Water

Considering the bent structure of water is was expected that v_1, v_2 and v_3 will be active both in IR and Raman spectra regions. These have been confirmed experimentally so water is a bent molecule.

(iv) Structure of Mercurous Salts

While recording the Raman spectra of aqueous solution of mercurous nitrate an additional line was obtained which is not present in Raman spectra of other metal nitrates. This line may be because of vibration of Hg-Hg convalent bond in diatomic Hg_2^{2+}.

(v) Structure of Chloro Complexes of Mercury

If a solution of mercuric chloride and ammonium chloride in a 1:2 molar ratio is prepared and then the Raman spectrum of it is recorded a strong line is observed at 269 cm^{-1} which compares will with a strong line at 273 cm^{-1} observed in spectrum of solid tetra-chloromercuriate (II). Thus this line (at 262 cm^{-1}) is indicative of the formation of $HgCl_4^{2-}$ in solution.

(vi) Nature of Bonding

In the IR spectra of tetrahedral complexes ML4 (*e.g.*. $ZnCl_4^{2-}$, $CdCl_4^{2-}$ etc.) and octahedral complexs ML_6 (*e.g.*, SiF_6^{2-}, SF_6 etc.) are recorded no bands appear but in the Raman spectra of such complexes strong lines are observed. Making use of metal-ligand bond stretching force constant useful information about the strength of metal-ligand bond can be obtained. The force constant value for oxy anions (*e.g.*, PO_4^{2-}, SO_4^{2-} etc.) is quite high and it is accepted as an evidence for $d\pi - p\pi$ bonding between the central atom and oxygen atom in addition to σ-bonding.

(vii) Hydrogen Cyanide

In the Raman spectrum of hydrogen cyanide two lines appear at 2062 cm^{-1} and 2094 cm^{-1}. The lines can be explained if we consider two isomers in a state of dynamic equilibrium.

(viii) Sulphuric Acid

The Raman frequency for 25% H_2SO_4 corresponds of the frequency of $KHSO_4$ but if H_2SO_4 is further diluted its frequency corresponds to that of K_2SO_4. From this the existence of HSO_4^{-1} ion is evident.

$$O_2S\begin{matrix} OH \\ OH \end{matrix} \rightleftharpoons H^+ + HSO_4^- \rightleftharpoons 2H^+ + SO_4^{2-}$$

Applications to Organic Compounds

Raman spectra like IR spectra is used for detection of functional groups. Raman spectra yields more information about certain type of organic compounds than do their IR counterparts. For example, the double-bond stretching vibrations for olifins results in weak and sometimes undetectable infrared absorption but the Raman band (at about 1600 cm^{-1}) is intense and its position is sensitive to the nature of substitutent as well as to its. geometry. Similarly in case of cycloparaffin derivatives Raman peak appears in the region of 700—1200 cm^{-1} and this is attributed to "breaking" vibration in which the nuclei move in and out with respect to the centre of the ring.

Thus the Raman spectra can be employed for :

(i) detection of presence of absence of a specific linkage in a molecule.

(ii) determination of molecular structure of simple organic molecules.

(iii) the study of isomers.

(iv) detection of impurities in the dyes.

(v) classification of organic compounds.

Table 9.2 lists the frequencies of Raman lines which lie in the region 500—3500 cm^{-1}.

Table 9.2

Group	*Wave number* cm^{-1}	*Group*	*Wave number* cm^{-1}
C—H (aromatic)	3050	S—H	2575
		O—H	3400
C—H (aliphatic)	2910—2980	C—Cl	650—720
C—O (anhydride)	1575	C—Br	630—670
C=O (ketone)	1600	C—I	500—550
C=O (aldehyde)	1710	C≡N	2150
C=O (acids)	1650	C—C	1000
		C=C	1900—2400

(i) Sturucture of Benzene

In the Raman spectrum of Benzene two lines at 1000 cm^{-1} and 1050 cm^{-1} are observed. These may be due to C—C and C=C linkages respectively. So this is considered to support the kekule-structure of benzene.

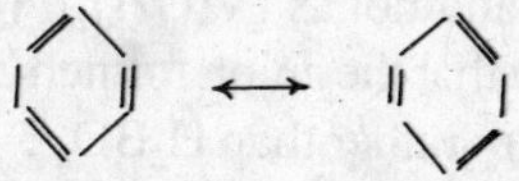

(ii) Location of Groups on Benzene Ring

In the Raman spectra of all the meta substituted benzenes an intense and strongly polarised line appears at 995 cm^{-1} but this line is not present in case of *ortho-* or *para-* substituted compounds. In *para-* substituted compounds a line appears of 625 cm^{-1} which is again absent in *ortho-* substituted compounds. Thus it is possible to distinguish between *meta, para* or *ortho-* substituted-compounds by using the Raman spectrum of the compound.

(3) Biological Applications

Raman spectroscopy has been used for the study of biological systems. Major advantages of using this technique are small sample requirements, minimal sensitivity towards interference by water, the spectral detail and the conformational and environmental sensitivity.

(4) Applications in Physical Chemistry

Raman spectroscopy has been used for study of electrolytic dissociation, hydrolysis, transitions of state etc.

(i) In electrolytic dissociation, the Raman lines have been used for determination of number and nature of ions formed. It has been utilised to know if the dissociation is complete or partial.

(ii) Raman spectra has also been used to know the degree of hydrolysis. This is achieved by measuring a set of lines which are characteristics of the base, acid or the salt.

(iii) The Raman spectra of amorphous salt gives a broad and diffused band where as that of a crystalline compounds gives fine sharp lines.

(iv) Ionic equilibrium in solutions such as the one for the reaction

$$HNO_3 + H_2O \rightleftharpoons H_2O + NO_3^-$$

Can be easily monitored by the intensity of NO_3^- ion and HNO_3 in the Raman spectrum.

Similarly in perchloric acid solutions vanadium (*iv*) appears to be present as VO^{2+}(aq) and not as $[V(OH)_2]^{2+}$(aq). Similar studies of boric acid solution show that the anion formed by the acid dissociation is the tetrahedral $B(OH)_4^-$ rather than $H_2BO_3^-$.

Raman spectroscopy has also been used to find the dissociation constants of strong acids such as H_2SO_4, H_2SeO_4, H_5IO_6 etc.

5. Applications in Polymer Chemistry

Polymers show the properties of liquids and crystals simultaneously in their spectra. This is attributed to the simultaneous existence of amorphous and crystalline regions in polymers. Stretched polymers show a close resemblance with molecular crystals. The spectrum of polymers can be interpreted in terms of unit cells containing a few elements of different polymer chains. Like single crystals, it is possible to examine stretched polymers with polarised radiations. It is also possible to draw a number or empirical conclusions regarding the structural units of macromolecules from their Raman spectra. *e.g.,* following groups/bonds how intense Raman bands but are very weak

C=C	1660 cm^{-1}
Benzene nuclei	1000, 1600, 3060 cm^{-1}
symmetric Si–O–Si vibrations in silicones	490 cm^{-1}
C—S vibrations in polysulphides	650 cm^{-1}
S—S vibrations in polysulphides	510 cm^{-1}.

6. Quantitative Applications

Raman spectra is found to be an easy, rapid and accurate method of analysis of such mixtures whose analysis by other methods is troublesome. This is because Raman spectra tend to be less cluttered with peaks than infrare spectra and so the peak overlap in mixtures is less likely and thus the quantitative measurements are likely to be simpler. The height of Raman peaks varies linearly with concentration. Moreover Raman spectrum instrumentation is not subject to attack by moisture and presence of small quantities of water in a sample do not interfere. Inspite of all these advantages this technique has not yet been exploited widely for quantitative analysis. D.E. Nicholson (1960) analysed a mixture of eight components by this method. The results of analysis agreed with preparatory data to about 1%.

Applications of other Types of Raman Spectroscopy

Two new Raman spectrosocopic methods have been developed. These are (*i*) Resonance Raman spectroscopy, (*ii*) Coherent Anti-Stoke's Raman spectroscopy.

(i) Resonance Raman Spectrosocpy

The basis of this type of Raman spectroscopy is the fact that a laser source with a frequency near or coincident with an electronic absorption peak of a substance produces a marked increase of Raman scattering associated with the chromophore. Due to this the sensitivity and selectivity increase.

(ii) Coherent Anti-Stoke's Raman Spectroscopy (CARS)

This technique does not suffer from the drawbacks of Raman

spectrosocpy, namely, low efficiency, its limitation to visible and near u.v. regions and susceptibility to interference from fluorescence.

The efficiency of coherent Anti-Stoke's emission is high, with as much as 1% excitation radiation being converted to new frequency. It has a very high sensitivity.

SOLVED PROBLEMS

1. What are the relative intensities of a Raman line excited by He—Ne laser at 632.8 nm and one excited by an argon laser at 488.0 nm?

[**Ans.** $(15{,}798/20{,}487)^4 = 3.35$]

2. By what factor is the scattered intensity of a given band reduced in changing the excitation frequency from an argon laser (488.0 nm) to neodymium-doped laser (1060 nm)?

(**Ans.** It is reduced by a factor of about 22.7.)

3. When excited by mercury line at 435.8 nm, the spectral trace of benzene contains Raman lines at 606, 850, 991, 1176, 1584, 1605, 3047 and 3063 cm^{-1}. At what wavelengths will these Raman lines appear when benzene is irradiated with (*a*) He—Ne laser (623.8 nm) (*b*) an argon ion laser (488.0 and 514.5 nm) and (*c*) a krypton laser (568.2 and 64.71 nm)?

[**Ans.**

Benzene Δcm–1	*Wavelength of Raman Lines (nm) for Exciting Lines*				
	623.8	*488.0*	*514.5*	*568.2*	*647.1*
606	6580	5029	5311	5885	6735
850	6888	5091	5380	5970	6848
991	6751	5128	5421	6021	6914
1176	6837	5177	5436	6089	7004
1584	7033	5289	5602	6244	7210
1605	7043	5295	5608	6252	7221
3047	7839	5732	6102	6872	8060
3063	7849	5738	6107	6879	8071

4. If unfiltered laser excitation from either the krypton or argon laser were used to excite the Raman spectrum of benzene in

previous problem, to what extent would each suite of Raman lines overlap?

(**Ans.** The C—H stretching frequencies at 3047 and 3063 cm^{-1} would overlop).

5. For each unknown mixture of the triethylbenzenes, compute the volume %age of each. The scattering coefficient of the pure compound at each analytical wave number is as under:

		1,2,3 – 620 cm^{-1}	1,2,4 – 716 cm^{-1}	1,3,5 – 570 cm^{-1}
Mixture	A	0.627	0.208	0.555
	B	0.209	0.069	0.185
	C	0.251	0.054	0.189
		0.157	0.077	0.211

[**Ans.**

	Volume%		
Unknown	*1,2,3—*	*1,2,4—*	*1,3,5—*
A	33.3	33.3	33.3
B	40.0	26.0	34.0
C	25.0	37.0	38.0]

6. For carbon disulphide, all vibrations that are Raman active are infrared active and *vice versa,* where as for Mitrous oxide (N_2O) the vibrations are simultaneously Raman and infrared active. What can we conclude about the structure N_2O and CS_2?

(**Ans.** N_2O has no centre of symmetry while CS_2 does have one. The structure must be N—N—O and S—C S).

7. For CCl_4, four principal Raman lines appear at 2/8, 3/4, 458, and 781 cm^{-1}. None of these infrared frequencies is absorbed. The depolarisation ratios are 0.86, 0.86, 0.046 and 0.83 respectively. What can be conclude about the symmetry of the molecule? If its spatial

configuration planar or tetrahedral?

(**Ans.** The molecule has centre of symmetry. If planar, the molecule would have three unpolarised Raman lines; If tetrahedral it would have but one. The spatial configuration is tetrahedral).

8. Deduce the structure of the compound whose spectra is shown in the accompanying figure. The molecular weight is 140.

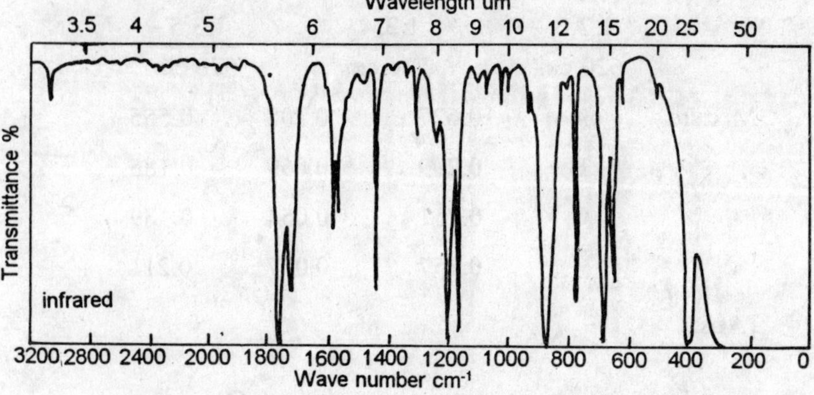

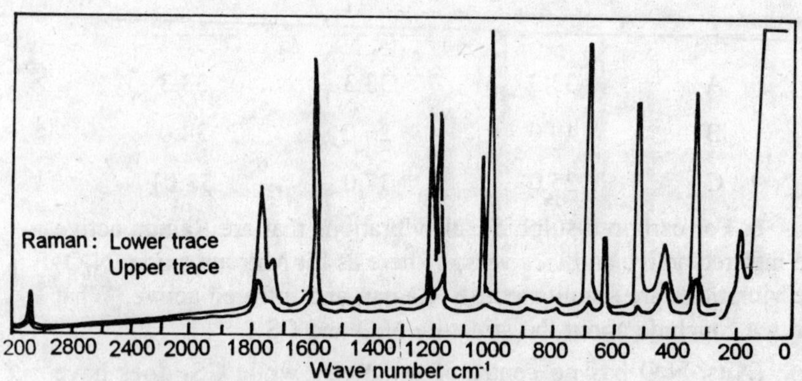

(**Ans.** Benzoyl chloride).

9. The molecular weight of the compound is 54. Using the spectra in the accompanying figure, determine the structure of the compound.

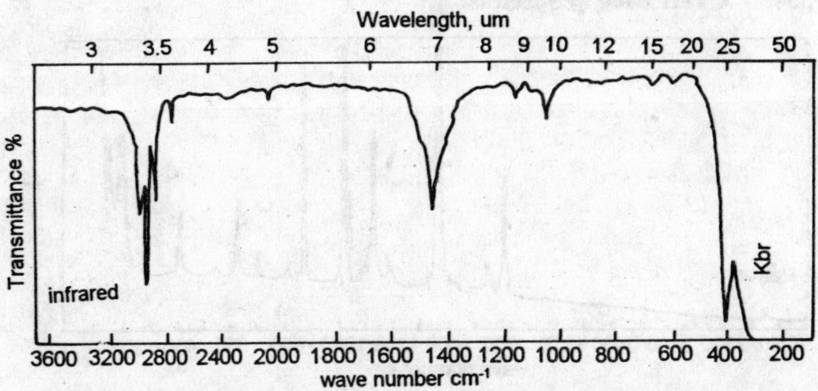

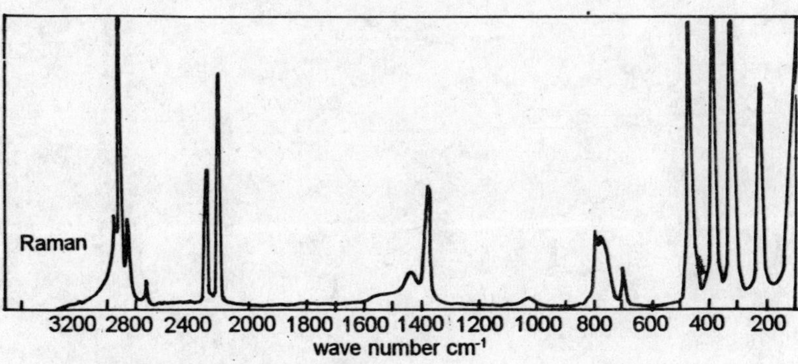

(**Ans.** Dimethyl acetylene).

10. From the infrared and Raman spectra shown, and the molecular weight of 123, deduce the structure of the compound.

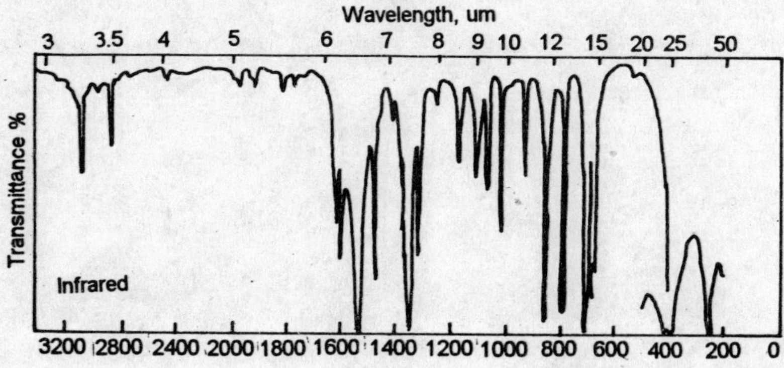

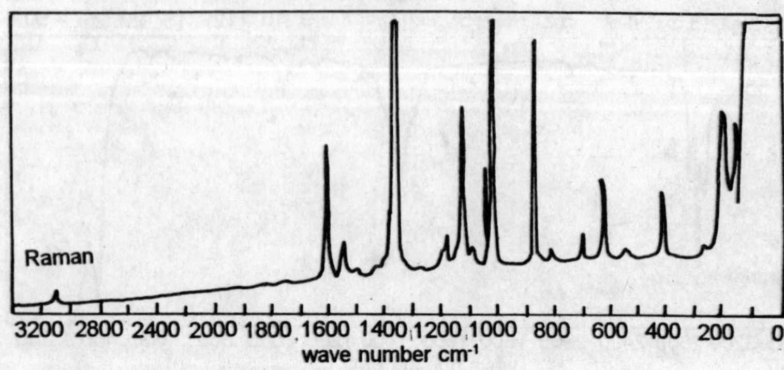

(**Ans.** Nitrobenzene).

10

Mass Spectrometry

Introduction

Mass spectrometry evolved from studies about the behaviour of positive ions in magnetic and electrostatic fields. Such studies were carried out in the beginning of 20th century and the next two decades method was refined and important informations were obtained from the use of mass spectrometry.

In mass spectrometry a mass spectrum is obtained by converting the compounds of a sample into rapidly moving ions (generally positive ions) and resolving these ions on the basis of their mass to charge ration. The spectrum produced, known as *mass spectrum* gives important information about various masses produced and also about their relative abundances. It is also possible to place certain functional groups into certain areas by detailed interpretation of mass spectrum.

The first true application of mass spectrometry was described in 1940 when reliable mass spectrometers became commercially available. In later years this technique was also used in quantitative analysis as well as in qualitative analysis.

It is one of the important analytical techniques that is used to get information about the molecular structure of organic as well as inorganic compounds. This technique is used for quantitative determination of components in complex hydrocarbon mixtures. It is widely used in petroleum industry.

Though this technique of analysis is based on a simple principle yet it is quite complex and need very expensive instruments. However analytical chemist is using this technique quite frequently because of its speed and reliability.

The Mass Sepctrometer

Mass spectrometer is an instrument used for separating individual atoms or molecules making use of difference in their masses.

Theory

Fig 10.1 is a diagram of a single focusing mass spectrometer. The sample is introduced into the instrument in such a way that its vapour is bombarded by electrons having an energy of about 70 eV. Positive ions formed in the ion source are accelerated between two plates by a potential difference of a few thousand volts (V). The ions pass through the source slit and are deflected by a magnetic field (H) according to their mass/charge ratios. They then pass through the exit of collector slit, impinge upon the collector, and the signal received is amplified and recorded (usually on photographic paper). The height or intensity of the resulting peak is proportional to the abundance, that is, the number of ions of identical mass received by the collector.

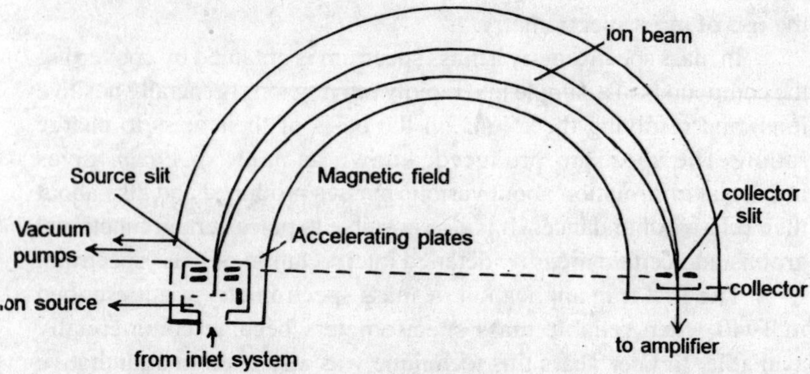

Fig. 10.1: Single focusing mass spectrometer

To understand the theory of mass spectrosocpy let us consider a molecule M. This molecule when bombarded with a beam of electrons ionises as under

$$M + e^- \longrightarrow M^+ + 2e^-$$

The ions produced are then accelerated in an electric field. Let,

V be the voltage of the electric field, then energy given to each particles is eV and so the kinetic energy $\frac{1}{2}$ mv² should be equal to this energy (ev) *i.e.*,

$$ev = \tfrac{1}{2}mv^2$$

or $$v^2 = \frac{2ev}{m}$$

or $$v = \sqrt{\frac{2ev}{m}}$$

where, v = velocity of particle
m = mass of particle
e = charge on an electron
V= accelerating voltage.

The same energy ev is possessed by all the particles but since the value of m changes from particle to particle so the velocity v changes in such a way that there is no change in $\frac{1}{2}mv^2$. For any particle with mass m_1 and velocity v_1 we have

$$\tfrac{1}{2}m_1v_1^{\ 2} = ev$$

Similarly for a particle with mass m_2, m_3.....etc. and velocity v_2, v_3, etc. we have

$$\tfrac{1}{2}m_2v_2^2 = ev$$

$$\tfrac{1}{2}m_3v_3^2 = eV$$

....

....

Form these we have

$$\tfrac{1}{2}m_1v_1^2 = \tfrac{1}{2}m_2v_2^2 = \tfrac{1}{2}m_3v_3^2$$

Thus we find that the velocities of various particles is different and it depends upon the mass of particle.

When these charged particles get accelerated due to applied voltage and enter a magnetic field H, they are attracted towards the field and they move in circles around it. The force of attraction on particles is given by Hev and it is balanced by the balancing centrifugal force which is equal to $\frac{mv^2}{r}$. When particles start moving uniformly around the circular path, these two opposing forces become equal *i.e.*,

$$\frac{mv^2}{r} = Hev$$

or $$\frac{1}{r} = \frac{Hev}{mv^2}$$

or $$r = \frac{mv}{eH}$$

where r = radius of the circular path adopted by the particle.

Since $$v = \sqrt{\frac{2evV}{m}} \quad \text{[Already derived]}$$

So, we get

$$r = e\frac{m}{H}\left(\frac{2ev}{m}\right)^{\frac{1}{2}}$$

Squaring both sides, we get

$$r^2 = \frac{m^2}{e^2H^2} \cdot \frac{2eV}{m}$$

or $$r = \left(\frac{2V}{H^2} \cdot \frac{m}{e}\right)^{\frac{1}{2}}$$

or $$\frac{m}{e} = \frac{H^2r^2}{2V}$$

Thus we find that radius of the circular path depends as voltage V, the magnetic field H and the ratio $\frac{m}{e}$. Since V, H and e are constant so the radius of path depends on mass m particle.

Thus the radius of the ion path may be changed by varying either the magnetic field (H) or the accelerating voltage (V). By either method, ions of different mass-to-charge ratios (m/e) can be made to impinge upon the collector in turn thus giving rise to a spectrum (Fig. 10.2). Magnetic field variation enables a wide range to be covered in a single sweep, but, for very rapid scanning a voltage sweep must be employed.

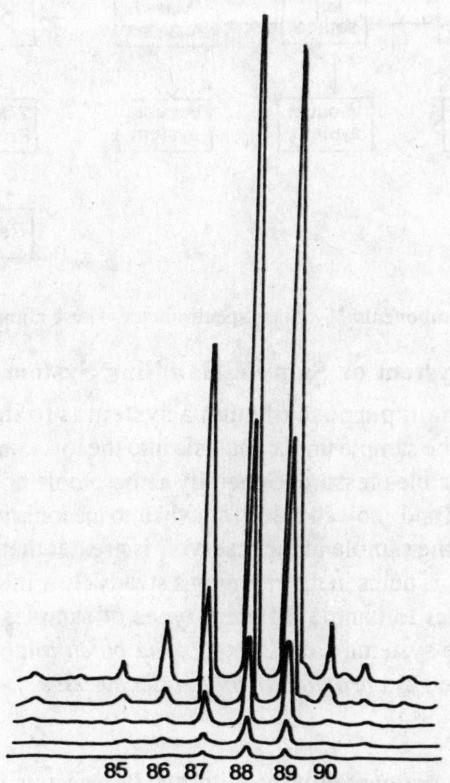

Fig. 10.2: A portion of a mass spectrum traced by a five-element galvanometer.

The ion-source, ion path and collector of the mass spectrometer must be under high vacuum (10^{-7} mm Hg) for optimum operation. Any extraneous materials in the source, including atmospheric gases, will themselves ionise in the electron beam and their spectra will be recorded.

Components of Spectrometer

The components of a mass spectrometer are shown in Fig. 10.3.

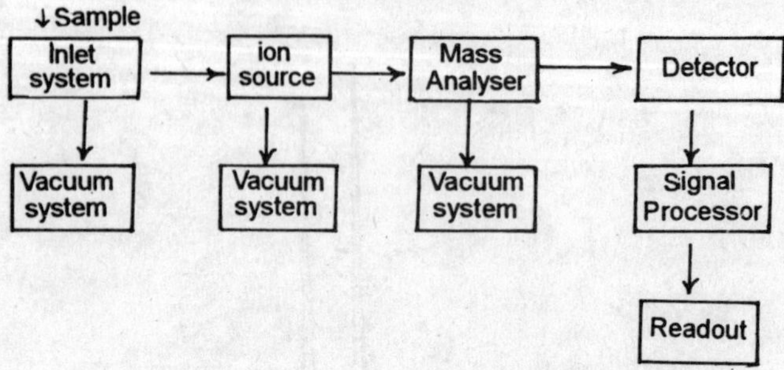

Fig. 10.3: Components of a mass spectrometer. These components.

(1) Inlet System or Sample Handling System

The main purpose of such a system is to introduce a small quantity of the sample under analysis into the ion source as a gas at low and reproducible pressure. Generally a micromole or less of the sample is volatilized and allowed to leak slowly into the ionisation chamber. The pressure of the sample in the reservoir is greater than in the ionisation chamber. This helps in maintaining a steady flow into the chamber.

In order to handle different types of samples various types of inlet sample systems are used, *e.g., the batch inlet, the direct probe inlet* and *the gas chromatographic inlet systems.*

Batch Inlet

These are generally used to handle gaseous samples. In these cases the gas is simply transferred from a gas bulb into the metering reservoir and then expanded into expansion reservoir.

Such system can be used for liquids boiling below 150°C. In their case a suitable quantity can be evaporated into evacuated reservoir at room temperature. In case of volatile liquids the sample may be introduced directly into ionisation chamber.

Direct Probe Inlet

Such an Inlet is suitable for non-volatile or thermally unstable materials. In this the sample is directly introduced into the ion source by means of a sample probe which consists of a holder for a small capillary tube or cup and is provided with a heater to volatilise the sample. The use of a probe permits the study of such non-volatile materials as carbohydrates, steroids and low molecular weight polymeric substances.

Gas Chromatographic Inlet Systems

In this case the effluent from a chromatographic column act as sample source.

2. Ion Sources

Various ion sources such as *electron impact source, chemical ionisation source, spark source, Field ionisation sources* etc. are used in mass spectroscopy.

Electron Impact Source

This is an important source for mass spectroscopy and it involves bombarding the sample with a beam of energetic electrons. (Fig. 10.4).

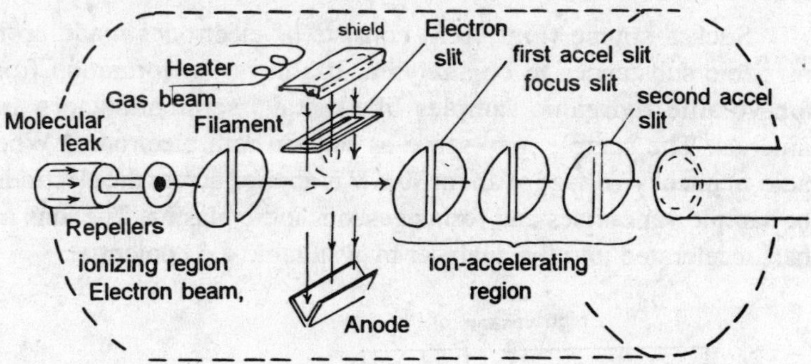

Fig. 10.4: An ion source

A typical ionisation source or *ion gun* is shown in Fig. 10.4. The positive ion formed are forced through the slit of first accelerating plate by small potential difference between this plate and the repeller. The final velocity of the particles is due to high potential difference between first and second accelerators. This type of source is quite energetic and helps to produce a large number of positive ions.

Such a source has the following disadvantages :

(i) The electron impact process is not efficient.

(ii) They require the sample in vapour state.

(iii) All background gases also get ionised along with the sample.

(iv) There is an increase in molecular ion intensity when the energy of impinging electrons is about the same as the ionisation energy. This greatly reduces the absolute ion intensity.

(v) Those molecules which are unstable with respect to electron impact do not exhibit molecular ion peaks.

Chemical Ionisation Source

In such a source methane is introduced into a modified electron beam source at a pressure of about 1 Torr. The sample is then introduced into it, the sample then ionises largely because of collision with reagent ions rather than electrons. Generally the ions produced are CH_5^+, CH_4^+, CH_2^+, H_2^+ and $C_2H_5^+$ etc. Such a source causes less fragmentation and so the mass spectra can be interpreted more easily.

It needs some modifications of the spectrometer so that it can sustain higher pressure in the ion source region.

Spark Source

Such a source (Fig. 10.5) consists of electrodes made up of inorganic substances to be analysed. It allows ion formation from non-volatile inorganic samples like metals, semiconductors and minerals. The sample may serve as one or both electrodes. When radio frequency voltage of about 30 KV is applied across the electrodes the sample vapourises and forms gaseous ionic plasma. The ions are then accelerated into the analyser by a suitable d.c., potential.

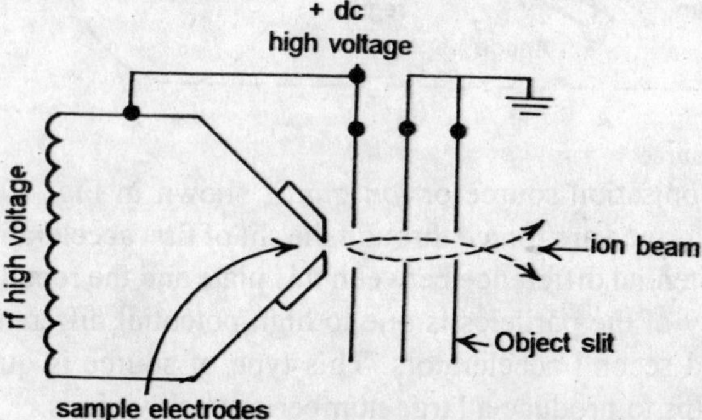

Fig. 10.5 : Spark source ionization.

Field Ionisation Source

Such a source consists of a metallic anode in the form of a blade with one or more fine points and a cathode that also acts as a slit. The distance between anode and cathode is about 0.5-2.0 mm. When 5.20 KV voltage is applied the ionisation of organic compounds take place.

This is considered to be a gentle technique as fragmentation is minimised. The major products being molecular ion and it provides information about the molecular weight of the substance.

3. Mass Analyser

This is the part of the mass spectrometer which separate ions in accordance with their masses. The important characteristics which mass analyser should possess are

(i) it should produce a high level of ion currents

(ii) it should distinguish between minute mass differences

(iii) it should passes a high resolving power.

Since all the above characteristics are not possible in any single arrangement because (*i*) and (*ii*) above are incompatible. Generally a compromise has to be made and main difference among various mass spectrometers lies in their system for separating ions.

Important mass analysers are :

(i) Single Focusing Analysers with Magnetic Deflection

In such analysers, which are commonly used, at any given voltage (V) all the singly charged ions are given the same K.E. and so $\frac{1}{2}mv^2 = eV$.

In any magnetic field (H) such a particle experiences a force (F), given by

$$F = HeV$$

The balancing centrifugal force being $\frac{mv^2}{r}$ so we have

$$HeV = \frac{mv^2}{r}$$

or

$$\frac{m}{e} = \frac{H^2r^2}{2V}$$

Since in most spectrometers H and r are fixed so $\frac{m}{e}$ ratio of the particles reaching the slit is found to be inversely proportional to accelerating voltages. So it is possible to focus any desired mass on exit slit by suitable adjustment of V.

(ii) Double Focusing Analyser

Such an analyser is shown in Fig. 10.6.

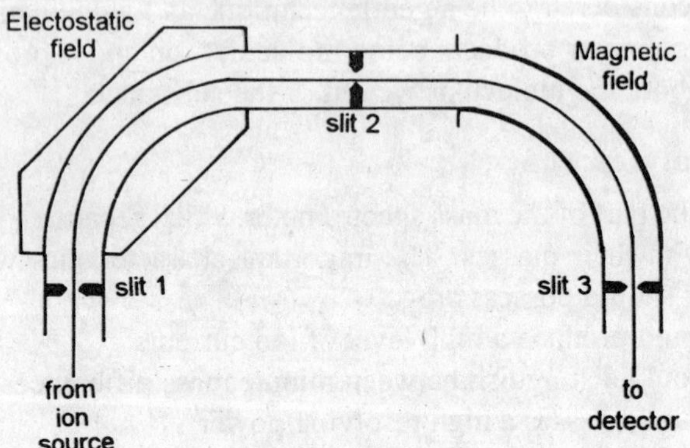

Fig. 10.6: Design of a double-focusing separator

Single focusing analyser fails to discriminate between small mass differences because of small variation in K.E. of particles of a given species. To achieve better focusing the energy spread has to be reduced before ions are allowed to enter the magnetic field. *Mattauch* and *Herzog* designed 'double focusing' instruments. In these instruments the beam is first passed through radial electrostatic field which focus only particles of the same K.E. on second slit (slit 2), which then serve as a sources for the magnetic separator. Such an arrangement can distinguish between particle differing slightly in masses. They have thus high resolving power but have slow ion currents.

(ii) Time of Flight Analysers

It has been noticed that all ions leave the accelerating field with different velocities which depend upon their mass. Such ions get separated while passing through a magnetic field. When such ions are allowed to travel in straight lines through a magnetic field free zone, they are found to take different time to travel any given distance. The measurement of this 'time of flight' is the basis of separation of ions in such instruments. Such instruments which use this type of analyser are found to be less satisfactory but have the following advantages :

(i) They have the ruggedness and are easily accessible to ion source.

(ii) They allow ready insertion of non-volatile or heat sensitive samples.

(iii) they are quite useful for study of short lived species because of their instantaneous display features.

(iv) They are smaller, more mobile and more convenient to use.

Quadrupole Analyser

Such an analyser is shown in Fig. 10.7.

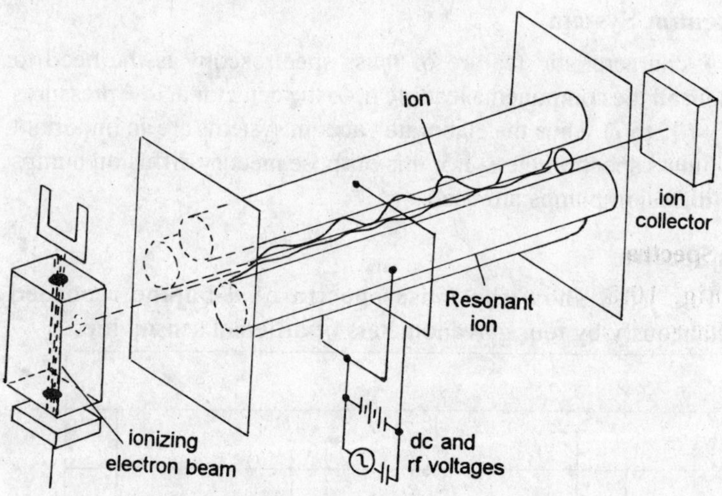

Fig. 10.7: A quadrupole mass spectrometer

It consists of four short, parallel metal rods arranged symmetrically around the beam. The opposed rods are connected to the positive of a *d.c.* source and the other pair to the negative. A radio frequency *a.c.* potential is applied to both pairs. None of these fields accelerate the positive ions but they cause particles to oscillate about their central axis of travel. By suitable adjustment it is possible that only particles which possess certain $\frac{m}{e}$ ratio can pass through the array without being removed by collision with one of the rods. Such analysers are less expensive and compact.

4. Detectors

The ion currents are generally of the order of 10^{-17} to 10^{-9} amperes (10^{-17} amperes corresponds to about 60 ions reaching the source per second). To record peaks of diverse sizes mass spectrometers are provided with several pens, with each corresponding to a different current range. Some times photographic plates. Faraday cylinders, electron multipliers and electrometers are used. Minicomputers and microprocessors from an integer part of modern mass spectrometer.

5. Vacuum System

A characteristic feature of mass spectroscopy is the need to maintain all the components leading upto the detector at low pressures (10^{-4}—10^{-8} torr). Thus the elaborate vacuum systems are an important part of mass spectrometers. For this purpose mecury diffusion pumps or oil diffusion pumps are used.

Mass Spectra

Fig. 10.8 show the mass spectra of *n*-butane recorded simultaneously by four galvanometers of different sensitivities.

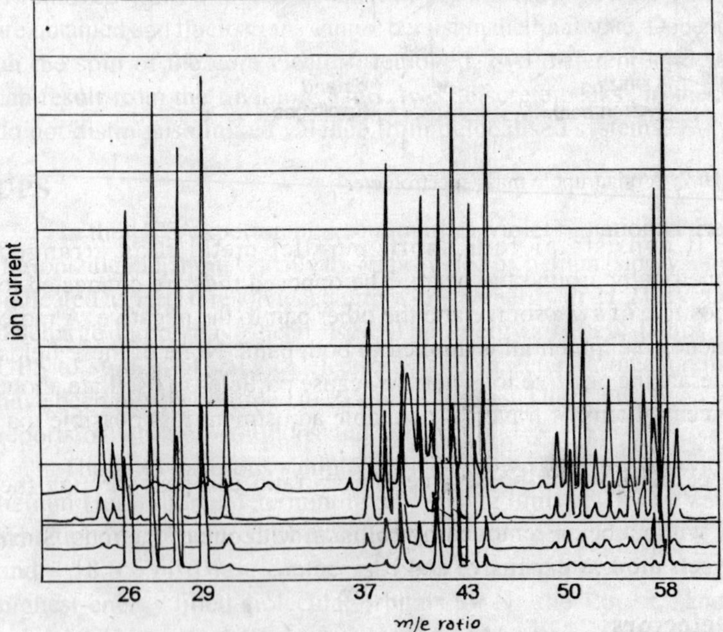

Fig. 10.8: Mass spectra of *n*-butane recorded simultaneously by four galvanometers of different sensitivities.

Mass spectrum generally contains an array of peaks of different heights and the detailed nature of spectrum depends upon the properties of the molecules as also the ionisation energy, the sample pressure and instrument design. Complete interpretation of a sample is seldom done. Often data are represented by a graph which plots relative abundance on the ordinate and m/e ratio on abscissa.

Resolutions of Mass Spectrometers

Resolutions means the ability of a mass spectrometer to differentiate between masses. It is usually denoted in terms of $\Delta m/m$ where m and $(m+\Delta m)$ are the masses of the two particles that give just appreciable peaks of equal size. Two peaks are considered to be separated when the height of the valley between them is more than 10% of their height.

Following factors generally decreases the resolution of mass spectrometer:

(i) Variation in magnetic field.
(ii) Poorly collimated ion beam.
(iii) Variation in accelerating voltage.
(iv) Distribution in K.E. produced in electron beam.
(v) Pressure in the spectrometer.
(vi) Space charge of the ion beam.
(vii) Width of the ion beam.

VARIOUS TYPES OF PARTICLES PRODUCED IN MASS SPECTROMETER

1. Molecular Ion or Parent Ion

Some minimum electron beam energy is needed to initiate the ionisation process for generation of mass spectrum.

$$M + e \rightarrow M^{+} + 2e$$

In this case M stands for the molecule and M^{+} represents the *molecular ion* or *parent ion*.

The minimum electron beam energy needed for most of the organic compounds is 9—15 eV. The simple mass spectrum produced in almost all cases contains only one peak which is known as *parent peak*. However in some cases small peaks also appear, these may be due to abundance of ^{12}C and ^{1}H in organic compounds. If there is any isotope in the compound another peak can also be observed.

The stability of the molecular ion increases if a molecule contains a π-electron system. This is so because a π-electron system can adjust more easily to the loss of electron in comparison to a σ-bond.

The decrease in the height of the parent peak is found to decrease in the following order.

Aromatic compounds > conjugated olifines > Alicyclic compounds > sulphides > straight chain hydrocarbons > ketons > amines > esters > ethers > carboxylic acids > branched chain hydrocarbons > alcohol etc.

The mass of molecular ion M^+ is identical to the molecular weight of the compound. Thus it is possible to calculate the molecular weight of the sample from the molecular ion peak. Under any given set of conditions, the intensity of molecular ion peak depends upon the stability of the ionised particle. Since the stability of particle is strongly affected by its structure so the size of molecular ion peak will show great variability.

2. Base Peak

When an electron beam having energy above the minimum electron beam energy a higher yield of molecular ions is produced. This is because of greater probability of ion producing collisions. Some of fragments with smaller or even larger masses than the parent molecule. This results is decrease of the molecular ion peak. The parent peak in mass spectrum is called *base peak.* The heights of all other peaks are measured with respect to parent peak.

3. Dissociation Process

When a sample molecule dissociates it forms molecular ions, some radicals and some stable molecules etc. In mass spectrum *m/e* is plotted against the intensity of peaks. The height of peak is indicative of the relative abundance of an isotope and its position gives the mass number of particle. For interpreting a mass spectrum the most intense peak is taken as 100 and other peaks are then recorded as %ages of this. Each peak is generally assigned a mass number for interpretation of mass spectrum.

Consider the mass spectrum of methane (Fig. 10.9).

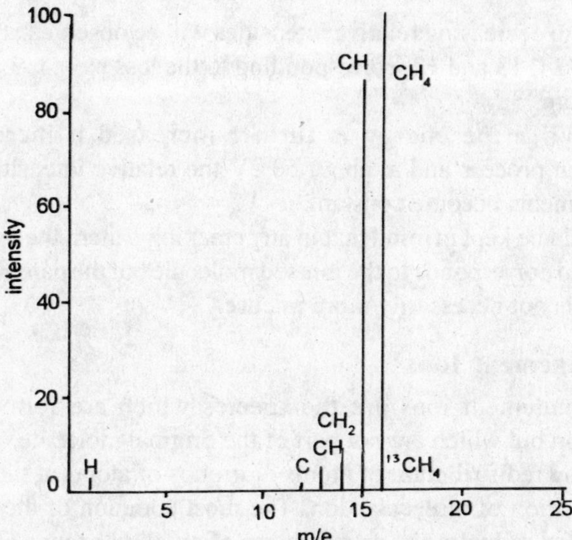

Fig. 10.9: Mass spectrum of methane.

The mass spectral lines of methane are as under :

m/e	*Relative abundance*	*Fragment*
1	3.1	H
12	1.0	C
13	3.9	CH
14	9.2	CH_2
15	85	CH_3
16	100	CH_4
17	1.1	$^{13}CH_4$

To under the various processes involved when methane is bombarded with electrons of increasingly higher energies, following may be considered:

(i) At lower energies (about 10 eV), the methane molecules ionises as

$$CH_4 \rightarrow CH_4^+ + e.$$

The mass of CH_4^+ is equal to the mass of CH_4.

(ii) When energy is further increased CH_4^+ gets fragmented and so the peaks to decreasing relative intensities will be observed at mass numbers 15, 14, 13 and 12, corresponding to the loss of 1, 2, 3 and 4 hydrogen atom.

(iii) When the energy is further increased it increases fragmentation process and at about 50 *e*V the relative intensities of various fragments become constant.

It should be kept in mind that in any cracking pattern the highest mass detected corresponds to the ionised molecule but the parent peak (base peak) is not necessarily more intense.

4. Rearrangement Ions

Rearrangement ions are the species which are formed in fragmentation but which are not part of the original molecule. These are formed by redistribution of atoms or groups of atoms at the time of decomposition of molecular ion. The most common of these are due to transfer of hydrogen or migration of an alkyl group. The re-arrangement is quite unpredictable and non-specific in case of hydro-carbons but the rearrangement is quite specific in case of molecules containing a hetero atom, *e.g.*

$$CH_3CH_2\overset{+}{X} \nearrow CH_3CH_2^{+} + X$$
$$\searrow CH_2{=}CH_2^{+} + HX$$

5. Multiple Charged Ions

Some times multiple charged ions are found in mass spectrum. In such cases a doubly charged ion is recorded at half ($\frac{1}{2}$) value and a triply charged ion is recorded at ($\frac{1}{3}$) one-third value of the *me/e* of a singly charged ion. Formation of such ions is quite common in heteroatomic molecules.

6. Negative Ions

Such ions may be formed when sample is bombarded with electrons. The various ways in which such ions are formed are as under:

(a) $AB + e^- \rightarrow A + B^-$ (Dissociation resonance capture)

(b) $AB + e^- \rightarrow AB^-$ (Resonance capture)

(c) $AB + e^- \rightarrow A + B^- + e$ (ion pair production)

To focus the negative ions the field is reversed. Some of the negative ions formed are O^-, OH^-, C_2H^- etc. However these ions are not useful in determination of structure of molecules.

7. Metastable Ions

Those ions which are produced from the decomposition taking place between the source region and the magnetic analyser are called metastable ions. Such ions appear in mass spectrum as broad peaks at *non-integral mass numbers.* These peaks are generally weak but are found to be quite useful to study the mechanism of fragmentation.

The relationship between the apparent *m/e* of the metastable ion and its parent is as under:

$$m_1^+ \rightarrow m_2^+ + m_0$$

where m_0 = mass of neutral fragment.

The metastable ion is observed at mass m_3 where m_3 is related to them by

$$m_3 = \frac{m_2^2}{m_1}$$

where m_1 is the mass of parent ion and m_2 the mass of metastable ion.

Interpretation of Mass Spectra

To interpret a mass spectrum it is essential to understand the ionisation process that occur before the ions reach the detector. Various methods used for determination of molecular formula from molecular weight determinations are as under :

1. The Exact Molecular Weight

It is quite possible to get the exact value of molecular weight of a given compound by observing its parent peak. The molecular formula of a compound can then be formulated from molecular weight.

2. The Isotopc Effects

For heavior isotopes peaks in a mass spectrum at *m/e* are observed one or more units higher than normal. such peaks occur at $(m+1)$ and $(m+2)$ etc. It is possible to make calculation of molecule weights from the heights of such peaks. So the molecular formula of the compound can be found.

3. The Nitrogen Rule

Beynon (1960) gave this rule, it states, "If the molecular weight of an organic compound is even, the number of nitrogen atoms in the molecule is even (may be zero) and if the molecular weight is odd, the number of nitrogen atoms in the molecule is odd." It further states that, "all fragment ions formed by cleavage of one bond have odd mass if they have an even number of nitrogen atoms and even mass if they have an odd number of nitrogen atoms."

This rule is found to be applicable in case of all compounds having covalent bonds and containing C, H, N, O, S, Si, As, P halogens and alkaline earth metals.

4. Ring Rule

This rule can be used to calculate the number of "unsaturated sites". In accordance with this rule, the number of unsaturated sites (R) is equal to the number of rings in the molecule plus the number of double bonds plus twice the number of triple bonds. For a compound $C_aH_bN_cO_d$, this rule may be stated as

$$R = a + 1 + \frac{c - d}{2}$$

In case of benzene, we get

$$R = 6 + 1 - 3 = 4$$

So it contains one ring and three double bonds.

Bonding

Bonding information may be frequently interpreted from fracture patterns. A few of the rules used in this regard are given below:

(a) Branched compounds tend to rupture at the branched carbon atom; the positive charge remains with the branched fragment. The greater the degree of branching the greater is the probability of rupture.

(b) Beta position cleavage is the most probable rupture for double and hetero bonds.

(c) Ring compounds usually exhibit mass numbers which are characteristic of the ring.

(d) Side chains are ruptured from saturated ring compounds

the alpha position.

(e) Carbonyl compounds break at the carbonyl bond. The positive charge remains with the carbonyl group.

Some Example of Mass Spectra

Mass spectra of some simple compounds is given in Fig. 10.10.

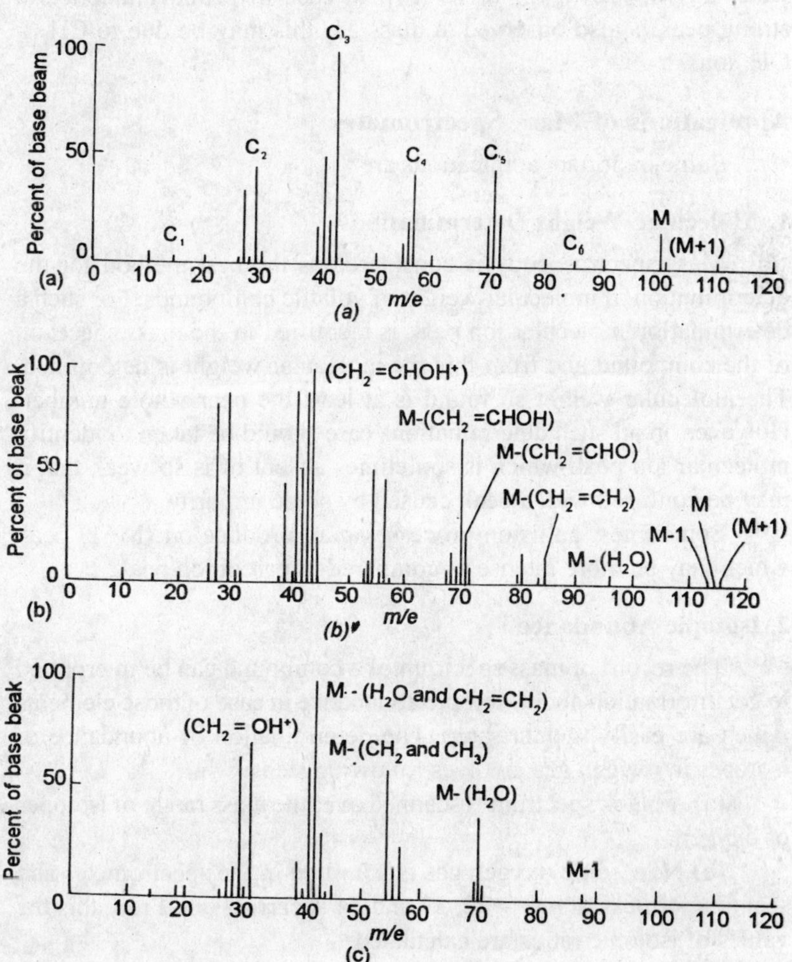

Fig. 10.10: Mass spectra of some simple compounds (*a*) *n*-Heptane; (*b*) 1-pentanol; (*c*) *n*-heptanal.

A close book at spectrum shown in Fig. 10.10 (*a*) reveals clusters of peaks differing in mass by 14. This type of pattern is typical straight chain paraffins. The observed mass decrease may be attributed to successive loss of methylene group.

Similar type of pattern can also be seen in left hand parts of Fig. 10 (*b*) and Fig. 10.10 (*a*). However, alcohols have a very weak or non-existent parent ion peak and give rise to a strong peak by loss of water at (M—18) [Fig. 10.10 (*c*)]. In case of primary alcohols a strong peak is also observed at mass 31, this may be due to $CH_2 = OH^+$ ion.

Applications of Mass Spectrometry

Some important applications are :

1. Molecular Weight Determination

Mass spectrosocopy is considered as the best method for the determination of molecular weight of volatile compounds. For such a determination molecular ion peak is identified in the mass spectrum of the compound and from this the molecular weight is determined. The molecular weight so found is at least the near whole number. However, in all such determinations care should be taken to identify molecular ion peak which is sometimes absent or is so weak that it may be confused with a peak caused by some impurity.

Sometimes, collision processes may produce an (M+1) peak which may be more intense as compared to parent ion peak.

2. Isotopic Abundance

The record of mass spectrum of a compound can be interpreted to get information about isotopic abundance in case of those elements which are easily vapourisable. The determination of abundance of isotopes in oxygen gas involves following steps:

(a) A blank spectrum is scanned over the mass range of isotopes of oxygen.

(b) Now some oxygen gas is admitted in the spectrometer and scanning of peak at $m/e = 32$, 33 and 34 is carried out. From this the values of isotopic ratios are calculated.

From these ratio of peak heights the relative abundances of various isotopes can be found out.

The isotopic ratio is found to be particularly useful for the detection and estimation of the number of sulphur, chlorine and bromine atoms is a compound. This is because of the large contribution made by them to (M+2) peak. Sometimes, we can identify combinations of chlorine and bromine atoms.

3. Isotopic Dillution Method

Mass spectroscopy has been adopted to the analysis of both inorganic and organic mixture of low volatility. By use of this method it has been possible to determine the amount of a component of a complicated mixtures from which it can not be separated quantitatively. For example to determine the amount of glycrine in a protein hydrolysate, if the weight of glycrine present in *xg* and the ratio $\frac{^{14}N}{^{15}N}$ is *p*. Now if *yg* of ^{15}N is enriched in synthetic sample of glycrine and the new ratio $\frac{^{14}N}{^{15}N}$ in this sample is *r*, which can be determined by of mass spectroscopy. Then the following relation can be written

$$\frac{x+v}{r+1} = \frac{x}{p+1} + \frac{y}{q+1}$$

Since all other quantities, in this relation, except *x* are known so we can calculated *x*.

This method has been found to be useful for determination of amount of caffeine in coffee and the determination of ^{18}O and ^{16}O ratio in CO_2 produced from rock carbonate samples. This study of CO_2 is used in geological survey of rocks.

4. Quantitative Analysis of Mixtures

For a successful mass spectroscopical analysis of organic mixtures the following conditions must be satisfied:

(i) Each component must exhibit at least one peak that differs markedly from others,

(ii) The contribution of each component to a peak must be linearly additive,

(iii) The sensitivity must be reproducible, and

(iv) Suitable standards for calibration must be available.

However for complex mixtures, it is seldom possible to find a mass peak that is unique for each component. A set of simultaneous equations must be solved in such a case, we can write in such cases as

$$I_1 = (i_1)_a p_a + (i_1)_b p_b + (i_1)_c p_c + \ldots\ldots$$

$$I_2 = (i_2)_a p_a + (i_2)_b p_b + (i_2)_c p_c + \ldots\ldots$$

Now, $(i_1)_a$, $(i_2)_a$ etc. can be obtained from spectra of pure compounds at known pressure which is equal to the sum of partial pressures, *i.e.*, $p = p_a + p_b + p_c \ldots\ldots$

If we solve these equations we can obtain the value of $p_a, p_b, \ldots$ and these values are proportional to the concentration of various components in the mixture.

A mixture of methane, ethane, propane, *n*-butane and iso-butane can be analysed by this method. In the spectrum of such a mixture both *n* butane and isobutane show peaks due to $C_4H_{10}^+$ $(\frac{m}{e} = 58)$ and $C_4H_9^+$ $(\frac{m}{e} = 57)$. This give rise to two equations and on solving these equations the amounts of these components can be found out. Making allowance for the contribution of butane to propane $(\frac{m}{e} = 44)$ it is possible to find propane content. Since the peak at 44 does not have any contribution from methane and ethane. Similarly ethane and methane content can be found out.

However, this method is not very reliable.

5. Distinction Between Cis- and Trans-Isomers

It is possible to make a distinction between cis- and trans-isomers by use of mass spectroscopy. The mass spectra is similar in both cases but there is a difference in molecular ion peak. The molecular ion peak is more intense for trans-isomer. However, the dehydration fragment is found to be stronger in case of cis-isomer. This method is seldom used but cis- and trans-isomers of hex-2-ene-1-ol have been characterised by this method.

6. Determination of Heat of Sublimation

When a sublime substance is placed in a sample reservoir

maintained at constant temperature an equilibrium gets established between the solid and vapour phase. Now, if a small amount of vapour is allowed to pass into the ionisation chamber from the reservoir and a mass spectrum is recorded a change in peak intensity occurs which is found to be directly proportional to the vapour pressure of the sample in the ion source. In the same way vapour pressure of the sublime substance are found out at various temperatures and to find the heat of sublimation a graph is plotted between log p *vs* $\frac{1}{T}$. The slope of such a graph gives the heat of sublimation.

This technique has been used to study the nature of species in the vapour phase over some substances having high boiling points, *e.g.*, Monomeric and dimeric species have been found in vapour phases over Na, L, K and Cs chlorides.

7. Determination of Ionisation Energy

Ionisation energy refers to the energy needed to remove an electron from a molecule or atom.

$$AB \longrightarrow AB^+ + e^-$$

$$A \longrightarrow A^+ + e^-$$

$$B \longrightarrow B^+ + e^-$$

According to **Frank Condon** principle the molecular separation in the ionised species remain the same as in a neutral atom or a molecule.

The ionisation energy can be easily determined by mass spectroscopy. For this purpose a plot of (peak height) ion current against electron beam energy (*ev*) is drawn.

The procedure adopted for determination of ionisation energy of nitrogen making use of argon as internal standard is as under :

(i) A blank mass spectrum is run upto $\frac{m}{e} = 50$ to confirm that peaks at $\frac{m}{e} = 28$ (N_2) and $\frac{m}{e} = 40$ (Ar) are not present in the back ground material.

(ii) Now a sample of nitrogen containing argon is introduced in the mass spectrometer and mass spectrum is recorded upto $\frac{m}{e} = 40$.

The electron beam energy is then reduced in steps of 5 eV and various values are recorded. This is repeated till the output current of ion $\frac{m}{e} = 40$ is equal to zero. Once again the beam energy is increased in steps of 5eV until full deflection on the original amplifier. Then a plot is drawn between the intensity of the ion beam and the electron beam energy for $\frac{m}{e} = 40$. Then the curve is extraploted. The value where the extrapolated curve meets the electron beam energy axis gives the I.E. of argon. This value is then compared with standard I.E. value of argon (15.76 eV) and error is found. Now the I.E. of nitrogen is calculated as under.

(iii) A mass spectrum is now run upto $\frac{m}{e} = 28$ (N_2^+) and then the ionisation efficiency curve is plotted. The curve is now extraploted to get the I.E. of nitrogen. The correction is then applied to get correct value.

8. Bonding

To get information about bonding the following rules are used :

(i) The rupture generally occurs at the branched carbon atom in branched compounds. The positive charge remains with the branched fragments. Greater the degree of branching more is the probability of rupture.

(ii) In case of double and hetero bond β-position cleavage is most probable.

(iii) The mass numbers which are characteristic of the ring are exhibited by ring compound.

(iv) The rupture of side chain occurs at α-position in saturated ring compounds.

(v) Carbonyl compound break at the carbonyl bond. The positive charge remains with the carbonyl group.

9. Determination of Bond Dissociation Energy

Using the concept of appearance potential of a fragment ion the bond dissociation energy of a molecule can be found out. The appearance potential of a fragment ion is the energy required to produce that ion in the mass spectrometer. Since the production of any ion other than molecular ion involves cleavage of bonds so the appearance potential of a fragment ion will furnish information on bond dissociation energies. For determination of ionisation potential from ionisation-efficiency curves is related to the energies of electron beam.

When a molecule AB is bombarded with electron beam, a positive molecule ion is produced.

$$AB + e^- \rightarrow AB^+ + 2e^-$$

Furthur dissociation of molecule ion AB^+ occurs as under

$$AB+ \rightarrow A^+ + B'$$

$$AB+ \rightarrow A' + B^+$$

The appearance potential of A^+ in the step $AB^+ \rightarrow A^+ + B'$ can be measured from the ionisation efficiency curve of ion A^+. Because A^+ is only produced after ionisation of AB so the appearance potential is equal to the sum of energies required to dissociate AB and to ionise the radical A°. The following expression can be written.

$$A(A^+) = I(A') + D(A\text{–}B)$$

where $A(A^+)$ = Appearance potential of A^+

$I(A°)$ = IE of A°

and $D(A\text{–}B)$ = Bond dissociation energy of AB.

The value $A(A^+)$ is measured from the ionisation efficiency curve of ion A^+ and I.E. of A° can be obtained from photoionisation method. Knowing the experimental value of $A(A^+)$ and $I(A°)$ it is possible to calculate the bond dissociation energy.

10. Reaction Kinetics

One of the important application of mass spectroscopy is in the study of reaction kinetics. It is used in the study of radical reactions. The basic principle involved is based on the fact that the concentration of many of the unstable species present in the reaction. To examine the time interval the reactive species are examined from the initiation of the reaction by changing the distance between the reaction vessel and the source. From the peak heights in spectrum it is possible to determine the concentration of different species. A graph is drawn between the time t and log peak height, a straight line of slope—2303/

K is evaluated. This graph is only valid for the first order reaction. For a second order reaction the graph is plotted between time and peak height. From the slope it is possible to calculate the value of K.

For calculating the energy of activation E by making use of the following equation

$$ln\frac{k_2}{k_1} = -\frac{E}{R}(\frac{1}{T_2} - \frac{1}{T_1})$$

where k_1 and k_2 are the rate constants at temperature T_1 and T_2 respectively.

11. Latent Heat of Vaporisation of Liquids

The Clausius-Clapeyron equation is used for the determination of latent heat of vaporisation of liquids which gives a relationship between the saturated vapour pressure of a liquid as a function of absolute temperature.

$$\log p = -\frac{\Delta H}{RT} + \text{Constant}$$

If p at a given temperature T is known them ΔH can be calculated. The height of peak in mass spectrum of that liquid whose ΔH value is to be measured is directly proportional to the saturation vapour pressure p at the temperature T. In the same way the values of p can be determined from the height of the peak at different temperatures. The value obtained from the slope of log p against 1/T.

12. Studies Involving Reaction Mechanism

For study of reaction mechanism by mass spectroscopy a particular atom in the reactant is labelled and then this atom is located in one of the products. The mechanism of esterification has been studied in this way by labelling oxygen atom ^{18}O. When the ester containing ^{18}O is saponified by OH^- the ^{18}O appears in carboxylic acid. From this it is confirmed that the process of saponification of ester involves cleavage of the acyl oxygen bond and not alkyl-oxygen bond.

13. Detection of Impurities

Impurities in a sample can be detected by mass spectroscopy.

Such a detection is possible only if the structures of impurities differ considerably from the structures of major products. In case impurities have larger molecular weights as compared to major products their detection becomes easier because their higher mass peaks are free from contributions of major components.

14. Identification of an Unknown Compound

To identify an unknown compound by mass spectroscopy following points are taken into consideration :

(i) A background spectrum is run before introducing the sample. This is done because weak peaks are generally present at *m/e*=41, 43, 55 and 57 due to hydrocarbon background. After this sample is introduced and spectrum is run.

(ii) It is then found if the peak at the highest mass is a molecular ion and then peaks immediately lower to it correspond to the loss of neutral molecules or not. Then from the data nature of bond, abundance of isotopes etc. is determined.

(iii) Fragmentation pattern is then used to get indication about functional groups and some information about the structure of the compound.

Finally a comparison is made between the mass spectrum of the unknown compound with the authenticated mass spectrum of a known compound.

The important advantages of identification by this technique are :

(i) Since the mass spectrum is always taken is gaseous state so complications arising due to solvent, change of state etc. have no effect.

(ii) Only a very small quality of the sample is needed.

The only disadvantage being that the spectrum gets complicated even if only small quantities of impurities are present and in such a case it is not easy to interpret the spectrum.

15. Characterisation of Polymers

Polymers can be easily characterised by mass sepstroscopy. For its characterisation polymer is pyrolised and the products are fed into mass spectrometer and identified. Information regarding the structure of polymer can be obtained and this information has been used to distinguish chlorinated polymers to know if the chlorine occurs randomly or occurs in blocks in the structure of the polymer.

16. Amino Acid Sequence Analysis in Peptides

The determination of the structures of peptides and protein is of major importance, and is often done by a process of stepwise hydrolysis and much column chromatography. A method for overcoming the lack of appreciable volatility of peptides, a disadvantage in mass spectrometry, is to reduce them with lithium aluminium hydride to the corresponding polyamino alcohols.

$$H_2N-\underset{R}{CH}-\underset{\|O}{C}-NH-\underset{R'}{CH}-\underset{\|O}{C}-OH \xrightarrow{LiAlH_4} H_2N-\underset{R}{CH}-CH_2-NH-\underset{R'}{CH}-CH_2OH$$

Polyamino alcohols are volatile and upon electron impact fragment in a characteristic manner. The problem of determining and amino acid sequence can be simplified further by the use of gas chromatography. If a polyamino alcohol mixture derived from a protein-hydrolysate peptide mixture is subjected to gas chromatographic separation, then the effluent fractions may be introduced into the mass spectrometer individually. The development of GC/MS has obviated the effluent isolation step and speeded up the process considerably. Derivatisation of amino alcohols also helps sequencing. Hydroxyl groups can be converted to trimethyslilyl ethers by treatment with N-(trimethylsilyl) diethylamine, a reagent that does not react with amino groups. Specific O-silyl derivatives show a considerable enhancement of fragment ions from the C-terminal amino acid.

Mass spectrometric examination of peptides can be carried out without prior reduction of the amide linkages. Compounds of sufficient volatility and stability can be obtained by esterification of the free carboxylic acid end-group and the conversion of the amine end-group to a characteristic derivative. For example, acetyl, trideuteroacetyl, trifluoroacetyl and carbobenzoxy.

Amino acid sequencing by MS is being used increasingly. The

interpretation of spectra, however, although based upon simple arithmetic summations, is tedious. Fortunately MS spectral data are amenable to computer analysis, thus speeding the sequencing process considerably.

(IX)

$C_{16}H_{13}N_2O_2Cl$

$C_{16}H_{13}N_2O_2Cl$ + O

$C_{16}H_{13}N_2O_2Cl$ + O

$C_{16}H_{13}N_2O_2Cl$ + 2O

$C_{16}H_{13}N_2O_2Cl$ + O - CH_2

Fig.

17. Drug Metabolism

Drug metabolites generally arise from relatively minor structural modifications of the parent molecule. Oxidation to hydroxyl derivatives or O-demethylation, generally renders a foreign molecule more readily eliminated from the body. Mass spectrometry is a technique which affords considerably structural information from a small sample and is consequently ideal for metabolism studies. The scope of the technique is extended further by its ability to detect drugs and metabolities that have been cold labelled, for example with ^{2}H, ^{13}C or ^{18}O.

The most obvious approach to identifying metabolites is to compare directly spectra of the pure drug with those of its biotransformation products. Direct insertion of metabolites, after isolation by column chromatography, preparative TLC or preparative GC, into the ion source of a high resolution instrument, followed by accurate mass measurement of each molecular ion, provides their molecular formulae. Differences in molecular formula show the transformation involved and the fragmentation pattern of a metabolite will generally yield information about the position of gain or loss of a molecular unit. Identification of the major metabolites of diazepam (IX) was established in this manner (Schwarts, Bommer and Vane 1967). Initial oxidation takes place at the positions indicated (*) in (X) and (XI). Nuclear magnetic resonance spectroscopy was used to assign unambiguously the positions of substitution. Secondary metabolism takes place by N-demethylation of (XI) to give (XIII) and by inserting a second oxygen (XII).

Combined GC/MS is becoming an increasingly common technique for the identification of organic compounds in admixture and in high dilution. It is now the method of choice in most metabolism studies.

For many years, 1-(2-acetylhydrazino) phathalazine (XV) was presumed to be a major metabolite of hydralazine (XIV). However, attempts to synthesise it resulted in the isolation of the triazolophthalazine (XVI) only. Paper chromatogrpahy of a metabolite, previously identified as the acetyl derivative (XV), was confirmed by a GC/MS study of the urine from three patients to be (XVI) (Zimmer, Kokosa and Garteiz, 1973).

$NH.NH_2$

Hydralazine (XIV)

$NH.NH.\overset{O}{\overset{\|}{C}}.CH_3$

(XV)

CH_3

(XVI)

A total ion current gas chromatogram is illustrated in Fig. 10.11 and Fig. 10.12 shows the mass spectral characterisation of compound (XVI).

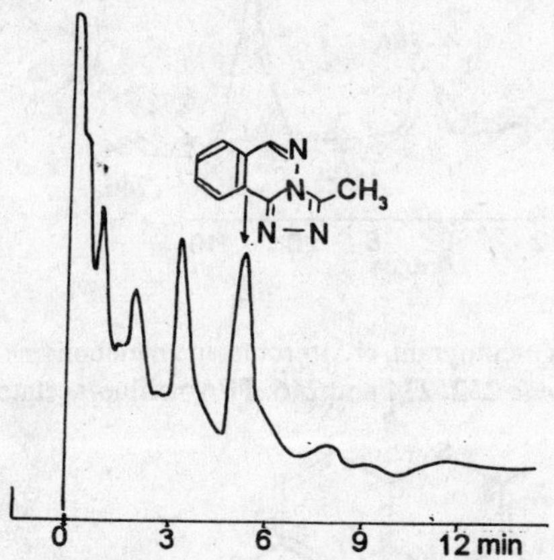

Fig. 10.11: Total ion current recording after injection of a 3 μl aliquot of a urine extract from a patient treated with hydralazine.

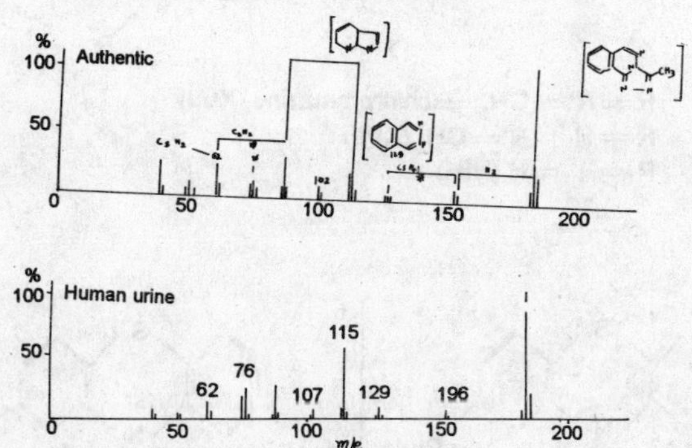

Fig. 10.12: Upper panel : Mass spectrum of authentic 3-methyl-s-triazolo 3, 4-a) phathalazine.
Lower panel : Mass spectrum of a compound with retention time of 5.5 min in a urine extract from a patient treated with hydralazine.

A fully computerised method for the quantitative determination of pictogram quantities of drugs by GC/MS has been described by Baczynskyj (1973).

The first application of mass fragmentography to the determination of drug metabolites was described for chlorpromazine (XVII) (Hammer, Holmstedt and Ryhage, 1968).

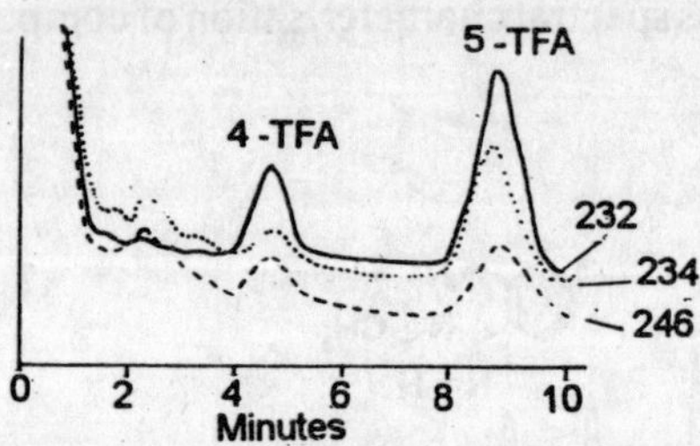

Fig. 10.13: Mass fragmentogram, chlorpromazine metabolism study. Fragments monitorer *m/e* 232, 234 and 246. TFA=trifluoracetate derivatives.

S
N
Cl
x
CH_2 — CH_2 — CH_2 — N
y
R
R'

$R = R' = CH_3$ = chlorpromazine (XVII)
$R = H$; $R' = CH_3$ (XVII)
$R = R' = H$ (HIH)

S
N
Cl
+
$A = {}^{35}Cl$
$A' = {}^{37}Cl$

S
N +
Cl
CH_2
$B = {}^{35}Cl$

A mass fragmentogram where the intensity of ions [*m/e* 232 (A), *m/e* 234 (A^{1}) and *m/e* 246 (B)] have been monitored with time is shown Fig. 10.13. Fragmentation at bond X gives rise to ions A, A^{1} and fragmentation at Y ion B. The intensity of these ions in the mass spectrum of a compound related to chlorpromazine will change according to the nature of the basic side-chain. Preparation of trifluoroacetate (TFA) derivatives of the mixture of metabolites increases the specificity of the method. By a consideration of their GC retention times and the relative intensities of the three ions monitored as seen in the mass fagmentogram, desmethylchloropromazine (XVIII) and didesmethylchlorpromazine (XIX) were identified as chloropromazine metabolites.

18. Clinical and Forensic Applications

A delicate balance of biochemical reactions is essential to ensure the health of the human organism. Specific disease states cause changes in body chemistry and excretion products and the latter may be detected in body fluids, expired air or sweet. Mass spectrometry and the GC/MS combination may be used to monitor these changes and aid diagnosis.

The health of an individual may also be impaired by accidential or deliberate abuse or overdosage of drugs or toxic chemicals. Forensic science and toxicology, these are areas where MS and GC/MS are finding increasing application.

Finkle (1973) for example, has reported the case of a male with a history of drug abuse found dead with narcotic overdosage being suspected as the cause.

Analysis by GC/MS showed that his blood and urine contained codeine, and that bile, and veins at the injection site, contained morphine, supporting the probable cause of death.

Chemical ionisation often gives peaks of much higher intensity in the molecular ion region than do corresponding electron impact spectra. In the identification of mixtures of drugs in body fluids this enhanced sensitivity coupled with spectrum simplicity is of special value.

Barbiturates have been prescribed so widely as sedatives that accidential or deliberate overdosage is relatively common. Because of the multiplicity of barbiturate analogues, identification by standard procedures is difficult, is not impossible. Fales, Milne and Axenrod (1970) have reported a method of barbiturate identification by GC/MS. The detection of quinalbarbitone and pentobarbitone in an extract of gastric contents is illustrated in Fig. 10.14.

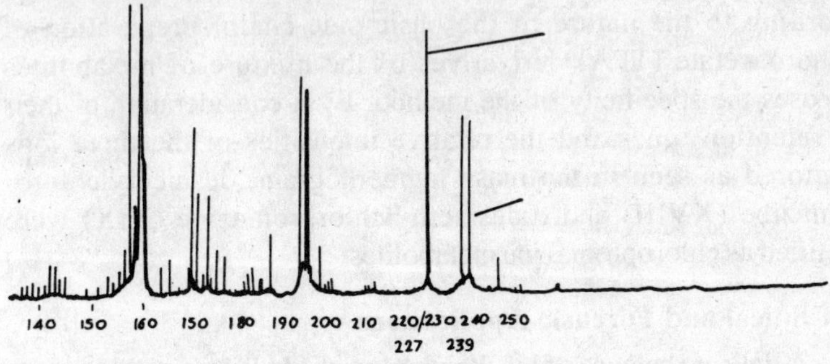

Fig. 10.14: Chemical ionisation mass spectrum of extract of gastric contents

Mass spectrometric techniques are advancing rapidly and are exploited in an increasing number of medical and related disciplines. Indentification and quantitation of drugs, metabolites and toxins will become more sophisticated, rapid medical diagnosis of MS will spread, and the monitoring of the environment will develop.

PROBLEMS

1. For a field strength of 2400 G in 180° magnetic-deflection spectometer, what electrostatic voltage range suffices for scanning from mass 18 to mass 200? The radius of curvature of the 180° analyzer tube is 12.7 cm.

2. For a drift length of 100 cm in a time-of flight mass spectrometer, what is the difference in arrival time between ion of $m/e = 44$ and $m/e = 43$ when the accelerating voltage is 2800 V?

3. The spectrum of cetyl palmitate (M.W. 480) was recorded on a linear scale in the presence of $C_{10}H_{16}$ (M.W. 438.970). On the spectrum the molecular ion peak of cetyl palmitate was separated from the M+1 peak by 200.5 nm, and from the perfluorocarbon molecular ion peak by 95.3 mm. What is the precise mass of the molecular ion of cetyl palmitate?

4. The present peak spectrum of tridecylbenzene (260.2504), phenyl undecylketone (260.2140), 1,2-dimethyl-4-benzoyl napthalene (260.1201), and 2,2 naphthyl benzothiophene (260.0922) would require that resolution for quantitative analysis based on the parent peak?

5. What resolving power is needed to separate (*a*) the CH_2N—C_2H_4 doublet at mass 200; (b) the N_2—CO doublet at mass 150; and (c) the CH_3—N doublet at mass 200?

6. The peptide was admitted to a high-resolution mass spectrometer and the parent peak mass was measured relative to the parent peak in the spectrum of dibromobenzene (236.8638). The measured ratio of unknown mass/reference mass 1.001197± 0.000002. Compute the exact weight of the peptide and deduce the molecular formula.

7. From the following exact molecular weights, estimate the empirical formulas assuming only C, H, O or N is present unless otherwise indicated : (a) 164.0473, (b) 120.0575, (c) 180.0939, (d) 4.0531, (e) 109.0528, (f) 190.9540 (contains Cl), (g) 181.0891, (h) 334.0873 (contains S), and (i) 177.0426.

8. (a) In the high-resolution spectrum of methionine, a quartet of peaks at appear nominal mass 88 : 88.0222, 88.0345, 88.0355, and 88.0267. Deduce the fragment ion responsible for each peak. (b) Methionine also gives a doublet at nominal mass 75. One line corresponds to $C_2H_4HO_2$; the other has *m/e* 75.0267. Outline the process leading to the fragment $C_2H_4NO_2$. Deduce the fragment ion of *m/e* 75.0267.

9. What is the probable composition of a molecule of mass 142 whose P_{M+1} peak is 1.1% of the parent peak.

10. Deduce the number of type of halogen atoms present in a molecule from the abundance of heavy isotopes and the intensity ratios given in the following table :

	P_M	P_{M+2}	P_{M+4}	P_{M+6}	P_{M+8}
Compound A	30	29	10	1	—
Compound B	13	30	19	6	1
Compound C	5	20	30	19	5
Compound D	23	30	7	—	—
Compound E	18	30	14	2	—

11. From the isotopic abundance information, what can be deduced concerning the empirical formula of each of the following

compound?

m/e	*% of base peak*	*m/e*	*% of base peak*	*m/e*	*% of base peak*
90(P)	100	89(P)	17.12	206(P)	25.90
91	5.61	90	0.58	207	3.24
92	4.69	91	5.36	208	2.48
		92	0.17		
230(P)	1.10	140(P)	14.8	151(P)	100
232	2.12	141	1.40	152	10.4
234	1.06	142	0.85	153	32.1
				154	2.9

12. The significant portion of the mass spectral data is given for the individual alcohols. Select appropriate analytical masses and write a series of four equations in terms of divisions of base peak due to each of the four alcohols.

	% of base peak				*Unknown mixtures*		
m/e	*Methyl*	*Ehthyl*	*n-Propyl*	*Isopropyl*	*A*	*B*	*C*
15	35.48	9.44	3.77	10.70			
19	0.29	3.13	0.90	6.51			
27	—	21.62	15.20	15.50			
29	58.80	21.24	14.14	9.49			
31	100	100	100	5.75			
32	68.03(P)	1.14	2.25	—	600	600	2350
33	0.98	—	—	—			
39	—	—	4.00	5.52	4800	3000	3000
43		7.45	3.18	16.76			
45		37.33	4.39	100			
46		16.23(P)	—	—	1000	1100	698
59			9.61	3.58	4000	2300	5000
60			6.36(P)	0.44(P)			
Sensitivity, divisions/10^{-3} tor	8.76	17.98	26.51	23.47			

13. The mixture peaks for three unknown mixtures of alcohols are shown in problem 12. For each mixture compute the mole percent

of each alochol.

14. A material containing only C, H and O, and in the form of leftlets melting at 40°C, possesses a rather simple mass spectrum with the parent peak at *m/e* 184 (10%), the base peak *m/e* 91, and small peaks at *m/e* 77 and 65. Metastable peaks appear at *m/e* 4.50 and 46.5. Deduce the structure of the compound.

15. The mass spectrum possesses a strong parent peak at *m/e* 122 (35%) plus peaks at *m/e* 92 (65%), *m/e* 91 (100%), and *m/e* 65 (15%). In addition there are metastable peaks at 46.5 and 69.4 mass units. Deduce the structure of the compound.

16. A solid, melting at 33°C, has the following mass spectrum : Parent ion at *m/e* 200 with P_{M+1} exhibiting a strength equal to 10.55% of the parent peak, and P_{M+2}, 5.77% of the parent peak. The major ion peaks occur at these mass units : 172 (18%), 155 (61%), 108 (10%), 107 (11%), 92 (30%), 91 (100%), and 65 (30%), Metastable peaks appear at 46.5, 53.5, 67.9, 106.3, 121.7, and 147.9 mass units. Deduce the structure of the compound insofar as possible.

17. Deduce the structural formula for each of these compounds from the mass spectral data :

m/e	$C_4H_8O_2$ % of base peak	*m/e*	$C_4H_6O_2$ % of base peak
27	39.3	15	27.7
29	19.8	26	22.4
39	14.8	27	68.1
41.	23.7	29	13.0
42	24.7	42	11.8
43	22.3	55	100
45	19.1	58	8.4
60	100	59	5.2
73	27.1	85	12.3
88(P)	1.6	86(P)	2.1

C		*D*	
m/e	*% of base peak*	*m/e*	*% of base peak*
29	18	63	22
39	23	64	20
51	29	95	33
65	18	92	82
78	50	93	18
91	100	120	100
105	41	121	34
134(P)	57.4	152(P)	45.0
135	5.80	153	4.1
136	0.41	154	0.4

18. Deduce the complete structural formula of the compound from the mass spectrum figure.

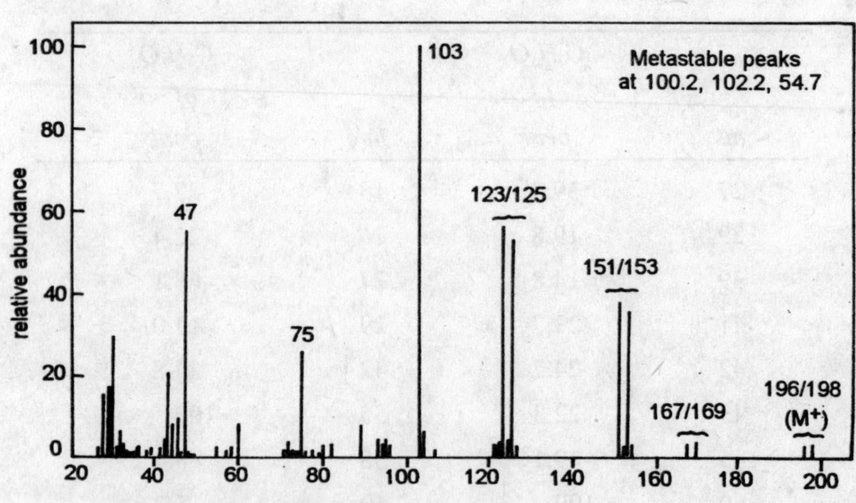

19. Deduce the structural formula of the compound (by 74°C) whose mass spectrum is shown.

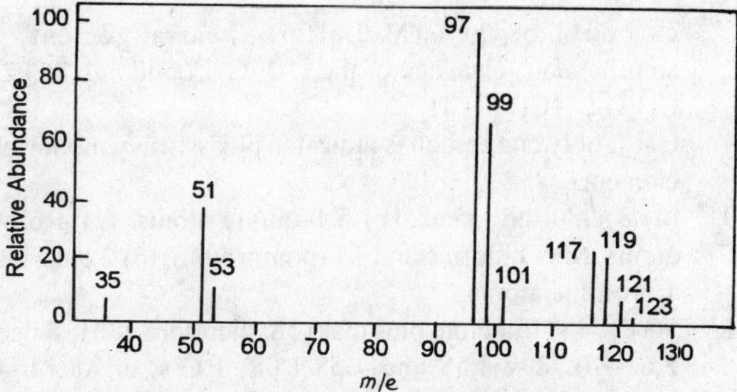

20. A low melting solid with molecular formula $C_5H_8O_4$ gave the mass spectrum shown. Deducc the structural formula of the compound.

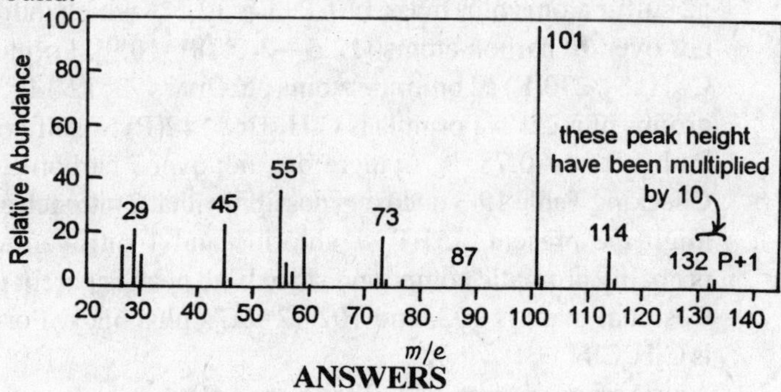

ANSWERS

1. 2490 V for mass 18 and 224 V for mass 200.
2. 9.03 μsec for mass 44 and 8.9 μsec for mass 43.
3. 480.493.
4. For the pair, tridecyl benzene and phenyl undecyl ketone, resolution required is 7140; for 1,2-dimethyl-4-benzoyl naphthalene and 2, 2-naphthyl benzothiophene, resolution required is 9320.
5. (a) 10,000, (b) 6500, (c) 10,000.
6. Mass =237.147±30.0005; $C_{12}H_{19}N_3O_2$.
7. (a) $C_9H_8O_3$. (b) C_8H_6O (c) $C_{14}H_{12}$. (d) $C_6I_6H_2$. (e) C_6H_7NO. (f) $C_6H_3Cl_2NO_2$. (g) $C_{13}H_1N$, (h) $C_{17}H_{18}O_5S$. (i) $C_9H_7NO_3$.
8. (a) C_3H_6NS; $C_4H_8SC_3H_6NO_2$; $^{15}CC_3H_7S$. (b) The hydrogen 3 carbon atoms from the carbonyl group is transferred to the

carbonyl oxygen (McLafferty rearrangement) with simultaneous cleavage of the C-2, C-3 bond *m/e* 75.0267 is CH_3—S—CH_2—CH_2^+.

9. CH_3I; only one carbon is indicated plus a heavy monoisotopic element.
10. (a) 3 chlorine atoms. (b) 5 bromine atoms. (c) 4 bromine atoms. (d) 1 chlorine and 1 bromine atom. (e) 2 chlorine and 1 bromine atoms,
11. 90(P) : 1 sulfur atom plus mass 58; therefore, $C_7H_{10}S$ because 5.61—0.78 = 4.58 and 4.58/1.08=4 C's, or 58/14=4CH_2 groups plus 2 H's. Compound is a dialkyl sulfide, 89(P) : 1 chlorine, nitrogen, plus residual mass 40. Since P+1 indicates not over 2 carbon atoms, probable formula is C_2ClNO. 206(P) : 2 sulfur atoms plus mass 140. P+1 is 12.5% which indicates not over 10 carbon atoms [12.5—2(0.78)=10.9]. Formula is $C_{10}H_{2}{}_2S_2$. 230(P) : 2 bromine atoms plus mass 72; 72/14=5CH_2 groups plus 2H's. Formula is $C_5H_{12}Br_2$. 142(P) : 1 sulfur atom P+1 is 9.54 – 0.78=8.76; therefore, not over 7 carbon atoms. Checking Table 19-3 uncovers possibility that 2 nitrogen atoms might be present. 151(P) 1 chlorine and 1 nitrogen atom; probably aromatic compound since base peak is parent peak. Residual mass is 102, and 102/12=8C's plus 6H's. Formula is C_8H_6ClN.
12. Appropriate masses : 32, 39, 46, and 59.
13.

UKNOWN	MeOH,%	EtOH,%	*n*-PrOH,%	*i*-PrOH, %
A	0.9	8.2	11.7	79.0
B	2.5	13.5	6.9	77.0
C	7.1	7.8	52.7	31.4

14. Metastable at 45.0 indicates the parent ion decomposes to mass 91 with loss of mass 93 neutral fragment(s). Metastable at 46.5 indicates that mass 91 decomposes further to mass 65 plus a neutral fragment of mass 26. These data, coupled with the peak intensities, indicate that we are dealing with a ring compound possessing the structure C_6H_5—CH_2 plus mass 93. Mass 93 is probably C_6H_5O. Molecule is tolyphenyl ether.
15. Meiastable at 69.4 indicates the decomposition route is from mass 122 to mass 92 plus a neutral fragment of mass 30. Metastable at 46.5 indicates the decompoisition of mass 91

to mass 65 plus a fragment of mass 26. Ion peak at mass 92, an even mass arising from an even mass molecular ion, suggests a rearrangement reaction. Coupled with the possible transition from mass 122 to 91, involving a loss of mass 31, this suggests the presence of a CH_2OH group which could participate in a McLafferty rearrangement to account for loss of CH_2O. Base peak at mass 91 suggests a tolyl group. Compound is 2-phenyl ethanol.

16. Metastable peaks indicate these transitions 147.9 : $200^+ \rightarrow 172^+ + 28$ (probable loss of $CH_2{=}CH_2$). 121.7 : $200^+ \rightarrow 156^+ + 44$ (probable loss of $CH_2{=}CH{-}OH$) 106.3 : $108^+ \rightarrow 107^+ + 1$ (loss of H). 67.9 : $173^+ \rightarrow 108^+64$ (probable loss of SO_2). 53.5 : $155^+ \rightarrow 91^+ + 64$ (probable loss of SO_2). 46.5 : $91^+ \rightarrow 65^+ + 26$ (loss of HC=CH from tolyl). One sulfur atom is indicated by the P+2 peak. The loss of $CH_2{=}CH_2$ and $CH_2{=}CHOH$, both as a result of rearrangement processes, plus the loss of mass 45 in the transition from mass 200 to mass 155, shows the presence of an ethoxy group which is linked to a sulfoxide group as shown by the subsequent loss of SO_2 from mass 155. Base peak at mass 91 suggests a tolyl group. Thes deduction account for the molecular weight of 200 and the presence of not more than 9 carbon atoms (P+1 abundance). The compound is $C_2H_5{-}OSO_2{-}C_6H_5{-}CH_3$; ring substitution cannot be ascertained.

17 88(P) : butyric acid. 86(P) : methyl acrylate. 134(P) : 3-phenyl propanaldehyde. 152 (P) : Formula is $C_8H_8O_8$; methyl salicylate.

18. Metastable peaks indicate these transitions. 102.2 : $153^+ \rightarrow 125^+ + 28$. 100.2 : $151+ \rightarrow 123^+ + 28$, 54.7 : $103^+ \rightarrow 75^+ + 28$. All mass 28 losses involve ethylene fragment from an ethoxy group.

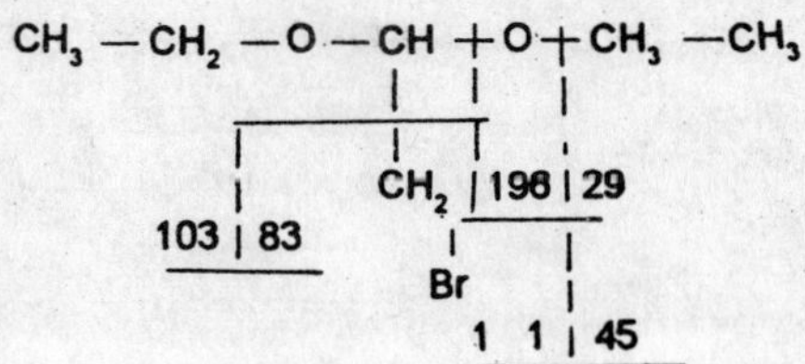

Fig.

19. CH_3CCl_3. Apparent loss of 20 mass units (117—7) cannot be corrleated with any known fragment except HF, mass 117 contains 3 chlorine atoms, and 117—3(35) leaves only 12 mass units, or 1 carbon atoms. Thus, HF is ruled out and mass 117 is CCl_3. Now mass 97 contains 2 chlorine atoms, yet cannot arise from mass 117 by loss of 1 chlorine atom. Mass 97 is CH_3CCl_2. Thus compound is CH_3CCl_3: no molecular ion peak is present.
20. From empirical formula, two usaturated bonds and/or rings are indicated. Mass 114, a rearrangement peak, arises from loss of water, indicating a COOH group. Mass 101 indicates loss of CH_3O or $HOCH_2$. Mass 59 suggests and ester, our second double bond. Pieces are : CH_3O + CO; COOH; and remainder of C_2H_4. Compound is methyl hydrogen succinate.

11

X-Ray Spectroscopy

Introduction

X-Ray photoelectron spectroscopy also called **ESCA** (or electron spectroscopy for chemical analysis) is a type of electron spectroscopy. ESCA is based upon the same processes that result in X-ray absorption and emission.

Principle

The absorption of an X-ray photon by an atom forms an excited ion and an electron.

$$A + h\nu_1 \rightarrow (A^+)^* + e_1^-$$

where $(A^+)^*$ is the excited ion produced.

By an interaction between the element A and X-ray photon $h\nu$, the relaxation of excited ion can take place in two ways.

$$(A^+)^* \rightarrow A^+ + h\nu_2 \quad \text{....(1)}$$

or

$$(A^+)^* \rightarrow A^{++} + e_2^- \quad \text{....(2)}$$

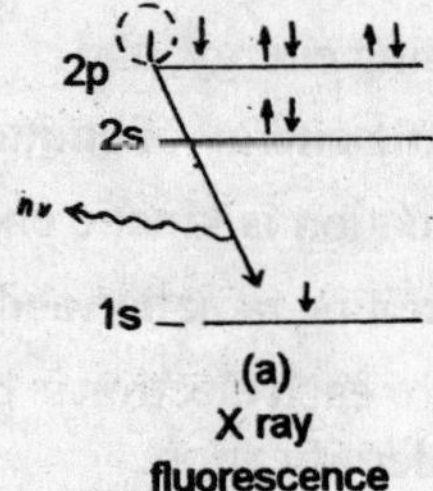

Fig. 11.1

The (1) process forms the basis of X-ray fluorescence methods (Fig. 11.1).

The two relaxation process (1) and (2) are competitive and the process (1) is favoured by high atomic number of the element involved. The X-ray fluorescence is not quite sensitive for detection of elements with atomic number less than 10.

Both X-ray absorption and X-ray photoelectron spectroscopy are based on the equation

$$A + h\nu_1 \rightarrow (A^+) + e_1^-$$

In the X-ray absorption spectroscopy an X-ray spectrometer is used for studying absorption as a function of wavelength or energy as the sample is irradiated with a polychromatic source.

In the photoelectron spectroscopy (PES), a monochromatic source is used and the energy of emitted electron is measured using an electron spectrometer (Fig. 11.2).

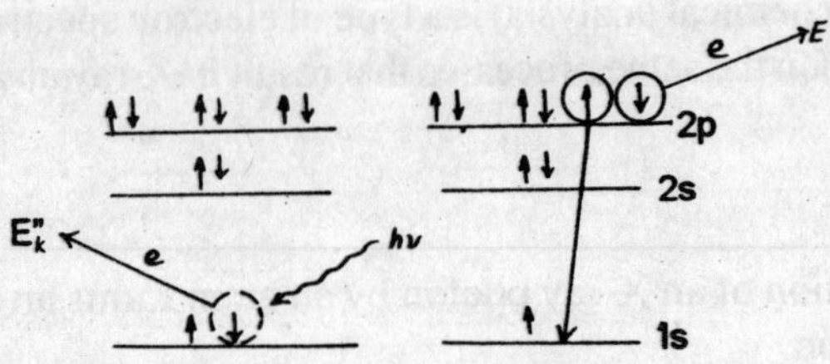

Fig. 11.2

Basically the method is equivalent to measuring the ionisation potential of both outer electrons and more firmly bound inner electrons. The kinetic energy of the emitted electrons is used for calculating the binding energy of electrons.

$$E_b = h\nu - E_k''$$

where E_b = binding energy, and

E''_k = kinetic energy of emitted electron

The more general equation is kinetic energy = $h\nu - I$

where $h\nu$ is the energy of exciting radiation and I is the binding energy.

The binding energy for each electron is unique for each element and so it can be used for identification.

Photoelectron Spectrosocopy

In X-ray photoelectron spectrosocopy the source generally used is an ordinary coolidge tube (Fig. 11.3). To get monochromatic radiations filters are used as shown in Fig. 11.3.

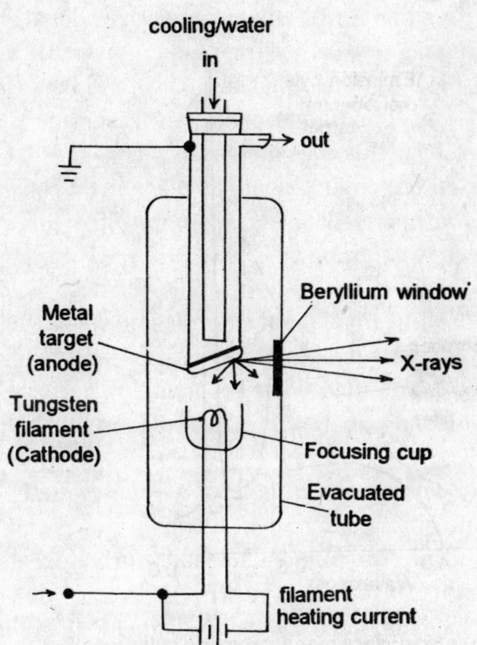

Fig. 11.3: Schematic diagram of the Coolidge tube.

Fig. 11.4 illustrates a common technique use for producing a monochromatic beam by using a filter. The Kβ line along with most of continuous radiations from emission of a molybdenum target is removed by a zirconium filter having a thickness of about 0.01 cm. The pure Kα line is then use for analysis.

Filtration of continuous radiations from a coolidge tube is done with a thing strip of metal as in case of molybdenum target. In this case, as with glass filters for visible radiation, relatively broad bands are produced and a significant loss of intensity also occurs.

The electrons produced by irradiation of the sample are passed into an electron spectrometer. Commonly used spectrometers use curved (one or more) magnetic fields for focussing electron beams on a detector. The trajectory of an electron depends upon its kinetic energy and the strength of the field.

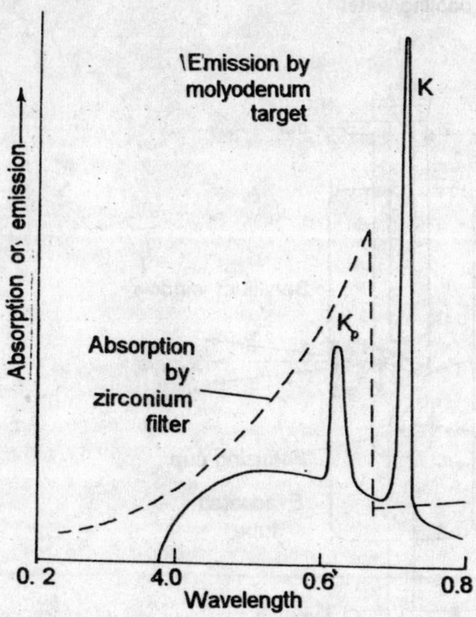

Fig. 11.4: Use of a filter to produce monochromatic radiations.

For estimation of kinetic energy or velocity of ejected electrons following are the main approaches.

In the first method shown in Fig. 11.5 the radiation is allowed to along the axis of a metal can that contains a concentrated metal grid and also holds the sample. Electrons which are ejected pass through the grid and are then collected by the can and the current is registered by a sensitive electrometer. When the grid is uncharged all the ejected electrons reach the can and maximum current is recorded. If grid is negatively charged slow moving electrons fail to reach the can as they are repelled and so the current falls. As the grid potential is increased from zero a curve of current versus grid potential can be plotted.

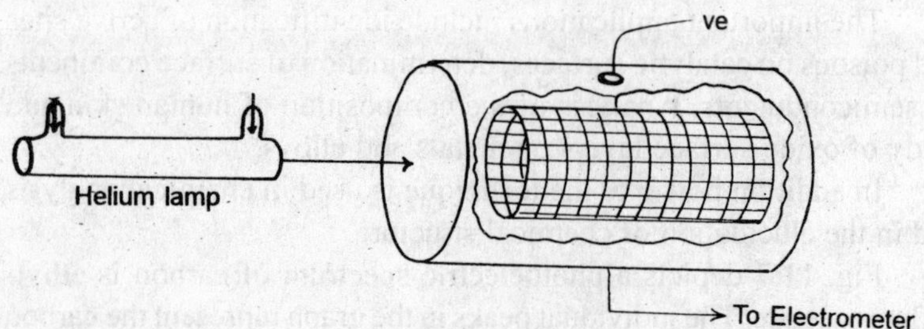

Fig. 11.5

Fig. 11.6 shows the second method.

In this method the ejected electrons are collimated into a beam by a slit in the can and then they can be sorted out by magnetic fields. Since the mass of electrons is same, the path of moving electrons is curved into a circle whose radius is directly proportional to the applied magnetic field.

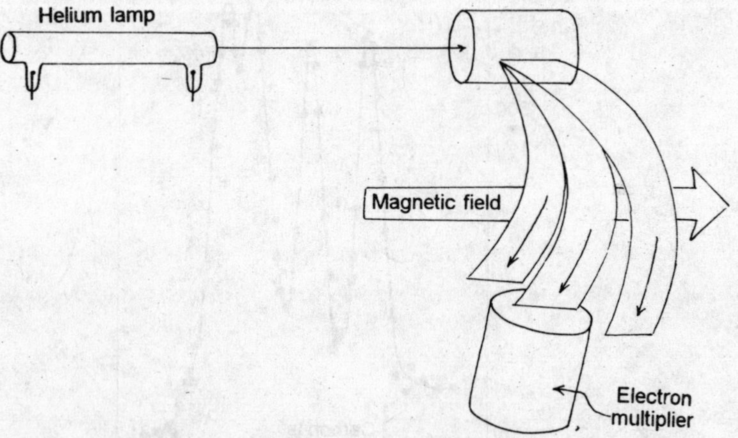

Fig. 11.6

Because the electron path in a spectrometer is influenced by the earth's magnetic field, shielding is done to reduce the effects of earth's magnetic field to zero. The avoid absorption of electrons by gaseous species the entire process is carried out at 10^{-5} to 10^{-9} atmosphere.

Applications

The important applications include identification of active sites and poisons on catalytic surfaces, determination of surface continents on semiconductors, Analysis of the composition of human skin and study of oxide surface layers on metals and alloys.

In addition however the technique is used in chemical analysis and in the elucidation of chemical structure.

Fig. 11.7 depicts a photoelectric spectrum of carbon is ethyl-trifluoroacetate. The individual peaks in the graph represent the carbon atoms shown above it. It therefore very well illustrates how this technique can be used for important structural information. The information obtained by this method is comparable to the information obtained from NMR or IR spectroscopy.

This technique can even distinguish among different oxidation states of an element.

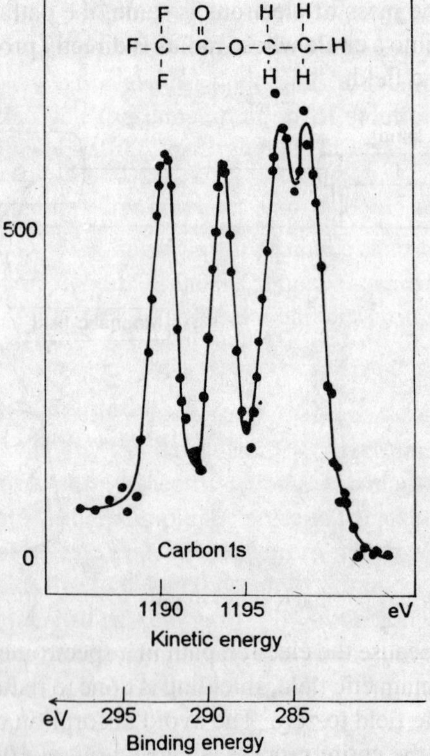

Fig. 11.7: Carbon is X-ray photoelectron spectrum for ethyl trifluoroacetate

Fig. 11.8 indicates the positions of peaks for sulphur in its, different oxidation states in various types of organic compounds.

From this figure and form the relevant data we find two peaks for $S_2O_3^{2-}$ ion which suggests that the two sulphur atoms present in this ion are in different oxidation states.

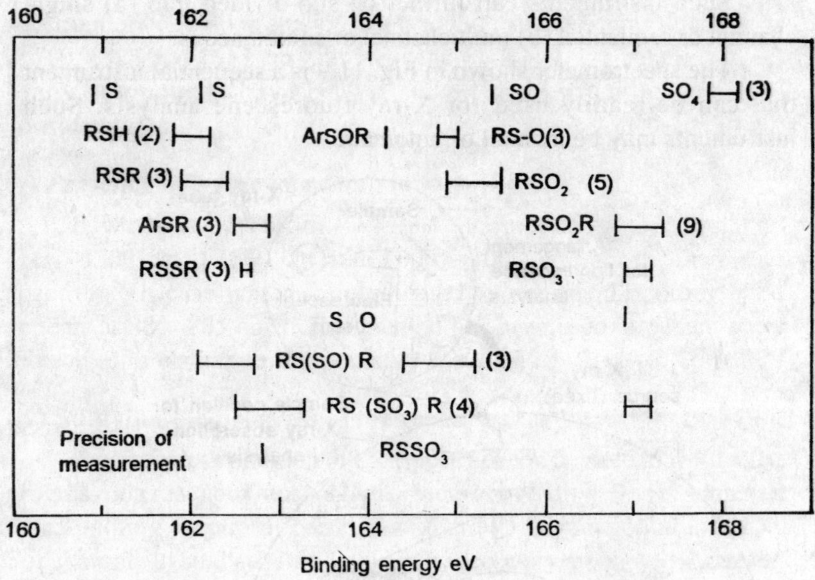

Fig. 11.8: Correlation chart for sulfur 2*s* electron binding energies.

Such information must also be present in the absorption edge of X-ray absorption spectrum also but ready extraction of information is not available because of insufficient resolution.

X-Ray Fluorescence Spectroscopy

When excitation is brought about by irradiation of the sample with a beam of x-rays from a coolidge tube or radioactive source, the elements in the sample are excited by absorption of the primary beam and emit their own characteristic fluorescent X-rays. It is the most widely used analytical method for the analysis and qualitative identification of elements having atomic numbers greater than eight.

Instruments

Of the various types of X-ray fluorescent instruments the three basic types are (i) wavelength dispersive (ii) energy dispersive and (iii) non-dispersive.

(i) Wavelength Dispersive Instruments

Such instruments can further be sub-divided into (a) single channel or sequential (b) multichannel or simultaneous.

The spectrometer shown in Fig. 11.9 is a sequential instrument that can be readily used for X-ray fluorescene analysis. Such instruments may be manual or automatic.

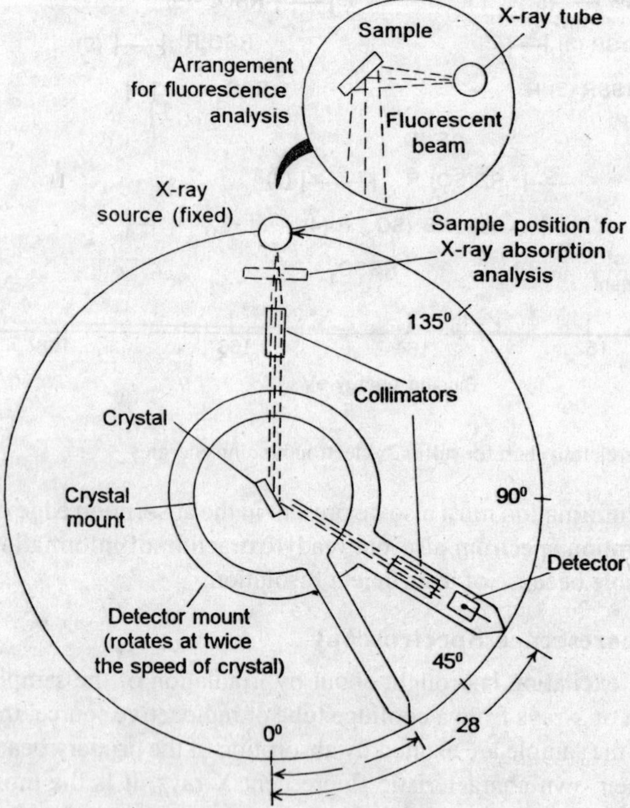

Fig. 11.9: An X-ray monochromator and detector. Note that the angle of the detector with respect to the beam (2θ) is twice that of the crystal face.

Modern single channel spectrometers are provided with two X-ray sources. One is a longer target for longer wavelengths and the other a tungsten target for shorter wavelength.

Multichannel instruments are large, expensive but permit the simultaneous detection and determination of as many as 24 elements. Such instruments are used for the determination of several components in materials of industry like steel, other alloy, cement, ores and petroleum products.

Both single channel and double channel instruments can handle samples in the form of metals, powdered solids, evaporated films, pure liquids or solutions.

(ii) Energy Dispersive Instruments

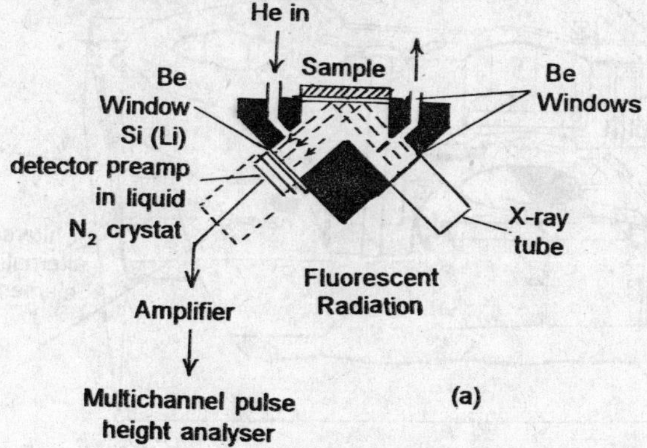

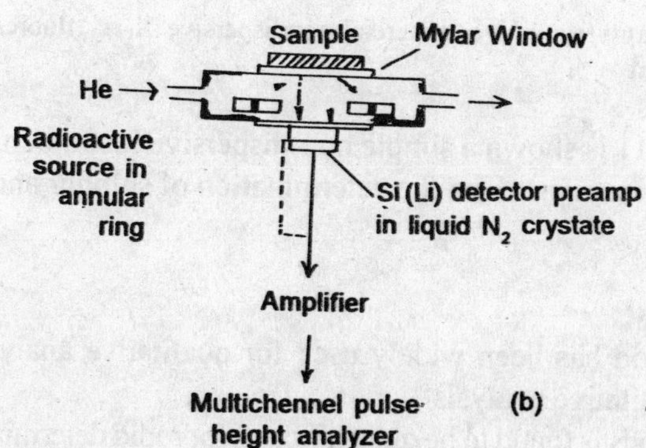

Fig. 11.10: Energy dispersive X-ray fluorescence spectrometer Excitation by X-rays from (a) a Coolidge tube and (b) a radioactive substance

An energy dispersive spectrometer is shown in Fig. 11.10. It consists of a polychromatic source such as a coolidge tube or a radioactive material, a sample holder, a silicon detector, and many other components needed for energy discrimination. Such sources are simple and permit the use of even weaker sources as radioactive materials or low power X-ray tubes. However they have lower resolution at wavelengths longer than about 1 A°.

(iii) Non-dispersive Instruments

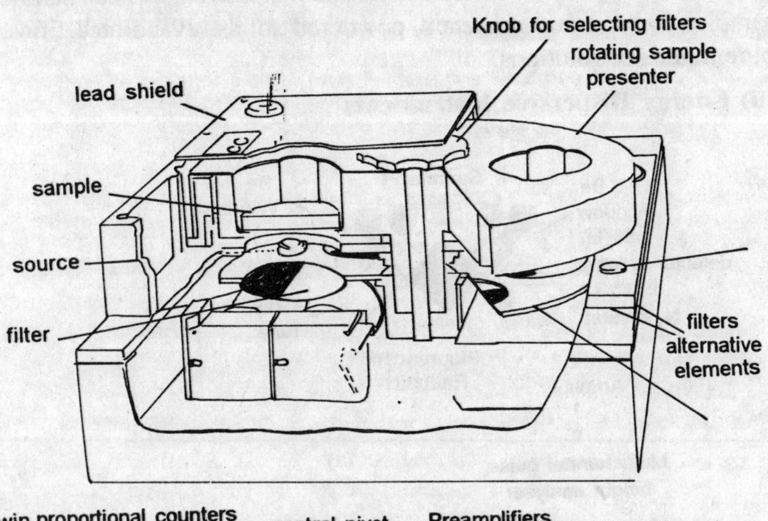

Fig. 11.11: Cutaway view of a commercial nondispersive X-ray fluorescence instrument

In Fig. 11.11 is shown a simple non-dispersive instrument. It is commonly used instrument for the determination of sulphur and lead in gasoline.

Applications

This method has been widely used for qualitative analysis as well as for quantitative analysis.

This methods is found to be quite efficient for rapid determination of all the elements with atomic number than eight. The fact that this method can be used even to analyse for oxygen furthur shows its importance.

These methods are so swift that they can be used for quality control in the manufacture of metals and alloys. Even direct determination of pigments in paint samples can be carried out by these methods.

X-ray fluorescence methods are also used widely for analysis of atmospheric pollutants.

Advantages of X-ray Fluorescence Methods

These have following advantages :

(i) The spectra are relatively simple.

(ii) They do not harm in any way the valuable objects like paintings, specimens, jewellery, coins etc. and so such object can be easily analysed.

(iii) Samples from visible speck to massive objects can be used.

(iv) These are quite convenient and quick methods.

Disadvantages of X-ray Fluorescence Methods

(i) They are not very sensitive.

(ii) They can not be used for lighter elements.

(iii) They are quite costly.

12

Electron Spin Resonance Spectroscopy

Introduction

Some atoms and molecules contain one or more electrons with unpaired spins and these are the substances which are expected to show *electron spin resonance (e.s.r.)* spectra; Since such substances show bulk paramagnetism this type of spectroscopy is often referred to as *electron paramagnetic resonance (e.p.r.)*.

E.S.R. or E.P.R. is a branch of absorption spectrosocopy in which the radiations of microwave frequency induces transitions between magnetic energy levels of electrons with unpaired spin.

Substances with unpaired electrons may either arise naturally or be produced artificially. The three simple molecules NO, O_2 and NO_2 come in first category which also include transition metals and their complexes *e.g.*, Fe^{3+}, $[Fe(CN)_6]^{3-}$ etc. These substances are stable and easily studied by *e.s.r.* unstable paramagnetic materials generally *free radicals* or *radical ions* may be produced either as intermediates in a chemical reaction or by irradiation of a normal molecule with u.v. or X-ray radiation or with a beam of nuclear particles. If the life time of such radicals is greater than about 10^{-6} second they may be studied by *e.s.r.*

The Position of E.S.R. Absorptions; The *g* Factor

We know that spin energy levels of an electron are separated in an applied magnetic field B_2 by an amount

$$\frac{\Delta E}{h} = \frac{g\beta B_2}{h} H_z \qquad .....(1)$$

Where b is Bohr magneton (9.273×10^{-24} J^{-1}) and g the Lande splitting factor. A resonance absorption will thus occur at a frequency $v = \frac{\Delta H}{h} H_z$. The position of absorption changes directly with the applied field (Eq.1). Since *e.s.r.* spectrometers operate at different fields it is convenient to refer to the absorption in terms of the observed 'g' value. Rearranging equation (1) we get

$$g = \frac{\Delta E}{\beta B_2} = \frac{hy}{\beta B_2} \qquad .....(2)$$

If resonance were observed at 8388.255 MHz in a field of 0.30 T, it would be reported as a resonance at a 'g' value of 2.0023. This very precise figure is the g factor for a free electron.

The g factor is a discussion less constant and the exact value of g-factor reflects the chemical environment, particularly when hetero atoms are involved because orbital angular momentum of the electron can have an effect on the value of transition $\Delta m_s = \pm 1$.

In many organic free radicals, the g-value of the odd electron is close to that of a free electron however in metal ions g-value are often greatly different from free electron value. The g-values of most of ionic crystals lie between 0.2 and 0.8 but many ionic crystals show a g-factor very close to free electron.

The measure the g-factor for free radicals it is convenient to measure the field separation between the centre of the unknown spectrum and that of a reference substance whose g-value is accurately known. The known may be a sample of finely powdered diphenyl-picrythydrazyl (DPPH) which is completely in free radical state. Two signals will be observed simultaneously with a field separation of ΔH. The g-factor for the unknown is given by

$$g = g_{std}\left(1 - \frac{\Delta H}{H}\right) \qquad .....(3)$$

where H is the resonance frequency, ΔH is positive if unknown has its centre at a higher field.

The Fine Structures of E.S.R. Absorption

In *e.s.r.* spectroscopy it is necessary to distinguish between two kinds of multiplet structures; there is the *fine structure* which occurs only in crystals containing more than one unpaired electron spin and the smaller *hyperfine structure*.

Electron Behaviour

The imposition of an external static magnetic field H_0 on an electron in a magnetic field establishes two energy levels. The difference between these two levels is given by

$$\Delta E = \frac{\mu e . H_0}{M_s} \qquad .....(1)$$

where μ_e is the electron magnetic moment and M_s the angular momentum quantum number which is either $+\frac{1}{2}$ or $-\frac{1}{2}$. Making susbstitution in (1) we get

$$v = \frac{\mu e . H_0}{h . M_s} = \frac{(9.285 \times 10^{-24}\ \text{JJ}^{-1}) H_0}{(6.626 \times 10^{-34}\ \text{J sec})\ (\frac{1}{2})}$$

$$= 2.083 \times 10^{10}\ \text{sec}^{-1}\text{T}^{-1}(H_0) \qquad .....(2)$$

$S=+\frac{1}{2}$ $I=+\frac{1}{2}$ $I=-\frac{1}{2}$

E_o No applied field $\Delta E = h\nu = g u_B H_o$

$I=-\frac{1}{2}$ $S=-\frac{1}{2}$ $I=\frac{1}{2}$

field H_o no proton

field H_o interaction with proton

(a)

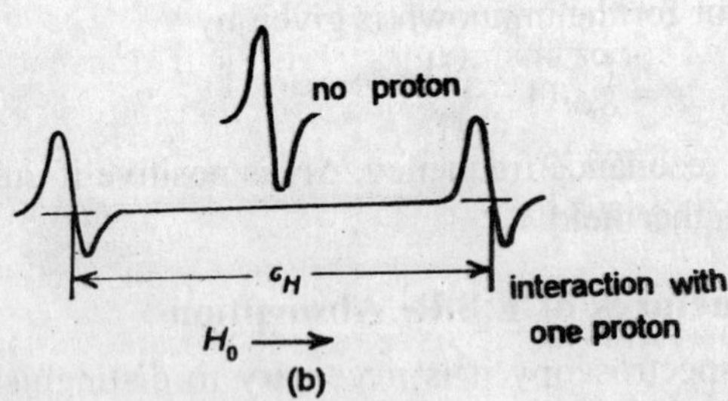

(b)

Fig. 12.1: Electron energy levels. (a) Splitting of energy levels by a magnetic field H_0 and by interaction of unpaired electron with one proton. (b) Splitting of spectral line.

The SI unit of magnetic field strength is a weber per meter squared, which is called a tesla (T). The relation between telsa and gauss is 10,000 G ≡ 1 T. When the magnetic field is expressed in kilo-gauss, the resonance frequency for transitions from one spin state to the other for a free electron (in MHz) is given by v=2803H_0. Most ESR measurements in solution are made with an exciting frequency of 9500 MHz and a field strength of 3400 G.

For a single spin system, the excess of population of the lower energy level to that in the upper energy level is very small. It is governed by the Boltzmann distribution, similar to the situation in NMR. At room temperature and for a free electron, the relative ratio of the population of upper-to-lower energy levels is about 0.998. This much greater population difference is what gives ESR its superior sensitivity in comparison to NMR. Even greater sensitivity can be obtained by working at low temperatures which was not always possible in NMR. Solids down to liquid helium temperatures have been studied by ESR.

Magnet and Modulation Coils

An electromagnet that is capable of producing steady fields from 50 to 5500 G is needed to handle samples whose *g*-factor ranges from 1.5 to 6. The *g*-factor is given by

$$g = \mu_e/\mu_B M_s$$

where *g*, called the spectroscopic splitting factor, has a value which is a function of the electron's environmental.

Microwave Bridge

The bridge, shown in Fig. 12.2, enables the microwave system to be operated as a balanced bridge with all the advantages of null methods in electrical circuits. The microwave bridge will not allow microwave power to pass in a straight line from one arm to the arm opposite. Power entering arm A will divide between arms B and C if the impedances of B and C are the same, so that no power will enter arm B. Under these conditions the bridge is said to be balanced, Arm C usually contains a balancing load. If the impedance of arm B (the sample cavity) changes because of some ESR resonance absorption by a sample in it, the bridge becomes unbalanced and some microwave power enters into arm D containing the detector—a semiconducting silicon-tungsten crystal which acts as a rectifier, covering the microwave power into direct current.

A set of coils mounted on the walls of the sample cavity and fed by a 0.1-MHz sweep generator provides modulation of the dc magnetic field at the sample position. As the main magnetic field is swept slowly through resonance over a period of several minutes, a dynamically recurring imbalance in the microwave bridge is detected and amplified for presentation on a recorder as the derivative of the microwave absorption spectrum against the magnetic field (Fig. 12.3). Due to instrumental considerations associated with the signal-to-noise ratio, ESR spectra are nearly always recorded as first-derivative spectra. For low-frequency modulation (400 Hz or less) the coils can be mounted outside the cavity and even on the magnet pole pieces. Higher modulation frequencies cannot penetrate metal effectively and the modulation coils must be mounted inside the sample cavity.

Sensitivity

The ultimate sensitivity at room temperature of practical X-bond (9500 MHz) ESR spectrometers is generally expressed as

$$N_{min} = 1\times10^{11}\frac{\Delta H}{\sqrt{\tau}}$$

where N_{min} is the minimum number of detectable spins per gauss, ΔH is the width between deflection points on the derivative absorption curve, and τ is the time constant of the detecting system and is inversely proportional to the bandwidth of the detection circuit. For a sample of very small dielectric loss, the minimum concentration is in the order of 10^{-9} M to barely see the line. For aqueous solutions, 10^{-7} M represents a reasonable lower limit. For structure determinations and quantitative analysis, the concentration should be about 10^{-6} M. Working at higher magnetic field strengths gives a higher sensitivity. For example, at 35,000 MHz (k-bond spectrometer) the sensitivity is 20 times greater.

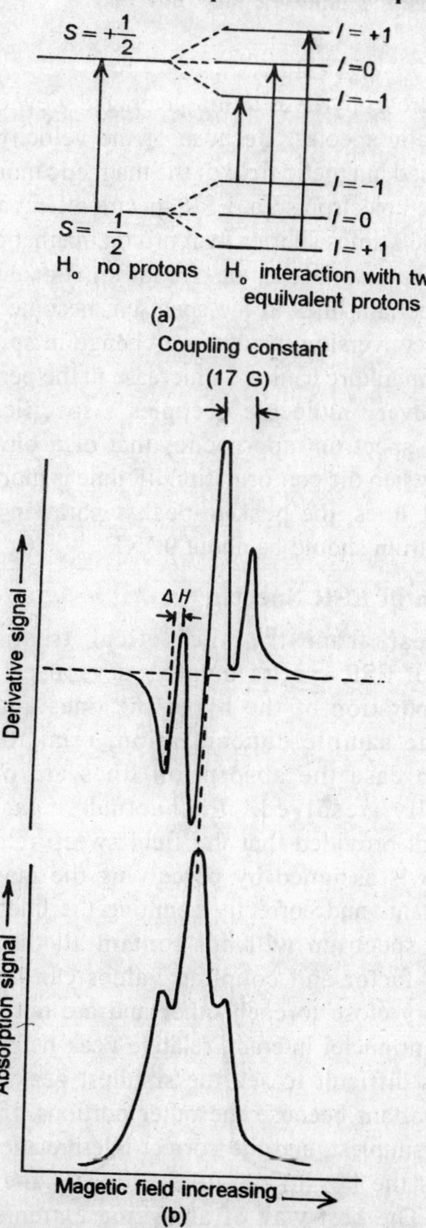

Fig. 12.2: Irradiated methanol. (a) Interaction of a unpaired electron with two equivalent protons. (b) Derivative and absorption signals of the triplet spectrum.

Linewidths

ESR spectroscopy fives a time-averaged view of the geometry of a paramagnetic species. Because of the velocity of precession around the applied magnetic field of the magnetic moment connected with the unpaired electron spin, ESR spectra reflect a time-averaged period about 1000 times shorter than proton magnetic resonance. As the temperature is lowered the rate of conformational interconversion decreases and certain lines in the spectrum become broader. Those emotional interconversion results in a change in spin state. Further lowering of temperature causes an increase in the percentage of time during which every molecule occupies a specific conformation. Eventually the spectrum approaches that of a blocked or frozen conformation (when the conformation lifetime is more than 10^{-6} sec). For sharp ESR lines, the peak-to-peak separation (ΔH) of a first derivative spectrum should be about 0.1 G.

Interpretation of ESR Spectra

It is clear from the theoretical treatment that the interpretation of ESR spectra involves several parameters; the *g*-factor, the separation of the hyperfine lines and their relative intensities, the sample concentration, relaxation times, and linewidths. In case the absorption lines are narrow and the spectrum fully resolved, the actual measurements are straightforward, provided that the field sweep rate is known. An ESR spectrum is assigned by perceiving the magnitudes of the coupling constants and correctly counting the lines. In most cases the measured spectrum will not contain all the lines expected because the *g*-factor and coupling values can be such that two lines come very close to each other and are not resolved. When many equivalent nuclei interact, relative peak heights become very large and it is difficult to see the smallest peaks. These smaller peaks are important because the outer portions of a spectrum are invariably the simplest, and the correct interpretation of these outer lines provides the key to the unravelling of the more complex central parts. The best way of analyzing extremely complicated spectra is to introduced approximate coupling constants in an electronic computer and compare the computedspectra with the experimental ones.

Spin Label

Biological and chemical compounds not containing an unpaired electron can, nevertheless, be studied when they are chemically bonded to a stable free radical. Because this radical, or *spin label,* produces a sharp and simple ESR spectrum that gives detailed information about the molecular environment of the label. Specific sites on a molecule can be tagged by choosing the appropriate spin label and reaction conditions. Five or six-membered heterocyclic molecules incorporating a nitroxyl group whose nitrogen atom is bonded to a tertiary carbon are excellent spin-labeling compound for biological macromolecules like proteins and nucleic acids.

If the unpaired electron in a nitroxide radical is totally localized on the oxygen atom then there will be no isotropic coupling to the nitrogen nucleus. The observe coupling arises from the charged resonance structure shown below:

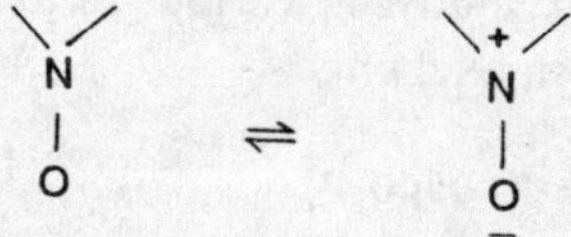

where the unpaired electron is on the nitrogen atom. This charged structure is stabilized to a larger extent when the nitroxide is in a polar medium of high dielectric constant, like water, than when it is in a nonpolar hydrocarbon environment. In a biological system, this solvent effect can be used to determine the polarity of the region surrounding a spin label.

ENDOR

Electron nuclear double resonance (ENDOR) is a method which is used for improving the effective resolution of an ESR spectrum. In this the sample is irradiated simultaneously with a microwave frequency that is suitable for electron resonance and a *rf* that is suitable for nuclear resonance. The *rf* is swept while on point of the, ESR spectrum is observed under conditions of microwave saturation. The ENDOR display is ESR signal height as a function of the swept nuclear rf. For systems with short nuclear relaxation times, like free radicals in solution, or with small nuclear magnetic moments, large *rf* fields are needed. However, for systems with long relaxation times, like solids, low power (5W) equipment is enough.

The ENDOR technique is useful when a lare variety of nuclear energy levels broaden the normal electron resonance line, masking the structure that contains important physical and chemical information. If the electron resonance can be held in a condition that is sensitive to nuclear transitions, the application of an rf field of the appropriate amplitude and frequency will expose the nuclear transitions responsible for the unresolved electron resonance.

ELDOR

In *electron double resonance* (ELDOR) the sample is irradiated simultaneously with two microwave frequencies. One of these is used to observes an ESR signal at some point of the spectrum, while the other is swept through other parts of the spectrum to display the ESR signal height as a function of the difference of the two microwave frequencies. The technique is useful for separating overlapping multiradical spectra and for studying various relaxation phenomena including chemical and spin exchange.

Applications of ESR spectroscopy

Electron spin has been widely applied to the study of chemical, photochemical and electrochemical reactions which proceed via free radical mechanism.

ESR studies are also used to obtain structural information about transition metals and their complexes. They are also used for the detection and semiquantitative determination of paramagnetic ions in biological systems in concentrations as low as a few parts per billion.

The interpretation of ESR spectrum of simple substances is very easy but in case of complex substances the ESR spectrum will not contain all lines expected because the *g*-factor and coupling values are such that two lines may overlap and thus the lines are not resolved. The best method to interpret the extremely complicated ESR spectrum is to introduce the approximate coupling constants in an electronic Computer and then compare the computed spectrum with the experimental spectrum. In case the two are same we may be sure of the structure of complex substance under investigation.

Some important applications of ESR are—

1. Study of Free Radicals

Fig 12.3 illustrates the application of ESR to the detection of a free radical intermediate in a chemical process. This spectrum was obtained after the addition of a base to a solution of hydroquinone in the presence of air. Oxidation of anion of hydroquinone to give quinone occurs. That is *Hydroquinone and quinone are not ESR active because neither contains an odd electron. The* postulated intermediate does

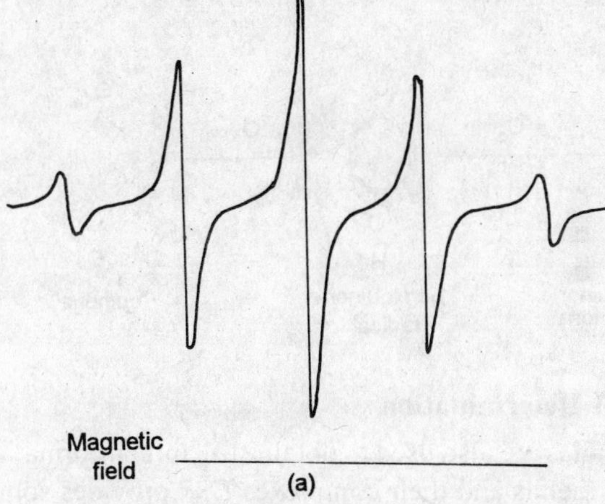

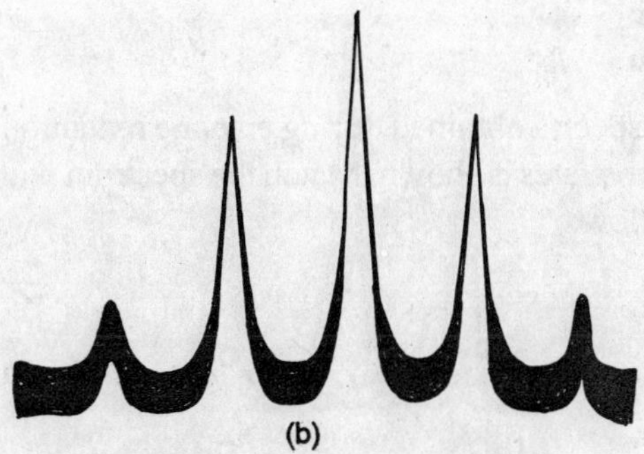

Fig. 12.3: The ESR spectrum for the semiquinone radical: (a) derivative spectrum; (b) absorption spectrum. (Reprint with permission from: B. Venkataraman and G.K. Fraenkel. J. Amer Chem. Soc. 1955, 77, 2711. Copyright 1955 American Chemical Society).

however, and the five peak hyperfine structure is consistent with a semiquinone radical. Here the four ring protons are assumed to be equivalent with respect to the interactions of their spins with that of the odd electron; five peaks result. This picture is reasonable since the position of the odd electron is not confined to the oxygen atom. In deed several resonance structures can be written in which the odd electron is associated with any of the several other atoms in the molecule.

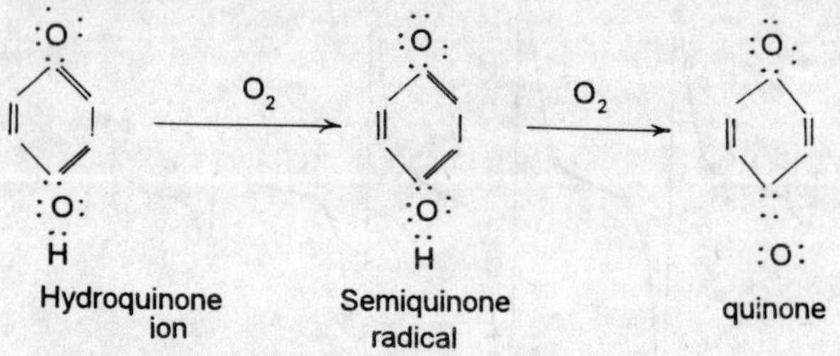

O_2 O_2

Hydroquinone ion   Semiquinone radical   quinone

2. Structural Determination

ESR studies are also used to obtain structural information about the transition metals and their complexes ESP provides some useful information about the shape of the raidcals and about the symmetry of the molecule but it does not tell us about the arrangement of atoms in the molecule.

Example 1

The ESR spectra obtained during enzyme oxidation by peroxide H_2O_2 of three substrates is shown. Match the spectrum with free radical arrangement below:

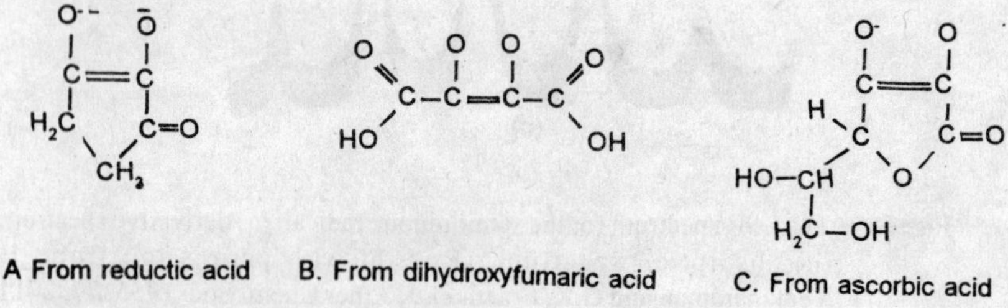

A From reductic acid   B. From dihydroxyfumaric acid   C. From ascorbic acid

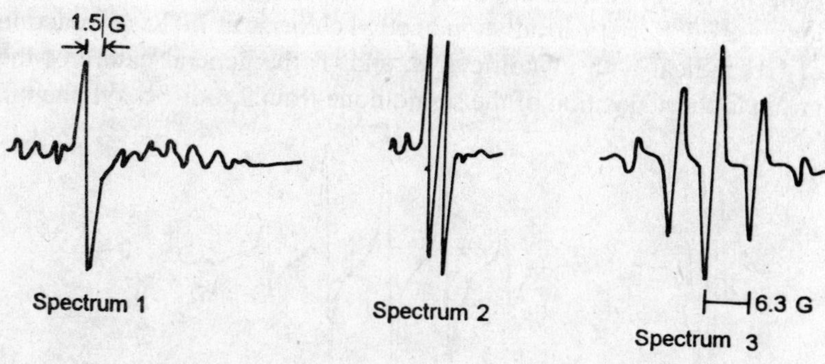

Solution

Spectrum 1 from dihydrofumasic acid spectrum 2 from ascorbic acid and spectrum 3 from reductic acid.

Example 2

For each spectrum shown, identify the semiquinone from which it arises. The five compounds are *p*-semiquinone itself and the following derivatives: monochloro-2,3-dichloro-, trichloro-, and tetrachloro. How does the sweptwing of each spectrum arise?

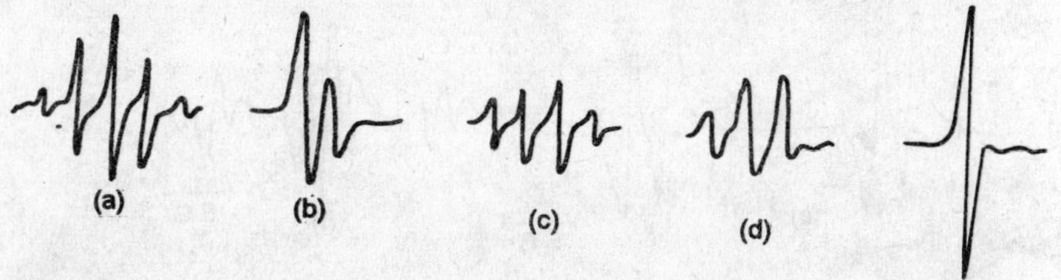

Solution

Spectrum (a) from semiquinone itself, (b) from trichloro-, (c) from monochloro-, (d) from 2,3-dichloro-, and (e) from tetrachloro derivative.

Example 3

For each spectrum in the accompanying diagram, deduce the structure of the free radical and diagram the energy levels. (a) γ-Irradiation of dimethyl sulfone at 77°Y, (b) alkali metal reduction of benzene in ethereal solution at —70°C, (c) γ-irradiation of *t*-butyl iodide at 77°K, (d) γ-irradiation of ethyl chloride at 77°K, (e) reaction of OH radical with glycollic acid, and (f) the general nature of the group in the 4-position of the semiquione from 2,6-di-*t*-butyl phenol.

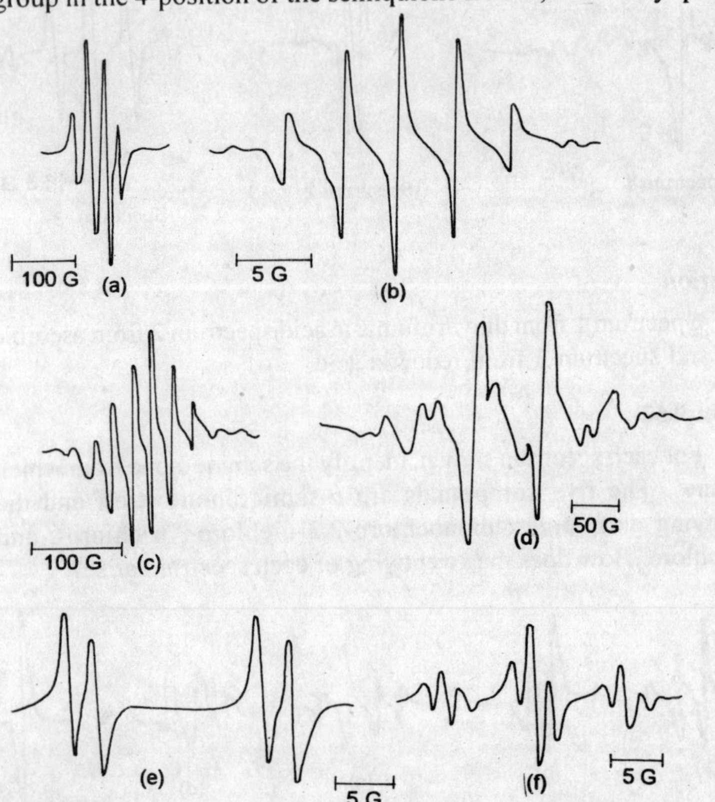

Solution

(a) CH_3, four peaks in 1 : 3 : 3 : 1 intensity ratio.

(b) $C_6H_6^-$, seven lines; 1 : 6 : 15 : 20 : 15 : 6 : 1.

(c) $(CH_3)_3C'$, ten lines, intensity ratio
1 : 9 : 36 : 81 : 126 : 126 : 81 : 36 : 9 : 1.

(d) a quartet from CH_3 with coupling constant of 65 G which is further split into triplets with coupling constants of 50 G due to CH_2; the CH_3CH_2 radical.

(e) $\dot{C}H(\dot{O}H)COOH$ radical with a^H_{CH} =15.3 G and a^H_{OH}=2.5 G.

(f) H = CH_2Y.

Example 4

Analyze the spectrum shown for the hypothetical radical

$H.X^{+}$ (I=I for X).

Solution

Lines 2 to 4 result from proton splitting; each line in turn is split into three components due to interaction with nucleus X; line 3 results from overlap of triplet wings.

3. Qualitative Analysis

The integrated intensity is usually related to concentration of paramagnetic species by comparison with a standard. The total area enclosed by either the absorption or dispersion signal is proportional to the number of unpaired electron spins in the sample. Comparison is made with a standard containing a known number of unpaired electrons and having the same line shape as the unknown (Gaussian or Lorentzian). A solid frequently used is 1,1-diphenyl-2'-picrylhydrazyl (DPPH) or solutions of peroxylamine disulfonate. DPPH contains 1.53×10^{21} unpaired spins per gram; substandards can be prepared by dilution with carbon black. Secondary standards include charred dextrose or synthetic ruby attached to the cavity. A dual sample cavity is used to minimize difficulties with actual physical interchange of standard and sample.

4. Study of Reaction Velocities and Mechanisms

Since a large number of organic reactions proceed by free radical mechanism so these can be studied by ESR spectroscopy the free radical. CH_2OH which contains one set of two equivalent protons has been characterised using this technique.

Very rapid electron exchange reactions can also be studied by ESR. For example when naphthalene is added to a solution naphthalene radical anion, an electron exchange reaction occurs which causes the broadening of the hyperfine component of ESR resonance line. The ratc constant for exchange of electron between naphthalene and naphthalene radical anion can be found by making use of this broadening.

5. Analytical Applications

ESR technique is used for determination of MN^{2+}, V^{4+} etc. This can also be used for determination of Cu^{2+}, Cr^{3+}, Gd^{3+}, Fe^{3+} etc.

For estimation of polynuclear hydrocarbons by ESR, these are first converted into radical cations and then absorbed in the surface of and an activated silica-alumina catalyst.

6. Biological Applications

ESR is used for the detection and semiquantitative determination of paramagnetic ions in biological systems in concentrations as low as a few parts per billion. These involve the use of *spin-label reagents.* These compounds are stable molecules that contain an odd electron, and in addition react with amino acids or other functional groups in biological systems. The ESR spectra of product then provide information about the environment such as structural features, polarity viscocity, phase changes and chemical reactivity.

7. Miscellaneous Applications

ESR spectroscopy is also used for confirming the fact that electrons in impurities (donors) are responsible for conducting properties of semiconductors. It has also been used to detect conduction electrons in solutions of alkali metals in liquid ammonia, alkaline earth metals, alloys etc.

UNSOLVED PROBLEMS

1. The ion $(SO_3)_2NO^{2-}$ in an aqueous solution of the potassium salt gives the ESR spectrum shown in the figure: three sharp equally spaced lines (13-G separation) of equal intensity. After 50-fold amplification two additional lines, denoted by an asterisk, appear in the spectrum. Comment on the origin of the spectrum [*Note* : ^{15}N ($I = \frac{1}{2}$) has a magnetic moment 0.6943 as large as that of ^{14}N].

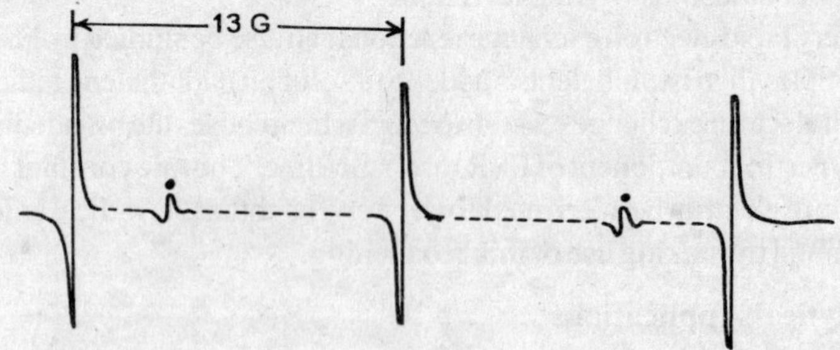

2. When 2,6-di-*t*-butyl-4-methyl-4-*t*-butyl peroxcyclohexa-2,5-diene-1-one is decomposed at 130°C in xylene within the sample cavity of an ESR spectrometer, the spectrum shown is obtained. Deduce the principal radical intermediate.

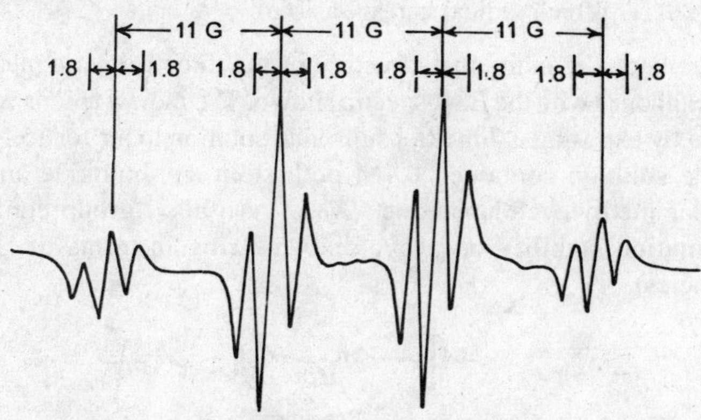

3. Diagram the splitting of the energy levels giving rise to the ESR spectrum of the seminaphtoquinone shown.

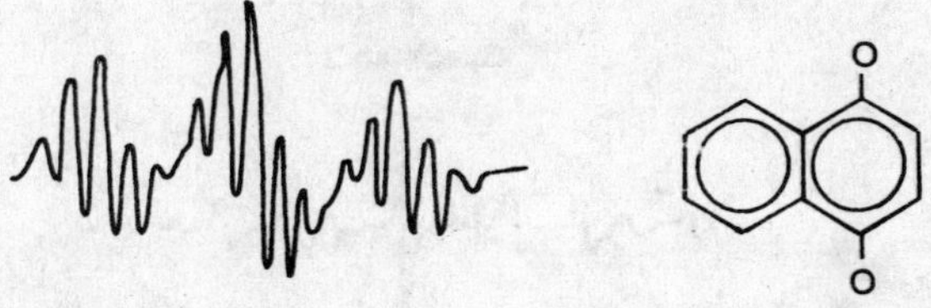

4. Assign the spectrum lines in the accompanying figure to the proper ring protons of naphthalene; the negative ion spectrum results from contact with alkali metal in an inert solvent.

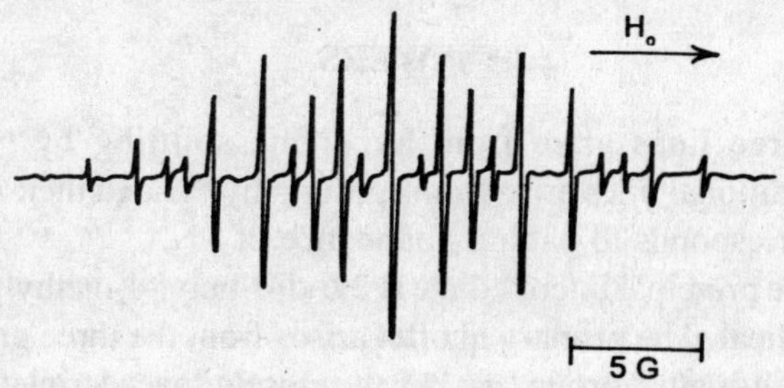

5. Photolysis of a solution of hydrogen peroxide in isopropyl alcohol at 110°K leads to free radical formation and an ESR spectrum consisting of seven lines with a hyperfine splitting of 20 G, a line-width of 10 G, and an approximate intensity distribution of 1 : 6 : 15 : 20 : 15 : 6 : 1. Which radical is responsible?

6. Match the individual structure of the three isomeric methyl cyclohexanones with the ESR spectra shown. The radical anions were detected by exposing a dimethyl sulfoxide solution to air for 15—25 sec. The solution contained 0.1M potassium *tert*-butoxide and a particular methylcyclohexanone. (*Note* : An alkyl group confers conformation stability on a cyclohexene ring in terms of ESR frequencies).

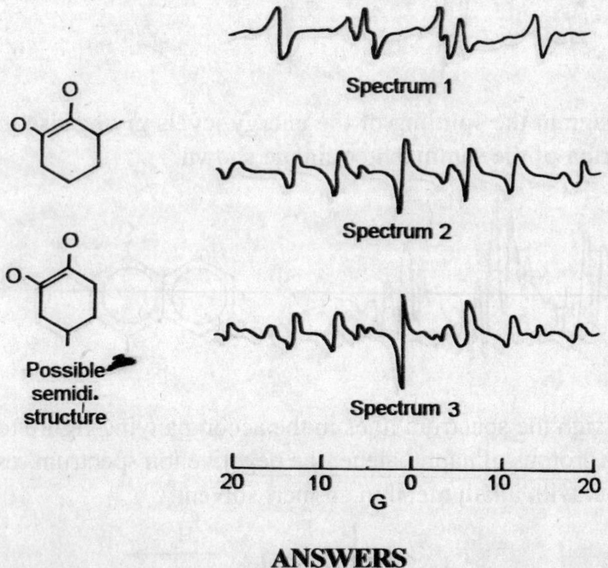

ANSWERS

1. Three lines arise from hyperfine splitting by ^{14}N; The additional lines arise from splitting by ^{15}N and their intensity corresponds to natural abundance of ^{15}N.
2. The principal intermediate is 2,6-di-*t*-butyl-4-methylphenoxy radical. The primary quartet arises from the three protons of the 4-methyl group (a_i=11); the closely spaced triplet (a_i=1.8) arises from weaker coupling of electrons with two *meta*-

protons on the ring.

3. Larger triplets arises from electron coupling with protons 2 and 3 (a_i=3); each triplet is split into five lines due to protons 5, 6, 7 and 8 (a_i=0.3).
4. A quintet of quintets with the coupling of the alpha protons about three times larger than coupling of β-protons.
5. The radical $(CH_3)_2$—C—OH. The small hydrogen splitting, likely to have considerable anisotropy, is not resolved but is responsible for line widths.
6. Spectrum A is from 2-methyl cyclohexanone spectrum B is from 4-methyl cyclohexanone spectrum C is a mixture of approximately three parts of B to one part of A, is from 3-methyl cyclohexanone.

13

Mössbauer Spectroscopy

Introduction

Mössbauer spectroscopy involves nuclear transition which result from the absorption of γ-rays by the sample. The condition for absorption depend upon the electron density about the nucleus, and the number of peaks obtained is related to the symmetry of the compound.

Mössbauer resonances have been observed in the following nuclei ^{197}Au, ^{161}Dy, ^{166}Er, ^{57}Fe, ^{155}Gd, ^{73}Ge, ^{177}Hf, ^{127}I, ^{129}I, ^{191}Ir, ^{193}Ir, ^{61}Ni, ^{195}Pt, ^{187}Re, ^{123}Sb, ^{181}Ta, ^{169}Tm, ^{182}W, ^{170}Yb, ^{67}Zn.

Mössbauer spectroscopy is concerned with transition between energy levels within the nucleus of the atom. Approximately one third of known elements, specially the heavier ones, when formed by the radioactive decay of an isotope of the same or different element, are initially produced in an excited nuclear state. After a very short delay (≃ microseconds) the micronucleus return to the ground state and emits energy of a very high frequency, generally in γ-rays region of the spectrum. It is the study of γ-ray emission and subsequent reabsorption which constitutes **Mössbauer** or **γ-ray** spectroscopy.

Principles of Mössbauer Spectroscopy

To understand the principles of this method, first consider a gaseous system consisting of radioactive source of γ-rays and sample capable of absorbing them. The source nucleus, on emission of γ-rays, decays to the ground state, and the energy of γ-radiations, $E\gamma$

(having a range of 10 to 100 KeV) is given by

$$E_\gamma = E_r + D - R \qquad ...(13.1)$$

where E_r is the difference in energy between the excited state and ground state of the source nucleus; D, the **Doppler shift,** is because of the translational motion of the nucleus, and R is the recoil energy of the nucleus. The recoil energy, similar to that occurring when a bullet leaves a gun, is generally 10^{-2} to 10^{-3} eV and is given by the equation:

$$R = E_\gamma^2/2mc_2 \qquad ...(13.2)$$

where m is the mass of the nucleus and c is the velocity of light. The **Doppler shift** accounts for the fact that the energy of a γ-ray emitted from a nucleus in a gas molecule moving in the same direction as the emitted ray is different from the energy of a γ-ray from a nucleus in a gas molecule moving in the opposite direction. The distribution of energies resulting from the translational motion of the source nuclei in many directions is referred to as **Doppler broadening.** The left-hand curve of Fig. 13.1 represents the distribution of energies of emitted γ-rays, E_γ, from this effect. The breadth of the curve results from Doppler broadening. The dashed line in Fig. 13.1 is taken as E_r, the energy difference between the nuclear ground and excited state of the source. The energy difference, R, between the dotted line and the average energy of the left-hand curve is the recoil energy transmitted to the source nucleus when a γ-ray is emitted.

In this spectroscopy the energy of the γ-ray absorbed for a transition in the sample is given by:

$$Er = Er - D - R \qquad ...(13.3)$$

Here, R is added because the exciting γ-ray must possess energy essential to bring about the transition and effect recoil of the absorbing nucleus. The quantity D has the same significance as before, and the value of E_r is assumed to be the same for the source and the sample. The curve in the right half of Fig. 13.1 shows the distribution of γ-ray energies essential for absorption. The relationship of the sample and source energies can be seen from the entire figure. As indicated by the shaded region, there is only a very slight probability that the γ-ray energy from the source will match that needed for absorption by the low probability that the γ-ray from the source will be absorbed to give a nuclear transition in the sample. The main cause for nonmatching of γ-ray energies is the recoil energy, with the distribution for emission centred about $E_r - R$, whereas that for absorption ($\sim 10^{-1}$ eV) is very much larger than the typical **Doppler energy.** The source would have to move with a velocity of 2×10^4 cm sec^{-1} to obtain a Doppler effect large enough to make the source and sample peaks overlap, and these velocities are not readily obtainable. However, it the quantity R could be reduced, or if conditions for a recoilless transition could be found, the sample would have a higher probability of absorbing γ-ray from the source. As indicated by equation (13.2), R can be decreased by increasing m, the mass. By placing the nucleus of the sample and source in a solid, the mass is effectively that of the solid and the recoil energy will be small as indicated by equation (13.2). For this reason, these spectra are generally obtained on solid samples using solid sources.

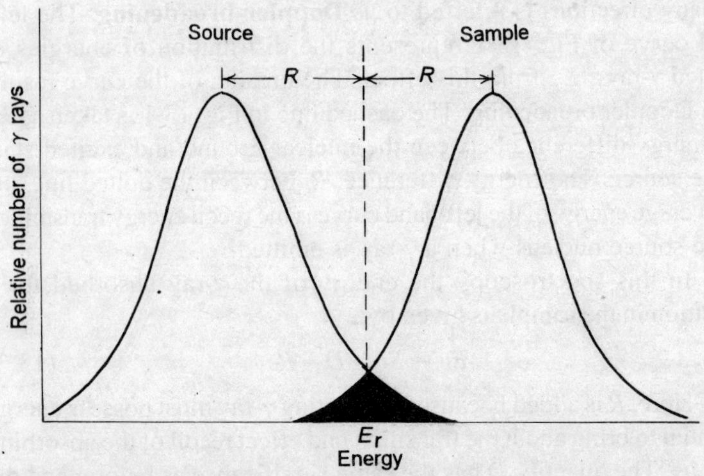

Fig. 13.1: Distribution of energies of emitted and absorbed γ-rays.

By placing the sources and sample in solid lattices, we have not effected recoilless transitions for all nuclei, but we have increased the probability of a recoilless transition. This is due to the fact that the energy of the γ-ray can cause excitation of lattice vibrational modes. This energy term would function in the same way as the recoil energy in the gas; *i.e.,* it would decrease the energy of the emitted particle and increase the energy required for absorption. Certain crystal properties and experimental conditions for emission or absorption will leave the lattice in its initial vibrational conditions for emission or absorption will leave the lattice in its initial state; *i.e.,* conditions for a recoilless transition will be satisfied. It must be emphasised that these conditions simply determine the intensity of the peaks obtained, for it is only the number of particles with matching energy that is determined by this effect. We shall not be concerned with the absolute intensity of band, so this aspect of MB spectroscopy will not be discussed. It should be mentioned, however, that for some materials (usually molecular solids), lattice and molecular vibrational modes are excited to such an extent that very few recoilless transitions occur at room temperature and no spectrum is obtained. Frequently, the spectrum can be obtained by lowering the temperature of the sample.

By going to the solid state we have very much reduced the widths of the resonance lines over that shown in Fig. 13.1. The Doppler broadening is now negligible, and R becomes ~ 10^{-4} eV for a 100 keV gamma ray and an emitting mass number of 100. The full width of a reconance line at half height is given by the Heisenberg uncertainty principle as $\Delta E = h/\tau = 4.56 \times 10^{-16}/0.977 \times 10^{-7} = 4.67 \times 10^{-9}$ eV or 0.097 mm sec^{-1} (for ^{57}Fe). The line widths are infinitesimal compared to the source energy of 1.4×10^{4} eV. The range of excited state lifetimes for Mössbauer nuclei is ~ 10^{-5} sec to 10^{-10} sec, and this leads to line widths of 10^{-11} eV to 10^{-6} eV for most nuclei.

Here we are concerned with the factors affecting the energy needed for γ-ray absorption by the sample. There are three main types of interaction of the nuclei with the chemical environment that result in small changes in the energy needed for absorption: (1) resonance line shift from changes in electron environment, (2) quadrupole interactions, and (3) magnetic interactions. These effects give us information of chemical significance.

Before discussing above factors, let us describe the procedure for obtaining spectra and the illustrate a typical mössbauer spectrum. The electron environment about the nucleus influences the energy of the γ-ray necessary to cause the *nuclear* transition from the ground to excited state, *i.e.*, E_r in the sample. The energy of γ-rays from the source can be varied over the range of the energy differences arising from electron environments in different samples by moving the source relative to the sample. The higher the velocity at which the source is moved toward the sample, the higher the average energy of the emitted γ-ray (by the Doppler effect) and vice versa. The energy ΔE_s of a photon associated with the source moving relative to the sample is given by:

$$\Delta E_s = \frac{v_0}{c} E_\gamma \cos\theta \qquad ...(13.4)$$

where E_γ is the stationary energy of the photon, v_0 is the velocity of the source, and θ is the angle between the velocity of the source and the line connecting the source and the sample. When the source is moving directly toward the sample, $\cos\theta = 1$. In order to obtain an mössbauer spectrum, the source is moved relative to the sample, and the source velocity at which maximum absorption of γ-rays occurs is determined.

For example in the Mössbauer spectrum of Fe^{3+} $Fe^{III}(CN)_6$ [where Fe^{3+} and Fe^{III} designate weak and strong field iron(III), respectively]. This substance contains iron in two different chemical environments, and γ-rays of two different energies are needed to cause transitions in the different nuclei. To obtain the Mössbaur spectrum, the source is moved relative to the fixed sample, and the absorption of γ-rays is plotted as a function of sources velocity as shown in Fig. 13.2. The peaks correspond to source velocities at which maximum γ-ray absorption by the sample occurs. Negative relative velocities correspond to moving the source away from the sample, and positive relative velocities correspond to moving the source toward the sample. The relative velocity at which the source is being moved is plotted along the abscissa of Fig. 13.2, and this quantity is related to the energy of the γ-rays. For a ^{57}Fe source emitting a 14.4 keV γ-ray, the energy is changed by 4.8×10^{-8} eV or 0.0011 cal mole^{-1} for every mm sec^{-1} of velocity imposed upon the source. This result can be calculated from equation (13.4):

$$\Delta E_s = \frac{1 \text{ mm sec}^{-1}}{3.00 \times 10^{11} \text{ mm sec}^{-1}} \times 14.4 \times 10^3 \text{eV} = 4.80 \times 10^{-8} \text{eV}$$

This energy is equivalent to a frequency of 11.6 MHz ($v = E/h$, where $h = 4.14 \times 10^{-15}$ eV sec). For other nuclei having a γ-ray energy of E_γ (in keV),

$$1 \text{ mm sec}^{-1} = 11.6 \times \frac{E_\gamma}{14.4} \text{MHz}$$

With reference to the abscissa of Fig. 13.2, we find that the energy difference between the nuclear transitions for Fe^{3+} and Fe^{III} in $FeFe(CN)_6$ is very small, corresponding to about 2×10^{-8} eV. The peak in the spectrum in Fig. 13.2 at 0.03 m sec^{-1} is assigned[6] to Fe^{III} and that at 0.53 to the cation Fe^{3+} by comparison of this spectrum with those for a large number of cyanide complexes of iron. Different line positions that result from different chemical environments are indicated by the values for the sources velocity in units of cm^{-1} or mm sec^{-1}, and are referred to as *isomer shifts, center shifts,* or *chemical shifts.* We shall now proceed with a discussion of the information contained in the parameters obtained from the spectrum.

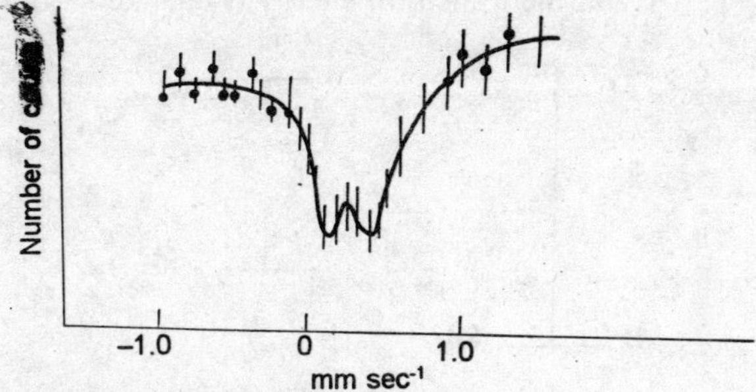

Fig. 13.2: MB spectrum of $FeFe(CN)_6$.

Interpretation of Isomer Shifts

The two different peaks in Fig. 13.2 arise due to the isomer shift differences of the two different iron atoms in octahedral sites. The isomer shift results from the electrostatic interaction of the charge distribution in the nucleus with the electron density that has a finite probability of existing at the nucleus. Only *s* electrons have a finite probability of overlapping the nuclear density, so the isomer shift can be evaluated by considering this interaction. It has to be remembered that *p, d,* and other electron densities can influence *s* electron density by screening the *s* density from the nuclear charge. Assuming the nucleus to be a uniformaly charged sphere of radius R and the s electron density over the nucleus to be a constant given by $\psi_s^2(0)$, the difference between the electrostatic interaction of a spherical distribution of electron density with a point nucleus and that for a nucleus with radius R is given by

$$\delta E = K[\psi_s^2(0)]R^2 \qquad ...(13.5)$$

where K is nuclear constant. Since R will have different values for the ground state and the excited state, the electron density at the nucleus will interact differently with the two states and thus will influence the energy of the transition; *i.e.*,

$$\delta E_e - \delta E_g = K[\psi_s^2(0)]R^2 \ (R_e^2 - R_g^2) \qquad ...(13.6)$$

where the subscript e refers to the excited state and g to the ground state. The influence of $\psi_s^2(0)$ on the energy of the transition is illustrated in Fig. 13.3 for ^{57}Fe, which has $I = 1/2$ for the ground state and $I = 3/2$ for the excited state. The energies of these two state are affected differently by $\psi_s^2(0)$, and the transition energy is changed.

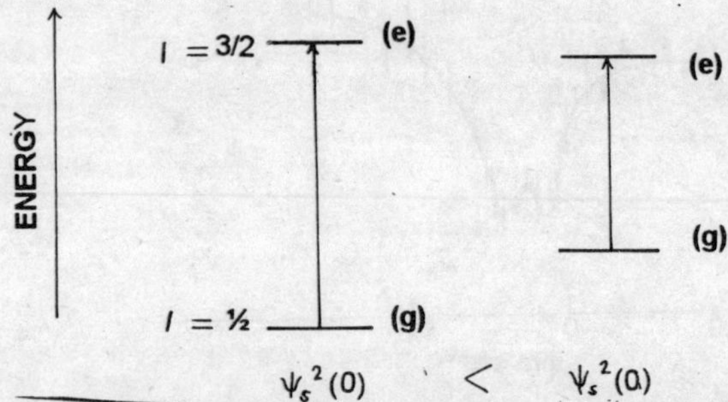

Fig. 13.3: Changes in the energy of the Mössbauer transition for different values of $\psi_s^2(0)$. This is a graphical illustration of equation (13.16a) with two different values of $\psi_s^2(0)$ for an ^{57}Fe nucleus. The differences in $\psi_s^2(0)$ must result from a cubic or spherical distribution of bonded atoms in order for this diagram to apply.

The R values are constant for a given nucleus, but $\psi_s^2(0)$ varies from compound to compound. The center shift in the Mossbauer spectrum is the difference between the energy of this transition in the sample (or absorber) and the energy of the same transition in the source. This difference is given by the difference in equations of the form of (13.6) for the source and sample, or:

$$C.S. = K(R_e^2 - R_g^2)\{[\psi_s^2(0)]_a - [\psi_s^2(0)]_b\} \qquad ...(13.6)$$

where the subscripts *a* and *b* refer to absorber and source, respectively. Standard sources are generally used (*e.g.*, ^{57}Co in Pd for iron Mössbauer spectra, or $BaSnO_3$ for tin spectra). The ^{57}Co decays to ^{57}Fe in an excited state *via* electron capture. The excited ^{57}Fe decays to stable ^{57}Fe by γ-ray emission. When standard sources are used $[\Psi_s^2(0)]_b$ is replaced by a constant, *C*. Moreover, the change in radius $R_e - R_g$ is very small, leading to the following commonly used expression for the center shift:

$$\text{C.S.} = K'\frac{\delta R}{R}[\psi_s^{\,2}(0)_a - C] \qquad ...(13.6)$$

where $\delta R = R_e - R_g$; *C* is a constant that is characteristic of the source; and K' is $2KR^2$. Both K' and $\delta R/R$ are constants for a given nucleus, so the center shit is directly proportional to the *s* electron density at the sample nucleus. The term *center shift* is used for the experimentally determined center of the peak; the term *isomer shift* is now used when the center shift has been corrected fro the small Doppler contribution from the thermal motion of the Mössbauer atom. The sign of δR depends upon the difference between the effective nuclear charge radius, *R,* of the excited and ground states $(R_e^2 - R_g^2)$. For the ^{57}Fe nucleus, the excited state is smaller than the ground state, and an increase in *s*-electron density produces a negative shift. In tin, the sign of δR is positive, so the opposite trend of shift with *s* electron density is observed.

The electron density in *p* or *d* orbitals can screen the electron density from the nuclear charge because of the fact that the electron density in *d* and *p* orbitals penetrates the s orbital. Hartree-Fock calculations show that a decrease in the number of *d* electrons causes a marked increase in the total *s* electron density at the iron nucleus. Accordingly, with comparable ligands and with negative $\delta R/R$, Fe^{2+} has an appreciably larger center shift than Fe^{3+}. When these ions are examined in a series of molecules, the interpretation becomes more difficult, for the *d, s* and *p* electron densities are modified by covalent bonding. For ^{57}Fe, for example, an increase in 4*s* density decreases the center shift, while an increase in 3*d* density increases the center shift. a series of high spin iron complexes have been interpreted on this basis. In the case of ^{119}Sn, the center shift increases with an increased *s* electron density and decreases with an increase in *p* electron density.

Quadrupole Interactions

The discussion of the center shift in the previous section applies to systems with a spherical or cubic distribution of electron density. The degeneracy of nuclear energy levels for nuclei with $I > 1/2$ is removed by a non-cubic electron or ligand distribution. For non-integral spins, the splitting does not remove the + or – degeneracy of the m_I levels, but we obtain a different level for each $\pm m_I$ set. Thus, the electric field gradient can lead to $I + 1/2$ different levels for half-integer values of I(*e.g.*, two for $I = 3/2$ corresponding to $\pm 1/2$ and $\pm 3/2$). For integer values of I we obtain $2I + 1$ levels (*e.g.*, five for $I = 2$ corresponding to 2, 1, 0, –1, –2). The influence of this splitting on the nuclear energy levels and the spectral appearance is illustrated[9] in Fig. (13.4) for ^{57}Fe. The ground state is not split but the excited state is split, leading to two peaks in the spectrum. The center shift is determined from the center of the two resulting peaks. When both the ground and excited states have large values for I, complex Mössbauer spectra result.

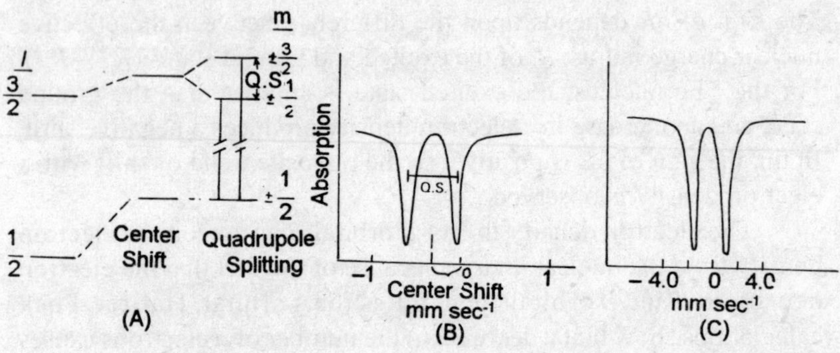

Fig. 13.4: The influence of a non-cubic electronic environment on (A) the nuclear energy states of ^{57}Fe and (B) the Mössbauer spectrum. (C) The iron MB spectrum of $Fe(CO)_5$ at liquid N_2 temperature.

The Hamiltonian for the quadrupole coupling is the same as for nqr:

$$\hat{H}_Q = \frac{e^2Qq}{4I(2I-I)}[3\hat{I}_z^{\,2} - I(I+1) + (\eta/2)(\hat{I}_+^2 + \hat{I}_-^2)]$$

For the $I = 3/2$ case (^{57}Fe and ^{119}Sn), the quadrupole splitting Q.S. is given by

$$\text{Q.S.} = \frac{1}{2} e^2 Qq(1 + \eta^2 / 3)^{1/2} \qquad \text{...(13.7)}$$

For ^{57}Fe, q and η cannot be determined from the quadrupole splitting. The sign of the quadrupole coupling constant is another quantity of interest. If $m_I = \pm 3/2$ is at high energy, the sign is positive; the sign is negative if $\pm 1/2$ from $I = 3/2$ is highest. From powder spectra, the intensities of the transitions to $\pm 1/2$ and $\pm 3/2$ are similar, and it becomes difficult to determine the sign. The sign can be obtained from spectra of ordered systems or from measurement of a polycrystalline sample in a magnetic field (*vide infra*). For systems in which the I values of the ground and excited states are larger than those for iron, the spectra are more complex and contain more information. The splitting of the excited state will not occur in a spherically symmetric or cubic field but will occur only when there is a field gradient at the nucleus caused by asymmetric p and d electron distribution in the compound. A field gradient exists in the trigonal bipyramidal molecule iron pentacarbonyl, so splitting of the nuclear excited state is expected, giving rise to a doublet in the spectrum as indicated in Fig. 13.4(C).

If the t_{2g} set and the e_g set of orbitals in octahedral transition metal ion complexes have equal populations in the component orbitals, the quadrupole splitting will be zero. Low spin iron(II) complexes (t^6_{2g}) will not give rise to quadrupole splitting unless the degeneracy is removed, and these orbitals can interact differently with the ligand molecular orbitals. On the other hand, high spin iron(II) ($t_{2g}{}^4 e_g{}^2$) has an imbalance in the t_{2g} set, and a large quadrupole splitting is often sen. If the ligand environment about iron(II) were perfectly octahedral, d_{xy}, d_{yz}, and d_{xz} would be degenerate and no splitting would be detected. However, this system is subject to Jahn-Teller distortion, which can lead to a large field gradient. when the energy separation of the t_{2g} orbitals from Jahn-Teller effects is of the order of magnitude of kT, a very temperature dependent quadrupole splitting is observed. The ground state in the distorted complex can be obtained if the sign of q is known. The sign can be obtained from oriented systems of from studies in a large magnetic field. Similar considerations apply to high spin and low spin iron(III) compounds.

Paramagnetic Mösssbauer Spectra

As shown in Fig. 13.4, a non-cubic electronic environment in an iron complex splits the degenerate $I = 3/2$ excited state of ^{57}Fe into m_I states $\pm 1/2$ and $\pm 3/2$. The Mössbauer spectrum consists of a doublet corresponding to the two transitions shown in Fig. 13.4. When an effective magnetic field, H_{eff}, acts on this system, the degeneracies of the $\pm 1/2$ ground state as well as the $\pm 3/2$ excited states are remove[10] as shown in Fig 13.5. The ordering of the levels reflects the fact that the ground-state moment is positive and the excited state moment is negative. The resulting six-line spectrum is shown in Fig. 13.5(B). Since the transitions are magnetic dipole in character, the selection rule is Δm= 0, ±1 as shown in Fig. 13.5(A). The Hamiltonian is

$$\hat{H}_m = -g_N \mu_N \hat{I} . H_{eff} \qquad ...(13.8)$$

where $g_N = 0.18$ and $I = 1/2$ for the the nuclear ground and $g = -0.10$ and $I = 3/2$ for the excited state. The intensity of the lines changes with θ, the orientation of the H_{eff} relative to the direction of the Mössbauer radiation. For an ordered sample, the $\Delta_m = 0$ transition has zero intensity for a field parallel to the radiation. When θ = 90°, the $\Delta m = 0$ transitions are the most intense of the six.

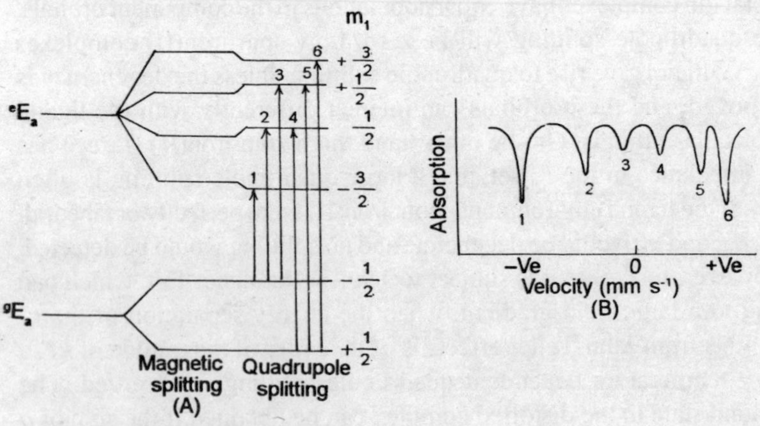

Fig. 13.5: Magnetic and quadrupole splitting in a ferromagnetic ^{57}Fe compound. (A) Energy level diagram. (B) Expected Mössbauer spectrum.

The effective field of equation 13.8 consists of the internal field and the applied field:

$$H_{eff} = H + H_{int} \quad ...(13.9)$$

and is given by

$$H_{int} = -\langle S\rangle.\mathbf{A}/g_N\mu_N \quad ...(13.10)$$

where $\langle S\rangle$ is the expectation value of the spin and **A** the hyperfine coupling tensor. The internal magnetic field controls the magnetic hyperfine and it depends in a complex way on the the zero field splitting parameters D and E, on the electronic g tensors and **A** tensors, and on the orientation of the molecule relative to the applied field.

Internal fields exist in some systems in the absence of an applied magnetic field. The $\langle S\rangle$ of equation (13.10) is zero for diamagnetic systems and for systems with integer values of S in the absence of an applied field. These spectra consist of quadrupole doublets. For compounds with half-integer spin (Kramers' doublest), two cases need be considered. With fast electron relaxation, the $\langle S\rangle$ averaged over all thermally accessible states is zero in zero magnetic field, leading to quadrupole doublets. In the slow relaxation limit, magnetic Mössbauer spectra can be obtained, generally at liquid helium temperatures. These spectra generally are studied at weak applied field to simplify them by decoupling hyperfine interactions with ligand nuclei.

In applied fields, magnetic Mössbauer splittings can be observed for both integer and half-integer spin systems. The applied field mixes the states producing a non-zero expectation value of the spin. The resulting spectra are interpreted using the spin Hamiltonian formalism given in equation(13.11).

$$\hat{H} = D[\hat{S}_z^2 - \frac{1}{3}S(S+1)] + E(\hat{S}_x^2 - \hat{S}_y^2) + \beta\hat{S}.g.\hat{\mathrm{H}} + \hat{S}.\mathrm{A}.I$$

$$-g_N\beta_N\hat{I}.\mathrm{H} + \frac{e^2Qq}{4I(2I-1)}\{3\hat{I}_z^2 - I(I+1) + \eta(\hat{I}_x^2 - \hat{I}_y^2)\}_{\mathrm{EFG}} \quad ...(13.11)$$

By standing the spectra as a function of temperature, field strength, and field direction, the various quantities in the spin Hamiltonaina can be determined. For Kramers' doublets complimentary information can be obtained from the epr. However for integer S system, *e.g.*, high spin iron(II), magnetic Mössbauer provides the sole means of obtaining electronic structural details.

Mössbauer Emission Spectroscopy

^{57}Co decays by an electron capture to ^{57}Fe ($T_{1/2}$ for ^{57}Fe is 0.1 μsec), populating and excited state of the iron nucleus. The emitted γ-rays can be absorbed by a standard single-line absorber to investigate the energy levels of the ^{57}Fe nuclei produced when the source decays. The cobalt-57 analogue of the compound to be studied is prepared and used as the source. Information regarding the short-lived iron complex in the source is obtained[15] from this experiment. One must be sure that the desired iron complex remains intact when the high energy iron atoms are formed in the cobalt decay process. The results obtained from some oxygenated complexes of ^{57}Co protoporphyrin IX dimethyl ester as a function of the axial base attached are shown in Table 13.1. These are to be compared with values of ΔE_Q and δ for oxygenated hemoglobin, obtained by absorption measurements, of 2.23 and 0.27 respectively.

Table 15.1: Mössbauer Emission Studies. (*Quadrupole splittings and isometric shifts for oxygenated complexes of 57$_{Co}$ protoporphyrin IX dimethyl ester. The ligand coordinated trans* to dioxygen are listed in the first column).

Ligand	ΔE_Q*(mm/sec)*	*δfe(mm/sec)*
1-methyl imidazole	2.17	0.29
1,2-dimethyl imidazole	2.32	0.30
pyridine	2.28	0.27
piperidine	2.25	0.30
ethylmethyl sulfide	2.27	0.30

Fig. 13.6 (A) shows the emission spectrum of the five-coordinate 1-methyl imidazole complex before oxygenation; Fig. 13.6(B) shows the spectrum of the same complex after oxygenation.

This technique is particularly important when the parent iron compound is difficult to prepare and isolate.

Applications

Some applications of Mössbuaer spectroscopy are :

Table 13.2 lists information about isotopes that have been studied by this technique.

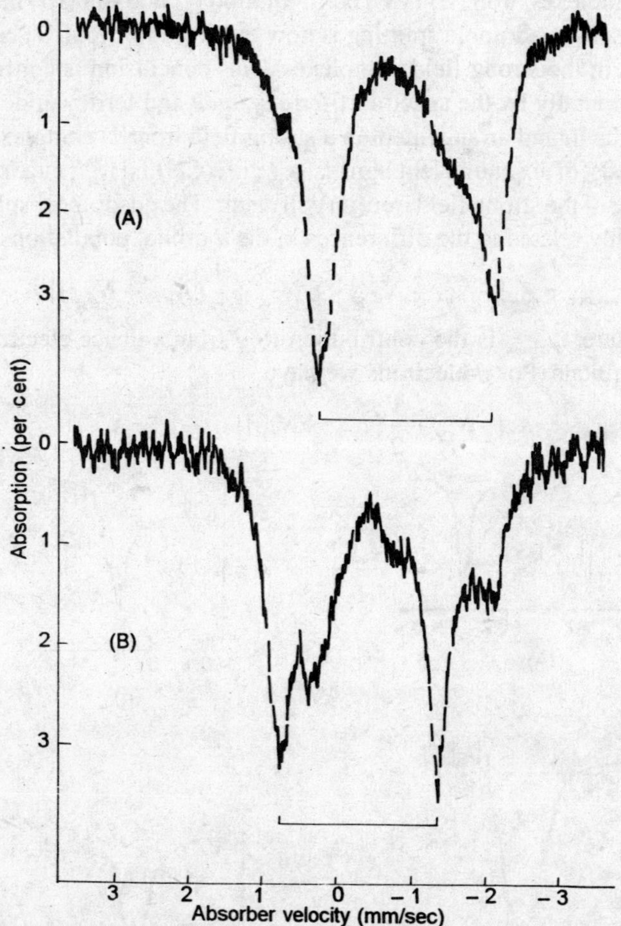

Fig. 13.6: Emission Mössbauer spectra of (A) the 1-methyl imidazole adduct of cobalt protoporphyrin IX dimethyl ester and (B) its O_2 adduct.

Facsimiles of spectra obtained on some iron complexes are given in Fig. 13.7. As already mentioned, for high spin iron complexes in which all six ligands are equivalent, a virtually spherical electric field at the nucleus is expected for $Fe^{3+}(d^5)$ $(t_2g^3e_g^2)$ but not for Fe^{2+}

$(d^6)(t_{2g}{}^4 e_g{}^2)$. Because of the field gradients at the nucleus, quadrupole splitting should be detected in the spectra of high spin iron(II) complexes but not for high spin iron(III) complexes. This is evident in spectra *A* and *B* of the complexes illustrated in Fig. 13.7. For low spin complexes, iron (II) has a configuration $t_{2g}{}^6$ and iron(III) has $t_{2g}{}^5$. As a result, quadrupole splitting is now expected for iron(III) but not iron(II) in the strong field complexes. This conclusion is confirmed experimentally by the spectra of ferrocyanide and ferricyanide ions. When the ligand arrangement in a strong field iron(II) complex does not consist of six equivalent lignads, *e.g.*, $[Fe(CN)_5NH_3]^{3-}$, quadrupole splitting of the strong field iron(II) will result. The quadrupole splitting is roughly related to the differences in the *d* orbital populations by

$$q_{\text{valence}} = K_d[-N_{d_z^2} + N_{d_{x^2-y^2}} + N_{d_{xy}} - \frac{1}{2}(N_{d_{xz}} + N_{d_{yz}})]$$

Here, q_{valence} is the contribution to q from valence electrons in the *d* orbitals. For *p*-electrons we have

$$q_{\text{valence}} = K_p[-N_{p_z} + \frac{1}{2}(N_{p_y} + N_{p_x})]..$$

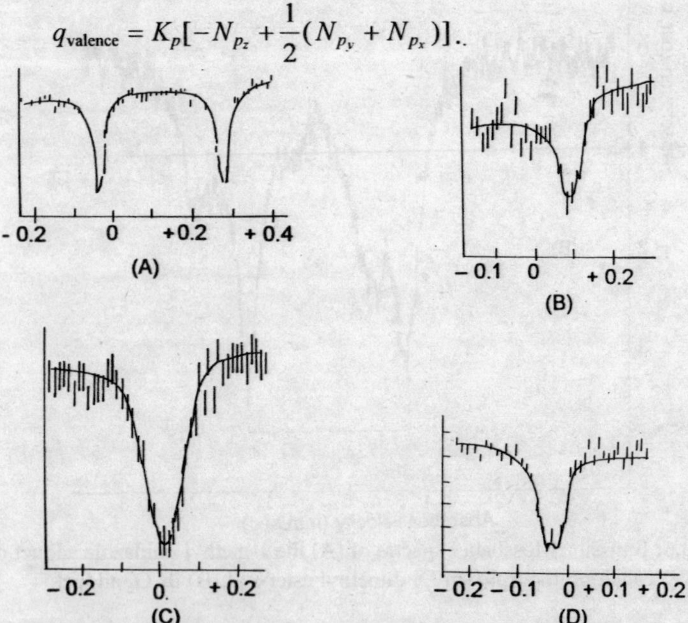

Fig. 13.7: Mössbauer spectra of some iron(II) and iron(II) complexes. (A) Spin-free iron(II)—$FeSO_4.7H_2O$. (B) Spin-free iron(III)—$FeCl_3$. (C) Spin-paired iron(II)—$K_4Fe(CN)_6.3H_2O$. (D) Spin-paired iron(III)—$K_3Fe(CN)_6$.

Table 13.2: Mossbauer Isotope of Chemical Interest[a].

Table 13.2: Mössbauer Isotope of Chemical Interest[a].

Isotope	Gamma energy (keV)	Half life[b] of precursor	Q[c] (mm sec^{-1})	δ_0[d] Spin	Natural abudnace (Barns)	$10^4 \frac{\delta R}{R}$ (10^{-19} cm^2)	(%)	
^{57}Fe	14.4	270 d	0.19	$\frac{1}{2} \to \frac{3}{2}$	+0.3	23.5	2.19	−18 ± 4
^{61}Ni	67.4	1.7 h; 3.3 h	0.78	$\frac{3}{2} \to \frac{5}{2}$	0.13	7.21	1.19	
^{67}Zn	93.3	60 h; 78 h	3.13×10^{-4}	$\frac{5}{2} \to \frac{3}{2}$	+0.17	1.18	4.11	
^{73}Ge	67.0	67 d	2.19	$\frac{9}{2} \to \frac{7}{2}$	−0.26	3.54	7.76	
^{83}Kr	9.3	83 d, 2.4 h	0.20	$\frac{9}{2} \to \frac{7}{2}$	+0.44	18.9	11.55	+4 ± 2
^{99}Ru	90	16.1 d	0.15	$\frac{5}{2} \to \frac{3}{2}$	>0.15	1.42	12.72	

Table (conti.)

^{107}Ag	93.1	6.6 h	6.68×10^{-11}	$\frac{1}{2} \to \frac{7}{2}$	—	0.54	51.35	
^{119}Sn	23.9	245 d	0.62	$\frac{1}{2} \to \frac{3}{2}$	0.07	13.8	8.58	$+3.3 \pm 1$
^{121}Sb	37.2	76 y	2.10	$\frac{5}{2} \to \frac{7}{2}$	–0.29	2.04	57.25	-8.5 ± 3
^{125}Te	35.5	58 d	4.94	$\frac{1}{2} \to \frac{3}{2}$	+0.19	2.72	6.99	+1
^{129}I	27.8	33 d; 70 m	0.63	$\frac{7}{2} \to \frac{5}{2}$	–0.55	3.97	0	+3
^{129}Xe	39.6	1.6×10^{7}y	6.84	$\frac{1}{2} \to \frac{3}{2}$	–0.41	1.95	26.44	0.3
^{133}Cs	81.0	7.2 y	0.54	$\frac{7}{2} \to \frac{5}{2}$	–0.003	1.02	100	
^{177}Hf	113.0	6.7 d, 56 h	4.66	$\frac{7}{2} \to \frac{9}{2}$	+3	1.20	18.50	
^{181}Ta	6.25	140 d, 45 d	$6.48 \times 10–3$	$\frac{7}{2} \to \frac{9}{2}$	+4.2	17.2	99.99	

Table (Conti.)

^{182}W	100.1	115 d	2.00	$0 \rightarrow 2$	−1.87	2.46	26.41	+1.3
^{187}Re	134.2	23.8 h	203.8	$\frac{5}{2} \rightarrow \frac{7}{2}$	+2.6	0.54	62.93	
^{186}Os	137.2	90 h	2.37	$0 \rightarrow 2$	+1.54	2.89	1.64	
^{193}Ir	73.1	32 h	0.59	$\frac{3}{2} \rightarrow \frac{1}{2}$	+1.5	0.30	62.7	+0.6
^{195}Pt	98.7	183 d	17.30	$\frac{1}{2} \rightarrow \frac{3}{2}$	—	0.63	33.8	
^{197}Au	77.3	65h, 20h	1.85	$\frac{3}{2} \rightarrow \frac{1}{2}$	+0.58	0.44	100	± 3

[b]d = days, h = hours, y = years, m = minutes.

[c]The ground-state quadrupole moment, where both ground and excited stated have $I > 1/2$.

[d]Cross-section for absorption of a Mössbauer gamma ray.

Values measured at room temperature for ΔE_Q and the isomer shift, δ, for a number of iron complexes have been collected and are listed in Table 13.3. For iron complexes, isomer shifts in a positive direction correspond to a decrease in electron density in the region of the nucleus. For high spin complexes, a correlation exists between isomer shift and *s* electron density. An increase in δ of 0.2 mm sec^{-1} is equivalent to a decrease in charge density of 8% at the nucleus. The negative values obtained for the low spin ferricyanides compared to high spin iron(III) complexes indicate more electron density at the nucleus in the ferricyanide ions. This has been explained as being due to extensive π bonding in the ferricyanides, which removes d electron density from the metal ion, which in turn decreases the shielding of the *s* electrons. This effect increases *s* electron density at the nucleus and decreases δ. Both strong σ donors and strong π acceptors decreases δ.

The Mössbauer spectrum of the material prepared from iron(III) sulfate and $K_4Fe(CN)_6$ is identical to the spectra for the compounds prepared either from iron(II) sulfate and $K_3Fe(CN)_6$ or by atmospheric oxidation of the compound from iron(II) sulfate and $K_4Fe(CN)_6$. The spectra of these materials indicate that the cation is high spin iron(III), while the anion is low spin iron(II).

The Mössbauer spectrum of sodium nitroprusside, $Na_2Fe(CN)_5NO$, has been investigated. This material has been formulated earlier as iron(II) and NO^+ because the complex is diamagnetic. The Mössbauer spectrum consists of a doublet with a ΔE_Q value of 1.76 mm sec^{-1} and a δ value of –0.165 mm sec^{-1}. Comparison of this value with reported results on a series of iron complexes led us to conclude that the iron δ value is close to that of iron(IV). The magnetism and Mössbauer spectrum are consistent with a structure in which there is extensive π bonding between the odd electron in the t_{2g} set of the orbitals of iron and the odd electron on nitrogen, as illustrated in Fig. 13.8. The filled π bonding orbital would a large contribution from the nitrogen atomic orbital, and the empty π antibonding orbital would have a larger contribution from the iron atomic orbital to produce iron(IV). More of the π electron density would be localized on nitrogen and the δ value for iron would approach that of iron(IV) due to decreased shielding of the *s* electrons by the *d* electrons. Since electron density is being placed in what was previously a π antibonding orbital of nitric oxide, a decrease in the N—O infrared stretching frequency is observed. The very large quadrupole splitting is consistent with very intensive π bonding in the Fe—N—O link.

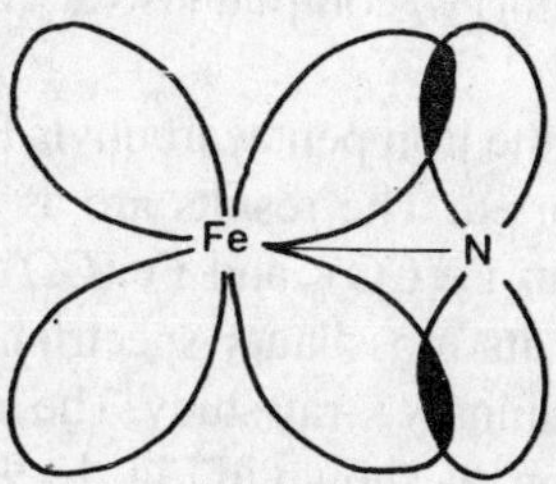

Fig. 13.8: Orbitals involved in the Fe—N π bonding to the NO group in $[Fe(CN)_5(NO)]^{2-}$.

Mössbauer spectra are generally useful in determining the oxidation state of atoms. It has been shown that the spectra originally assigned to some high spin iron(II) chelates of salicyladoxime and other chelates were, in effect, those of an iron(III) oxidation product. The oxidized materials gave a center shift [relative to $Na_2Fe(CN)_5NO.2H_2O$] of ~ 0.6 mm sec^{-1} and a small quadrupole splitting. This smaller-than-expected quadrupole splitting for iron(II) was rationalized in terms of π-backbonding. The authentic material prepared under conditions the at were rigorously air-free gave normal iron(II) centre shifts and large quadrupole splittings, ~1.4 Values for the center shifts of various oxidation states of high spin iron compounds are shown in Table 13.4.

Table 13.4: Isomer Shifts for High Spin Iron Compounds [Shifts Relative to $Na_2Fe(CN)_5NO.2H_2O$].

Oxidation state	+1	+2	+3	+4	+6
I.S.	~ + 2.2	~ + 1.4	~ + 0.7	~ + 0.2	~ – 0.6

The effect of pressure on the Mössbauer spectra of a wide variety of iron compounds is reported. Upon application of pressure (~165 kbar) to trisacetylacetaonato iron(III), a new species forms that is attributed to iron(II). The change is reversed upon removal of the pressure.

The determination of the oxidation states of tin compounds from

Mössbauer spectra is not as clear-cut as in the case of iron. Values of δ below 2.65 mm sec^{-1} are often due to tin(IV), and those above that value to tin(II). Exceptions are known. The isomer shifts of some four- and six-coordinate tin(IV) compounds vary considerably as the attached anion is varied.

The MB spectra of the iron pentacarbonyls $Fe(CO)_5$, $Fe_2(CO)_9$, and $Fe_3(CO)_{12}$ have reported. The results are as expected from the known structures for both $Fe(CO)_5$ and $Fe_2(CO)_9$. The structure of $Fe_3(CO)_{12}$ deduced from its mossbauer spectrum was at odds with infrared results and a preliminary x-ray study. The mössbuaer indicated more than one type of iron, as shown in Fig. 13.9(A). The outer two lines are assigned to one type of iron and the inner two to a second type. In general, the areas are roughly proportional to the number of a particular type of iron present. A definitive crystal structure study has supported the structure shown in Fig. (13.9)(B), which was predicted from the Mossbauer results. The spectra of $Fe(II)X_2(CO)_2P_2$ (where X = Cl, Br, and I, and P = phosphines and phosphites) have been interpreted in terms of the five different isomers that exist.

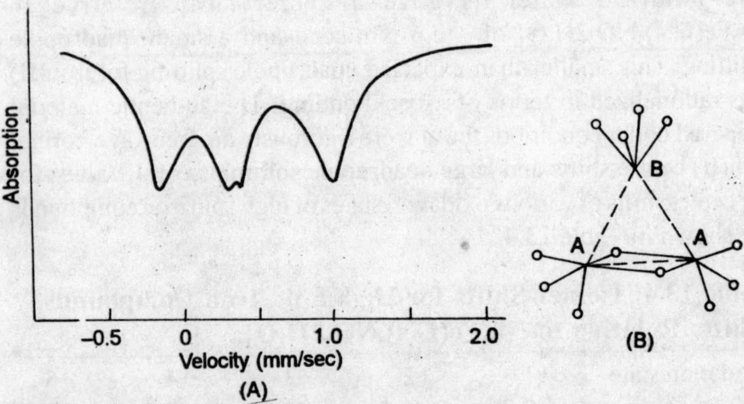

Fig. 13.9: The Mössbauer spectrum (A) and structure (B) of $Fe_3(CO)_{12}$.

Several systems involving spin equilibrium between high spin and low spin iron(II) complexes have been studied by Mossbauer spectroscopy. A typical result involves the hexadentate ligand {4-[(6-R)-2-pyridyl]-3-azabutenyl}$_3$ amine. The spectra obtained when two or three of the R groups are methyl are characteristic of low spin iron(II) ($^1A_1$) at 77 K, while at 294 K the large isomer shift and quadrupole spitting found are characteristic of high spin iron (II) ($^5T_2$). At intermediate temperatures, both forms are observed in the spectrum. this established that it is a true equilibrium ($^5T_2 \rightleftarrows {}^1A_1$) and that these states are long-lived on the Mossbauer time scale; *i.e.*, the lifetime must be 10^{-9} sec or greater.

There are several interesting biological applications of Mossbauer spectroscopy. Horseradish peroxidase is an iron(III) heme protein. It can be oxidized in two one-electron steps, producing red and green compounds. The Mössbauer spectrum changes upon oxidation to either form and in both cases is interpreted in terms of iron(IV). The removal of the second electron leading to the green form is believed to come from the protein or porphyrin.

Mossbauer studies have been of considerable utility in the study of the redox centers that exist in several classes of iron-sulfur proteins. A crystal structure investigation of the feredoxin HP_{red} photosynthetic (high potential) protein from *Chromatium* has been carried out, and it has been shown to contain an $Fe_4S_4(SCys)_4$ (SCys refers to cystein) core with the cubane-like structure shown in Fig. (13.10). Similar units are present in other proteins. The compound HP_{red} is readily oxidized to form HP_{ox}. The crystal structure does not provide information about the oxidation states of the iron or the charge on the cluster. The Mössbauer spectra of several proteins containing this unit consist of a quadrupole split doublet[(24)]. By comparing the isomer shift in HP_{red} with those of other iron systems, it was concluded that the core consists of an average of two Fe^{3+} and two Fe^{2+} ions. The irons are antiferromagnetically coupled, leading to a diamagnetic material, and the metal electrons in the system are delocalized so that only one kind of iron atom exists. The sensitivity of isomer shift to oxidation state is indicated in Table 13.5.

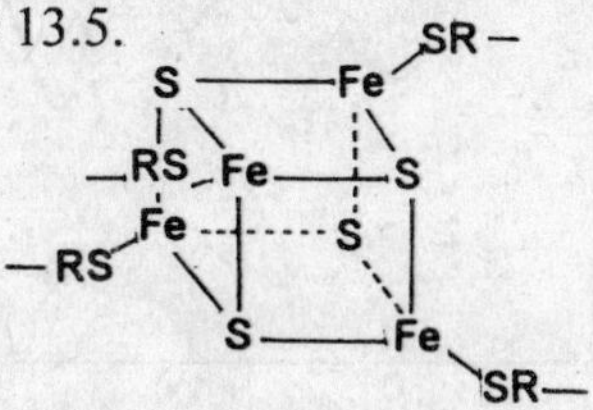

Fig. 13.10: Cubane-like structure of the Fe_4S_4(S—R—)$_4$ core of some ferredoxins.

The model compound $[C_2H_5)_4N]_2[Fe_4S_4(SCH_2C_6H_5)_4]$ has been prepared. The charge on this anion is known to be minus two, so it must be a $2Fe^{2+} + 2Fe^{3+}$ case. The Mössbauer spectrum shown in Fig. 13.11 and the crystal structure are similar to those of HP_{red} or oxidized ferredoxin with equivalent iron atoms and an isomer shift of 0.36 mm sec^{-1}. This supports the assignment of the corresponding ferredoxins as $2Fe^{2+} + 2Fe^{3+}$ systems.

Table 13.5: Centre Shifts for Various Iron-Sulfur Compounds.

Species[a]	δ
Fe^{3+} (rubredoxin)	0.25
$3Fe^{3+} + 1Fe^{2+}$ (*Chromatium* HP_{ox})	0.32
$2Fe^{3+} + 2Fe^{2+}$ (*Chromatium* HP_{red}) and ox. ferredoxin)	0.42
$1Fe^{3+} + 3Fe^{2+}$ (red. ferredoxin)	0.57
Fe^{2+} (rubredoxin)	0.65

[a]An entry such as $3Fe^{3+} + 1Fe^{2+}$ is meant to imply an average oxidation state for the iron atoms corresponding to the given combination.

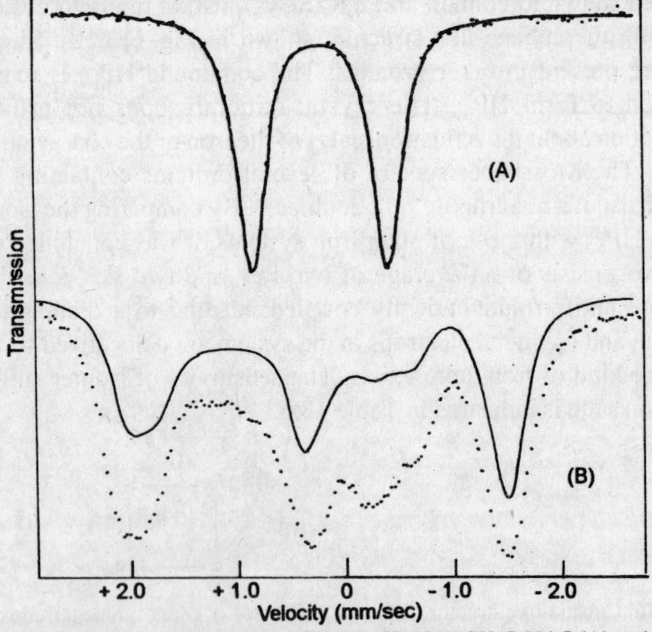

Fig. 13.11: Mössbauer spectrum of $(Et_4N)_2$ $[Fe_4S_4(SCH_2C_6H_4)_4]$ (A) at 1.5 K and $H_0 = 0$, and (B) at 4.7 K and $H_0 = 80$ koersteds.

The Mössbauer spectrum of $(Et_4N)_2$ $[Fe_4S_4(SCH_2C_6H_5)_4]$ shown in Fig. 13.11(A) is a quadrupole split doublet arising from the low symmetry about the iron center. The lines are further split by a magnetic field, as shown in (B). The solid line in (A) is a least squares fit of the data using Lorentzain line shapes. The solid line in (B) is a computer-generated spectrum employing H = 80 kilogauss, $\eta = 0$, and DE_Q = 1.26 mm sec^{-1}. The sign of the principal component of the field gradient is found to be positive from the spectral fit.

It has already been mentioned several times that it is difficult to interpret field gradients. However, it has been found possible to parametrize ions and groups attached to a central metal ion and to use these parameters, called *additive partial quadrupole splittings,* to predict the quadrupole coupling. The basic model is one of a point charge. In a diagonal electric field gradient coordinate system, the contributions to V_{xx}, V_{yy}, and V_{zz} from a charge Z are given by

$$V_{xx} = Zer^{-3}(3 \sin^2 \theta \cos^2 \varphi - 1)$$

$$V_{yy} = Zer^{-3}(3 \sin^2 \theta \sin^2 \varphi - 1)$$

$$V_{zz} = Zer^{-3}(3 \cos^2 \theta - 1)$$

where θ and φ have their usual definitions in the polar coordinate system. For a tetragonal complex (Fig. 13.12), for example, we must take each ligand (a point charge) and sum the individual contributions to V_{xx}, V_{yy}, and V_{zz}. Table 13.6 contains a summary of this calculation for the *cis* and *trans* complexes in Fig. 13.12.

Summing the contributions in Table 13.6, we obtain for the *trans* complex

$$V_{xx} = V_{yy} = (-2[A] + 2[B])e$$

$$V_{zz} = (4[A] - 4[B])e$$

where [A] and [B] are unspecified contributions from ligands A and B, respectively. For the *cis* complex we obtain

$$V_{xx} = V_{yy} = ([A] - [B])e$$

$$V_{zz} = (-2[A] + 2[B])e$$

Table 13.6: Individual Point Charge Contributions to the EFG Tensor in *trans*- and *cis*-MA,B4. [a]The Quantity [A] Equals $Z_A e/r_A^3$.

trans-MA_2B_4									
Ligand (Fig. 15–12)	θ	φ	*sin* θ	*cos* θ	*sin* φ	*cos* φ	V_{xx}/e	V_{yy}/e	V_{zz}/e
A_1	0	0	0	1	0	1	–[A]	–[A]	+2[A]
A_2	180	0	0	–1	0	1	–[A]	–[A]	+2[A]
B_1	90	0	1	0	0	1	+2[B]	–[B]	–[B]
B_2	90	90	1	0	1	0	–[B]	+2[B]	–[B]
B_3	90	180	1	0	0	–1	+2[B]	–[B]	–[B]
B_4	90	270	1	0	–1	0	–[B]	+2[B]	–[B]

cis-MA_2B_4									
Ligand	θ	φ	*sin* θ	*cos* θ	*sin* φ	*cos* φ	V_{xx}/e	V_{yy}/e	V_{zz}/e
A_1	90	0	1	0	0	1	+2[A]	–[A]	–[A]
A_2	90	90	1	0	1	0	–[A]	+2[A]	–[A]
B_1	0	0	0	1	0	1	–[B]	–[B]	+2[B]
B_2	180	0	0	–1	0	1	–[B]	–[B]	+2[B]
B_3	90	180	1	0	0	–1	+2[B]	–[B]	–[B]
B_4	90	270	1	0	–1	0	–[B]	+2[B]	–[B]

The ratio of the *trans* to cis quadrupole splittings is

$$\frac{(4[A]-4[B])e}{(-2[A]+2[B])e}=-2$$

If both isomers are available, Mossbauer spectra provide a ready means of distingushing them, with this ratio. In the partial quadrupole coupling approach, magnitudes are empirically assigned to [A] and [B], and the field gradients are predicted. A *cis* or *trans* geometry is readily determined with the above equations. For other geometries,

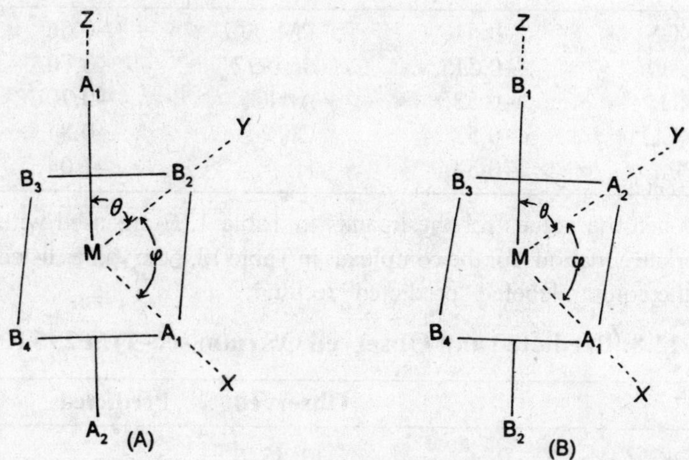

Fig. 13.12: Geometries and coordinates for (A) trans and (B) *cis* MB_4A_2.

the appropriate equations must be derived. The compound $Fe(CO_2)_2(PMe_3)_2I_2$ can exists as five different isomers whose calculated quadrupole splittings differ considerably. Two isomers can be prepared with observed quadrupole splittings of –0.90 and +1.31 mm sec^{-1}. These are readily identified as the (*trans* P, *cis* I, *cis* CO) and *all trans* isomers, respectively. Partial quadrupole splitting parameters for various ligands bonded to iron(II) are listed in Table 13.7

Table 13.7: Ligand Partial Quadrupole Splitting (PQS) Parameters for Iron(II)[a].

Ligand	PQS value	Ligand	PQS value
1	2	3	4
NO^{+}[b]	+0.01	$P(OPh)_3$	–0.55
X^-	–0.30	CO	–0.55
N_2	–0.37	PPh_2Et	–0.58
N_3^-	–0.38	PPh_2Me	–0.58
CH_3CN	–0.43	depb/2	–0.59
$SnCl_3^-$	–0.43	$P(OEt)_3$	–0.63
H_2O[b]	–0.45	depe/2	–0.65
$SnPh_3$	–0.50	$P(OMe)_3$	–0.65

1	2	3	4
NCS^-	–0.51	PMe_3	–0.66
$AsPh_3$	–0.51	dmpe/2	–0.70
NH_3[b]	–0.52	ArNC	–0.70
NCO^-	–0.52	CN^-[b]	–0.84
PPh_3	–0.53	H^-	–1.04

When the values for the ligands in Table 13.7 are used with appropriate equations for the complexes in Table 13.8 the values listed under the column lebeled "predicted" result.

Table 13.8: Predicted and Observed QS (mm sec–1) at 295 K[a]

	Observed	Predicted
trans-$FeCl_2(ArNC)_4$	+1.55	
cis-$FeCl_2(ArNC)_4$	–0.78	–0.78
$[FeC](ArNC)_5]ClO_4$	0.73	+0.78
trans-$Fe(SnCl_3)_2(ArNC)_4$	+1.05	
cis-$Fe(SnCl_3)_2(ArNC)_4$	0.50	–0.52
cis-$FeClSnCl_3(ArNC)_4$	0.61	–0.69($\eta = 0.60$)
trans-$FeH_2(depe)_2$	–1.84	
trans-$FeHCl(depb)_2$	<0.12	–0.20
trans-$Fe(EtNC)_4(CN)_2$	–0.60	
cis-$Fe(EtNC)_4(CN)_2$	0.29	+0.30
trans-$[FeH(ArNC)(depe)_2]BPh_4$	–1.14	–0.98
trans-$[FeH(CO)(depe)2]BPh_4$	1.00	–0.46

Other parameters have been reported for use with tin compounds. The approach has been applied to a large number of systems with a high degree of success. There clearly are some shortcomings in the point charge model, but more work is required to enable one to predict when it will break down in empirical type applications.

The sign and magnitude of the field gradient can be used to provide information about the electronic ground state of a transition metal ion complex. The *approximate* value of the field gradient for different ground states can be estimated by adding the *d* orbital contributions of the populated orbitals, employing Table 13.9.

The quantity $\langle r^{-3} \rangle$ is the expectation value of $1/r^3$ for the appropriate orbital function. This table is constructed by using the various orbital functions to evaluate the matrix elements:

$$q_{zz} = q = -\left\langle \psi \left| \frac{3\cos^2\theta - 1}{r^3} \right| \psi \right\rangle$$

$$\eta q = \left\langle \psi \left| \frac{3\sin^2\theta\cos 2\varphi}{r^3} \right| \psi \right\rangle$$

In a typical application, the $a_{1g}{}^2(d_{z^2})$, $e_{2}g^4(d_{xy}, d_{x^2-y^2})$ ground state for ferrocene is predicted to have a q_{zz} value of $2(-\frac{4}{7}\langle r^{-3}\rangle) + 4(\frac{4}{7}\langle r^{-3}\rangle) = \frac{8}{7}\langle r^{-3}\rangle$. A large positive field gradient is observed. In forming the ferricenium cation, a decrease in the quadrupole splitting is observed, which is consistent with the loss of an e_{2g} electron.[30]

Table 13.9: Magnitude of *q* and η for Various Atomic Orbitals.

Orbital	q	η
p_z	$-\frac{4}{5}\langle r^{-3}\rangle$	0
p_x	$+\frac{2}{5}\langle r^{-3}\rangle$	−3
p_y	$+\frac{2}{5}\langle r^{-3}\rangle$	+3
$d_{x^2-y^2}$	$+\frac{4}{7}\langle r^{-3}\rangle$	0
d_{z^2}	$-\frac{4}{7}\langle t^{-3}\rangle$	0
d_{xy}	$+\frac{4}{7}\langle r^{-3}\rangle$	0
d_{xz}	$-\frac{2}{7}\langle r^{-3}\rangle$	+3
d_{yz}	$-\frac{2}{7}\langle r^{-3}\rangle$	−3

14

Photoelectron Spectrosocopy

Introduction

Photoelectron spectroscopy has its basis in the photoelectric effect. There is a threshold frequency, υ_1, required in order for radiation impinging on a solid to be able to eject an electron from solid into vacuum. The work function ϕ is defined as the corresponding minimum energy required,

$$e\phi = h\upsilon_1 \qquad ...(1)$$

where e is the charge of the electron and h is Planck's constant. At higher frequencies of incident radiation, the emitted electrons have a kinetic energy corresponding to this excess frequencies *i.e.,*

$$E_{kin} = h\upsilon - e\phi \qquad ...(2)$$

However, if only ionisation process is concerned, it is found that the energy of the ejected electron E is given by the following equation.

$$E = h\upsilon - I \qquad ...(3)$$

where $h\upsilon$ in the energy of the incident radiation and I the ionisation potential, or the binding energy of the electron.

By Photoelectron spectroscopy, *the energies of both outer and inner orbitals of atoms and molecules can be determined.* For such studies high-energy quanta for *u.v.* (ups *i.e.,* ultra-violet-photoelectron spectroscopy or PES) or of *X*-ray region are required.

Photoelectron spectroscopy. uses the *far-u.v.* radiation from a helium discharge tube to expel an electron from the atom or molecule under study. This discharge tube provides an intense source of 58.4 nm radiation. Such radiation has quanta of energy 2046 KJ mol^{-1} or

21.21 eV. With such an ionising beam, the electron of the outermost electron shell of most atoms and molecules can be expelled. These electrons are most directly involved in the chemical, or bonding characteristics of the molecules.

Another techniques known as **(XPS)** or *electron spectroscopy for chemical analysis* **(ESCA)** makes use of low energy of "soft" X-rays. Various X-ray targets provide radiations with energies of several thousand electron volts. Such radiations can expel even the inner-most electrons of atoms or molecules. Such electrons are bound with an energy that depends primarily on the nuclear charge of the atom to which they are held. The 1s electrons of the carbon atoms in any type of molecule, for example have an ionisation energy of about 1200 eV. However this value also depends upon the chemical environment of the carbon atoms. Measurement of ionisation energies of such inner electrons thus reveals the various chemical environments of the atoms of a given type the molecules of sample under study. These results are therefore a guide to the nature of the molecule and so this technique has many analytical applications.

Making use of these techniques and the use of equation 14.3, a large number of molecules have been studied. When high-energy u.v. or X-ray ionising radiation is used, atoms and molecules have a variety of ionisation energies. Ionisation that require a large energy leave the expelled electrons with little K.E. and those requiring a smaller amount of energy leave the expelled electrons with more K.E. The pattern of electron kinetic energies thus yields a display of the various ionisation energies of the atoms or molecules of the sample.

The spectrum is a plot of the number of electrons emitted versus the kinetic energy of the electrons (Fig. 14.1).

Orbital Energies of Atoms. A variety of ionization energies are found for the atoms of the sample.

Each ionization energy is a measure of the energy required to expel an electron from one of the orbitals of the atom. Or we can think of it as the energy released when an electron with zero kinetic energy is "dropped" into the empty orbital of the ion of the atom. Thus the *ionization energies can be taken to be measures of orbital energies.*

The ionization spectrum of argon given in Fig. 14.2 illustrates the results obtained for free atoms. The ionizations can be described by the atomic symbols for the Ar^+ ions that are formed. More helpful are the designations of the orbitals of the parent arogn atoms from which an electron has been expelled.

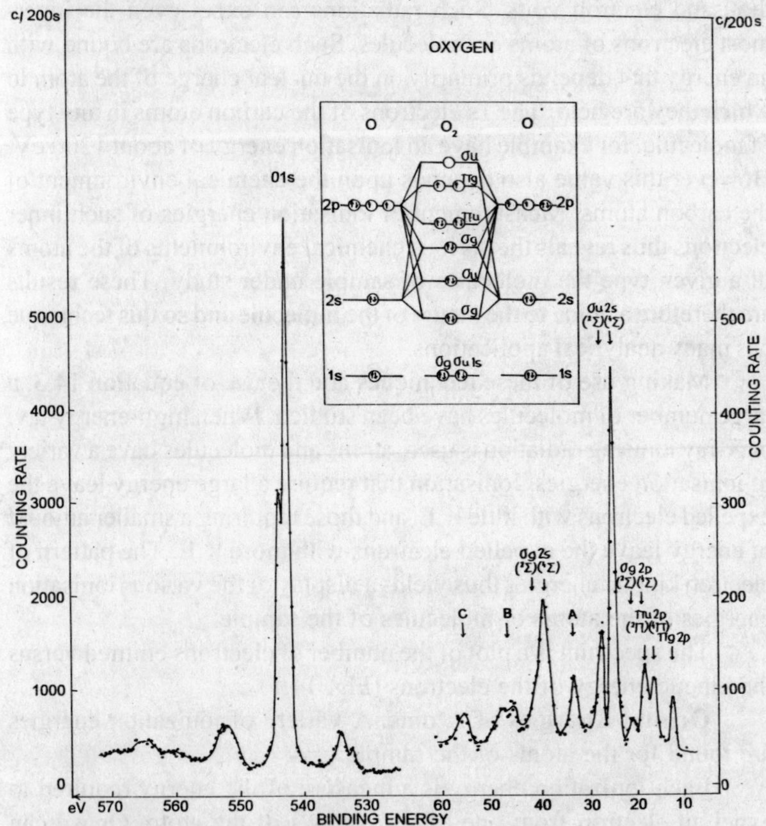

Fig. 14.1: XPS spectrum of core and valence electrons in O_2 excited by Mg Ka X-radiation. Two peaks are obtained for each fully occupied orbital, including the oxygen *1s* core orbital. The peaks A, B, and C, and the peaks toward higher binding energy from the O_{1s} peak, are satellite peaks (*vide infra*).

Orbital Energies of Molecules. Now let us discuss the results obtained when the ionization energies of molecules are measured. We must be prepared for the ionized molecule, A, to be produced in the ground vibrational state or various excited vibrational states. These possibilities can again be treated in terms of the potential-energy curves for the electronic states involved and the *Frank-Condon principle.*

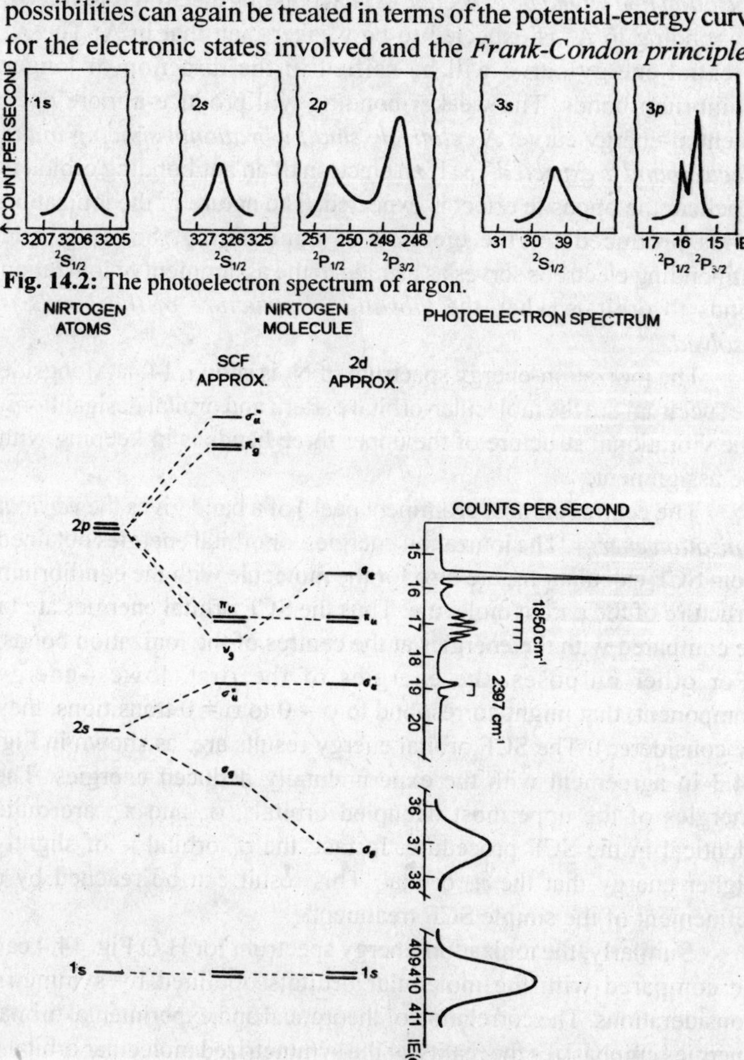

Fig. 14.2: The photoelectron spectrum of argon.

Fig. 14.3: Comparison of the molecular-orbital energy pattern of N_2 deduced from the photoelectron spectrum with that calculated by other methods.

Three types of situation can be expected. (1) If the ionized electron comes from a nonbinding or an inner orbital, the potential-energy curve for the molecular ion A^+ should be little different from that for A. We expect $\upsilon = 0$ to $\upsilon = 0$ transitions to occur. *A single ionization line would be expected.* (2) If a bonding electron is removed, the bonding in A^+ is expected to be weaker than that in A. The A^+ potential-energy curve will be shifted in the direction of longer equilibrium bonds. The weaker bonding will produce a more open potential-energy curve. A *relatively small vibrational spacing in the spectra band is expected.* (3) If an electron of an antibonding orbital is expelled, the opposite effect is expected. The nature of the ionization bands produced for the ejection of nonbonding, bonding, and antibonding electrons serves as a guide to the assignment of ionization bonds to orbitals when the *vibrational structure of the bonds is resolved.*

The ionization-energy spectrum of N_2 is shown 14.4. Alongside the spectrum are the molecular-orbital pattern and orbital desigantions. The vibrational structure of the upper three bands is in keeping with the assignment.

The centre (or most prominent peak) of a band gives the *vertical ionization energy*. The ionization energies, or orbital energies obtained from SCF calculations, are also for the molecule with the equilibrium structure of the parent molecule. Thus the SCF orbital energies are to be compared with the energies at the centres of the ionization bonds. (For other purposes, the energies of the first, lowest-energy component, that might correspond to $\upsilon = 0$ to $\upsilon = 0$ transitions, may be considered). The SCF orbital energy results are, as shown in Fig. 14.3 in agreement with the experimentally deduced energies. The energies of the uppermost occupied orbitals, σ_g and π_u, are quite identical in the SCF procedure. In fact, the σ_g orbital is of slightly higher energy that the π_u orbital. This result can be reached by a refinement of the simple SCF treatment.

Similarly, the ionization-energy spectrum for H_2O Fig. 14.4 can be compared with the molecular orbitals obtained by symmetry considerations. The correlation of theoretical and experimental orbital energies emphasizes the reality of the symmetrized molecular orbitals. Similar experimental results for the energies of molecular orbitals of a great variety of molecules are now available.

One illustration (Fig. 14.5) is shown to suggest the use of inner-orbital-ionization energies in providing a "fingerprint" of a molecule. Please note that each carbon atom with a different environment reveals itself through the chemical shift of its *1s* ionization energy.

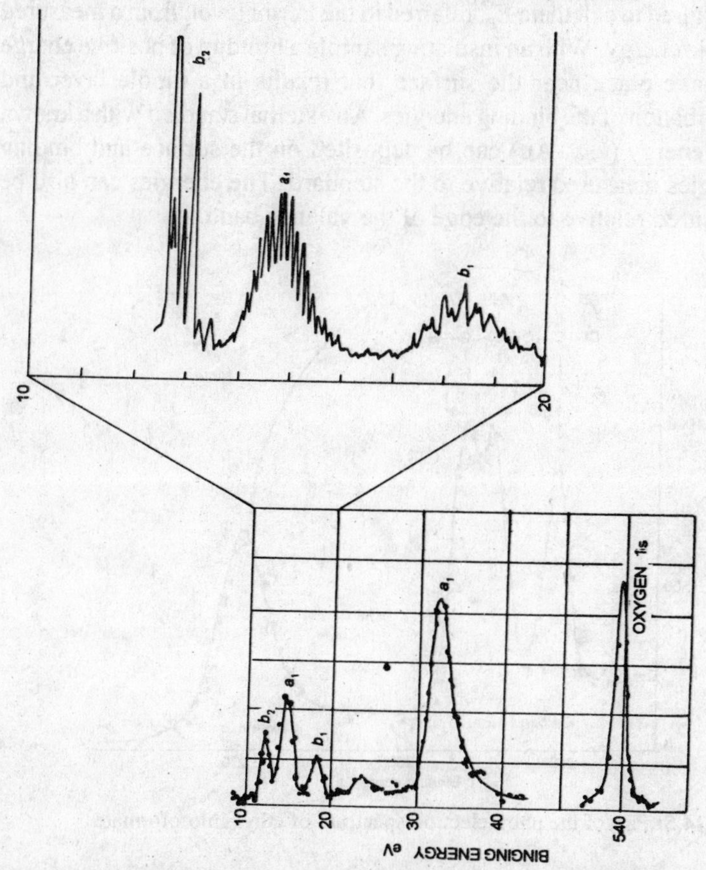

Fig. 14.4: The photoelectron-spectrum pattern for the H_2O molecule (*a*) attained with 58.4-nm helium discharge radiation; (*b*) obtained with Mg K_{α} X-rays. The symmetries of the orbitals, from which the electron is removed are indicated.

XPS

Solids (including frozen solutions), gases, liquids, and solutions have been studied by XPS. Binding energies, E_B, of both core and valence electrons can be measured as seen in equation ($E_{kin} = h\upsilon - e\phi$

– E_B). In the XPS experiment, the binding energies are expressed relative to a reference level. As the solid samples is in electrical contact with the spectrometer, the Fermi level, $e\phi$, of the sample and spectrometer are the same. Thus, only the spectrometer work function is required to calculate E_B, referred to the Fermi level, from a measured kinetic energy. With an insulating sample a buildup of positive charge can take place near the surface that results in a dipole layer and perturbation of the binding energies. An external standard with a known core energy (*e.g.*,, Au) can be deposited on the surface and binding energies measured relative to the standard. The energies can also be measured relative to the edge of the valence band.

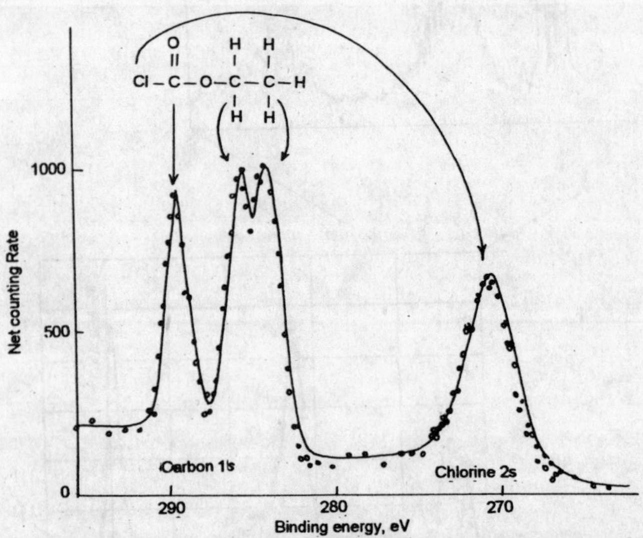

Fig. 14.5: Part of the photoelectron spectrum of ethyl chloroformate.

The binding energies of atoms change with oxidation state partial charges on the atoms. Compilations are available that enable us to use binding energies to determine these quantities by a fingerprint type of application. Generally, *the binding energies of cations increase with increasing oxidation state.* Opposite behaviour is observed for PbO in comparison to PbO_2 and this is attributed to differences in the *Madelung potential in the two structures.* Very small differences in binding energy are found for Ag_2O compared to Ag and for Cu_2O

compared to Cu. On the other hand, the E_B (1s) for Cu and CuO differ by 4.4 eV.

Please note that linear correlations have been found between core-electron binding energies and **Mössbauer isomer shifts** for compounds of tin and iron. Also, a correlation of chlorine core binding energies with nuclear quadrupole resonance frequencies has been found.

Attempts have been made to correlate the binding energies to orbital energies from molecular orbital calculations. The basis for this approach is **Koopman's theorem,** that states "*the vertical ionization energy for removal of an electron from a molecular orbital corresponds to the negative of the eigenvalue obtained from the molecular orbital calculation*". It is thus assumed that the molecular orbtials for the parent molecule and the ionized molecule are the same. Any electronic relaxation or change in correlation energies causes Koopman's theorem to break down leading to limited quantitative success for this approach.

The photoemission process is made up of the following events:

1. Absorption of the photon by the bound electron, thus increasing the kinetic energy of the electron.

2. Ejection of the electron from the atom using some of its kinetic energy to overcome the Coulomb attraction of the nucleus.

3. Simultaneous with 2, the electrons in the outer orbitals readjust (intra-atomic relaxation) to a lower-energy final state, transferring this energy to the outgoing electron.

Relaxation does not always result in the formation of the ground state of the ion. Simultaneous with the photoionization of an electron, there can be excitation of one of the remaining electrons to an initially unoccupied orbital. This phenomenon is known as a "**shakeup**" process. The higher energy, low intensity structure on XPS core-electron pekes can then be used to study various electronic transitions taking place simultaneously with the photoionization. These satellite features are found in the range from zero to 50 eV toward higher binding energy than the main peak. Thus, an electronic absorption at 50 eV (=404,000 cm^{-1} = 25 nm) is a very high energy absorption in the vacuum ultraviolet. Photoemission from states belonging to a shell with an empty or partially filled subshell will excite all states accessible by redistributing the electrons within that shell that have the same symmetry as the state produced by the one electron transition.

An example of such satellite XPS features appears in Fig. 14.1.

Broad features are seen toward higher binding energy than the two O_{1s} peaks for O_2. The features marked A, B and C are "shakeup" peaks appearing as satellites on the valence-electron peaks.

In ions with a partially filled outer shell, the coupling of the outer shell to the core hole (all spin and angular moment) will give a set of final states with distinct energies. The spectrum will exhibit fine structure (*vide infra*).

The "mixed-valent" species $[(NH_3)_5 Ru\text{-(pyrazine)}Ru(NH_3)_5]^{5+}$, has also been studied by XPS. The XPS spectrum has been reported to contain ionization peaks from two non-equivalent transition metal ions. The carbon *1s* peaks occur in the same spectral region as the metal ionization peaks, and the conclusion rests on the ability to subtract these peaks from the spectrum. The XPS technique has a very short time scale, on the order of 10^{-17} sec. However, when a core electron is removed from a mixed-valent compound, only discrete valances are obtained and fluctuations cannot persist in the final state. Depending on the spin of the core electron removed, two different spin states can result from the final ion. Thus, two "discrete peaks" in the XPS do not distinguish mixed valence from delocalised systems.

UPS

In the UPS experiment, vacuum ultra-violet radiation is used as the ionizing radiation; Usually this is provided by helium [singly ionised, indicated as He(I)] resonance lamp with an energy of 21.21 eV. Other discharge lamps have been used. The energy of these lamps limit UPS to studies of valence electrons, and in general, measurements have been mostly confined to gaseous samples. There have been some reports of work on solutions and solids.

The UPS gasesous sample of N_2 is shown in Fig. 14.6 with the helium He(I) source determining an ionising limit of 21.21 eV, three vibrationally structured photoionizations are seen (~ 15, 6, ~ 17.0, and ~ 18.8 eV). These can be assigned to ionisation from the three highest-energy filled molecular orbitals for N_2, the $2\sigma_U$, π_U, and $3\sigma_g$ orbitals. The peaks have been assigned on the basis of the observed vibrational structure. Please note that an π PS spectrum has there peaks (vibrational structure not seen due to lower resolution) in addition to a peak at 37.3 eV for ionisation from the $2\sigma_g$ level level as well as a single peak at 409.9 ev for both the $1\sigma_g$ and $1\sigma_U$ levels.

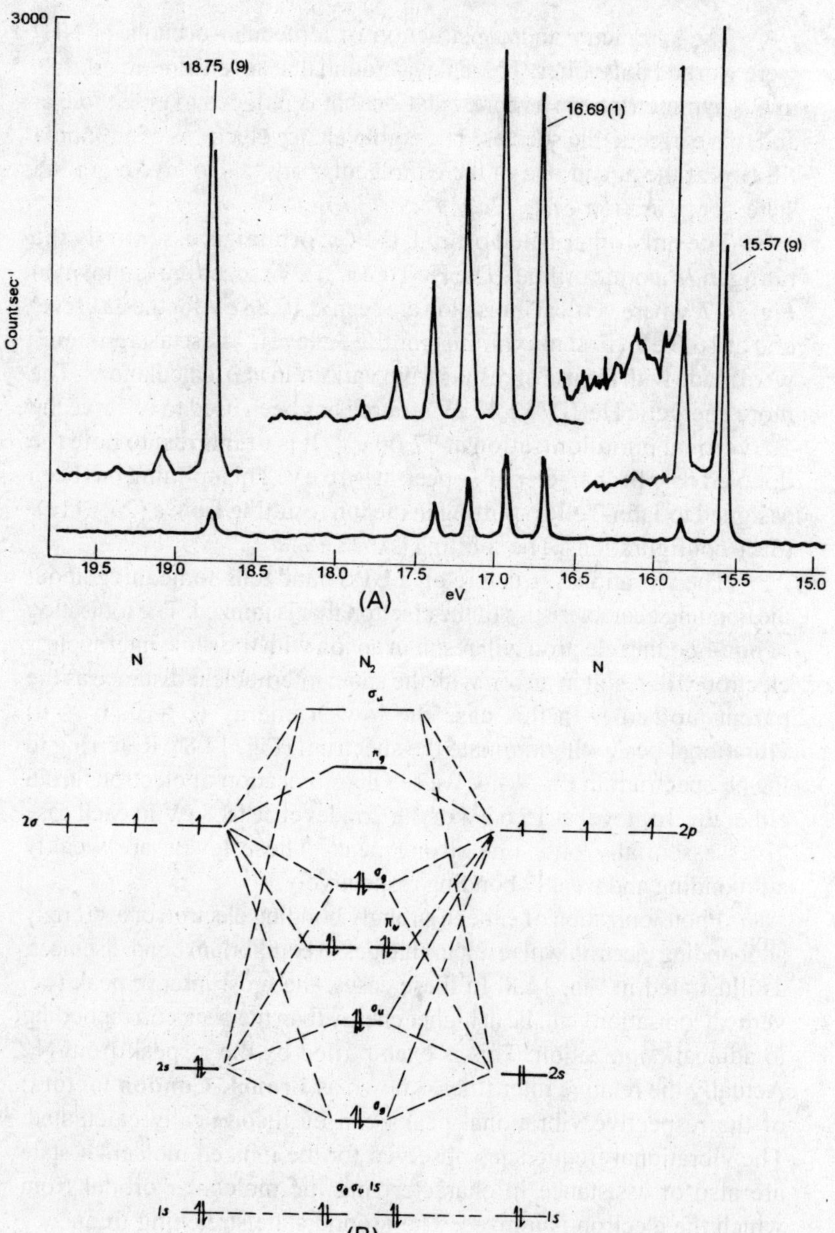

Fig. 14.6: PES spectrum for a gaseous nitrogen sample.

The symmetry and construction of molecular orbitals of NH_3 were worked out earlier. There it was found that seven atomic orbitals in C_{3V} symmetry form a representation that is reduced to give three a_1 and two e irreducible species. The eight valence electrons of ammonia fill two of the a_1 and one of the e molecular orbitals to give a ground state configuration of.... $(2a_1)^2\ (1e)^4\ (3a_1)^2$.

The only other filled orbital, the *1a*, orbital, is essentially the nitrogen *1s* atomic orbital. The He(I) (21.21 eV) spectrum is shown in Fig. 14.7, where vertical ionisation are seen at 10.88 eV for the $3a_1$, level and at 16.0 eV (first maximum) for the *1e* level. These assignments were made with the use of results from various in *m.o.* calculations. The more energetic He(II) (42.42 eV) source has been used to observe the *2a*, vertical photoionisation at 27.00 eV. It is of interest to note the doubled or split character of *1e* peak at ~ 16 eV. This splitting has been assigned to Jahn-Teller splitting in the ion resulting from a $(2a_1)^2\ (1e)^3\ (3a_1)^2$ configurations. The splitting is 0.78 eV.

The vibrational structure on a **UPS** band tells some thing about the bonding characteristics of the electron that is ionized. The ionisation of non-bonding electron will result in an ion with the same internuclear electron will result in an ion with the same internuclear distance as the parent molecule. In this case the lowest energy ($\upsilon = 0, \upsilon' = 0$) vibrational peak will dominate the spectrum (Fig. 14.8). Referring to the N_2 spectrum in Fig. 14.6, we see that ionisation of electrons from either the $3\sigma_g$ level at 15.6 eV or the $2\sigma_U$ level at 18.8 eV in each case gives essentially only one strong peak. These levels are weakly antibonding and weakly bonding respectively.

Photoionization of either a strongly bonding electron or a strongly antibonding electron will result in changes in equilibrium bond distances, as illustrated in Fig. 14.8. In these cases, the most intense peak (i.e. vertical ionisation) will be at higher energy than the peak corresponding to adiabatic ionization. This is exemplified by the π_u peak from N_2. Actually the relative intensities (known as **Franck-Condon factors**) of the respective vibrational peaks can be theoretically calculated. The vibrational frequencies observed for the ionized molecular state are also of assistance in characterizing the molecular orbital from which the electron is ionized. The ground state stretching frequency for N_2 is 2345 cm^{-1}. Table 14.1 gives the stretching frequencies seen for various N_2^+ ionized states. A comparison of each of these values with the 2345 cm^{-1} N_2 frequency leads to the implied m.o. characters of the lost electron given in Table 14.1.

Table 14.1: Vibrational Structure in the N_2UPS Spectrum.

Peak	*ν(N—N)*	*Implied M.O. Character*
$\sigma_g(2p)$	2150 cm^{-1}	weakly bonding
$\pi_u(2p)$	1810 cm^{-1}	strongly bonding
$\sigma_u(2s)$	2390 cm^{-1}	weakly antibonding

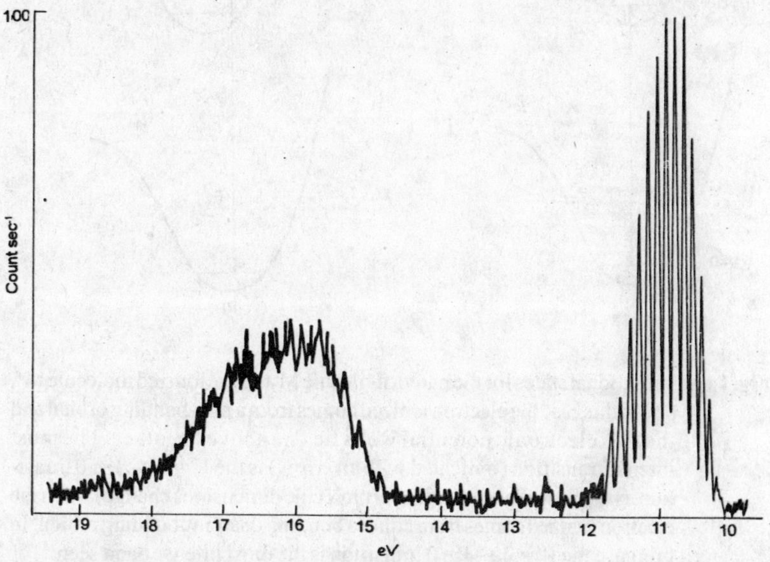

Fig. 14.7: PES spectrum of a gaseous sample of ammonia.

In addition to vibrational structure, other fine structure is seen in UPS spectra. The first (*i.e.*, lowest binding energy) peak in the spectra of CH_3Cl, CH_3Br, and CH_3I is, obviously, from the highest occupied molecular orbital, and this is largely localized in the halogen atom (with some X—H anti-bonding and C—H bonding character. Spin-orbit splitting is seen in this peak, where the separation varies in the series: CH_3I^+, 0.62 eV; CH_3Br^+, 0.31 eV; and CH_3Cl^+, ~0 eV. These spin-orbit interactions agree with Mulliken's predications of 0.625, 0.32, and 0.08 eV, respectively. Many other such observations of spin-orbit interactions have been noted. In fact, recent work at the high resolution of 10 meV has detected spin-orbit splittings in NO of 0.012 to 0.016 eV.

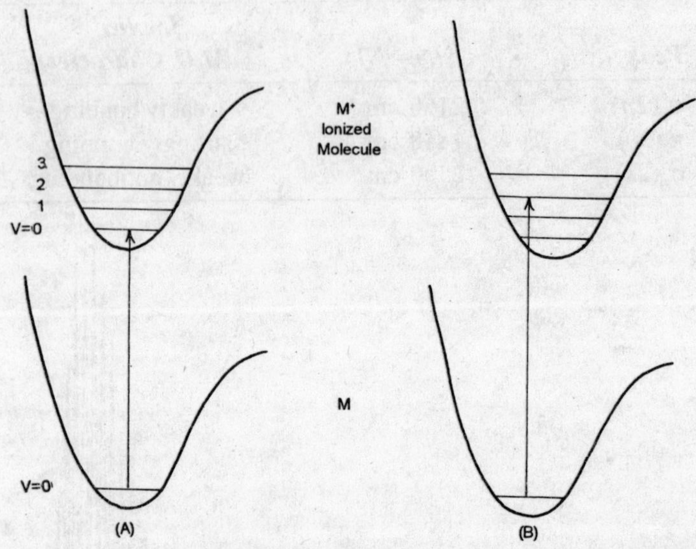

Fig. 14.8: Vibrational states for the parent molecule M and the ionized molecule M^+. (A) In this case the electron ionized comes from a non-bonding orbital and the two electronic potential wells lie one above the other. The most intense transition (indicated with an arrow) is the $V'=0 \leftarrow V=0$ transition. (B) The wells are displaced (molecule dimensions change) when the electron ionized comes from either a bonding or an antibonding orbital. In this case the $V'=2 \leftarrow V=0$ transition is the most intense band seen.

A large number of small gaseous molecules have been studied with UPS and, for that matter, XPS. Some radical and excited state species have also been studied. The three expected state (^2P, ^4S, and ^2D) of O^+ have been detected. An electrodeless microwave discharge was used to produce the excited state species $O_2(^1\Delta_g)$, and the photoelectron spectrum of this species shows a vibrationally strcutured peak because of the formation of O_2 $(2\Pi_g)$. Comparison with the peak corresponding to ionization of ground state $O_2(^3\Sigma_g^-)$ to the same ionic state gives a value of 11.09 ± 0.005 eV for the adiabatic ionization potential of $O_2(^1\Delta_g)$. Several other such species (*e.g.*, SO and NF_2) have also been studied.

As compared with diamagnetic compounds, paramagnetic species show additional complexity in the UPS (and XPS) spectra. The oxygen molecule has two unpaired π_g electrons. The UPS spectrum of O_2 is given in Fig. 14.9. The photoionization of an electron from the partly filled, antibonding π_g(2p) molecular orbital appears as the first peak in the UPS spectrum. Only one ionic state is realized. On the other hand, photoionization of an electron from one of the other filled molecular orbitals leads, in each case, to *two* electronic states of the ion O_2^+. Thus, if an electron is ionized from the filled bonding π_u level, the *unpaired electron remaining* in the π_u orbital can be aligned either with or *against* the two unpaired electrons in the antibonding π_g level. When the electron is aligned with the two π_g electrons, there are three unpaired electrons, the total spin $S = 3/2$, and the electronic state of O_2^+ molecule is $^4\Pi_u$. The other alignment gives a $^2\Pi_u$ electronic state for the O_2^+ molecule. The $^4\Pi_u$ and $^2\Pi_u$ states of O_2^+ are at different energies, and thus there is a splitting of the π_u orbital ionization peak. In Table 14.2, the observed features for the O_2 molecules from both UPS and XPS spectra are given.

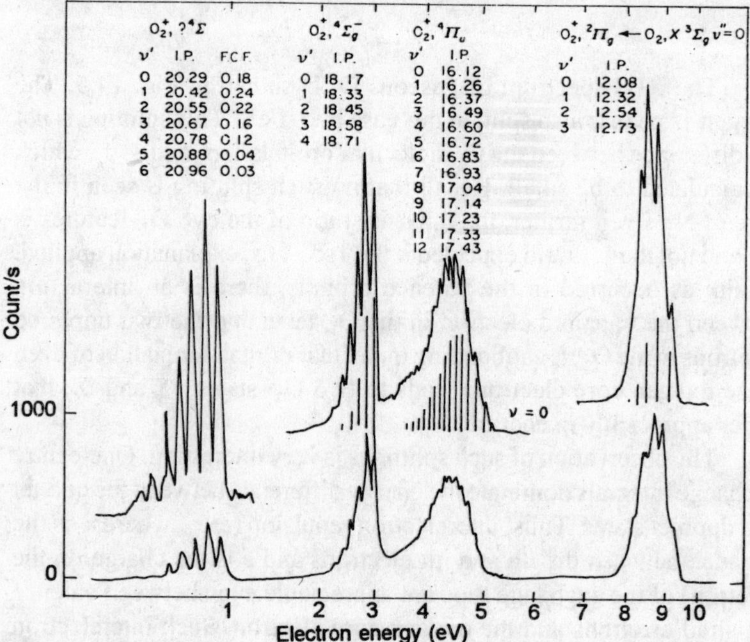

Fig. 14.9: Oxygen PES spectrum.

Table 14.2: Vertical O_2 Ionization Data.

One-Electron Molecular Orbital	Ionic Electronic State	XPS, eV	PES, eV
$\pi_g 2p$	$^2\pi_g$	13.1	12.10
$\pi_u 2p$	$^4\pi_u$	17.0	16.26
	$^2\pi_u$	?	?
$\sigma_g 2p$	$^4\Sigma_g$	18.8	18.18
	$^2\Sigma_g$	21.1	20.31
$\sigma_u 2s$	$^4\Sigma_u$	25.3	24.5 [*via* He(II)]
	$^2\Sigma_u$	27.9	
$\sigma_g 2s$	$^4\Sigma_g$	39.6	
	$^2\Sigma_g$	41.6	
Oxygen 1*s*	$^4\Sigma$	543.1	
	$^2\Sigma$	544.2	

The XPS spectrum of gaseous O_2 is shown in Fig. 14.5. The oxygen 1*s* peak is also split, in this case by 1.1 eV. This splitting is *not* the difference between the O_2 molecular orbitals $1s\sigma_u$ and $1s\sigma_g$, which is calculated to be small. Recall that no such splitting is seen in the case of N_2. Even further, the intensity ratio of the two O_{1s} features is 2:1 and not the 1:1 ratio expected if the $1s\sigma_u/1s\sigma_g$ explanation applied. Again, as occurred in the valence orbitals, there is an interaction between the unpaired electron in the O_{1s} level and the two unpaired electrons in the $O_2^+\pi_g$ antibonding molecular orbital. Ionization of even these oxygen core electrons leads to two O_2^+ states, $^2\Sigma$ and $^4\Sigma$, that differ appreciably in energy.

The observation of such splittings is very interesting. One-centre exchange integrals dominate the energy difference between the quartet and doublet states. Thus, interelectron repulsion (e^2/r_{ij}, where r_{ij} is the distance between the *i*th and *j*th electrons and *e* is the charge on the electron) of the exchange type are appreciable even between valence unpaired electrons and the oxygen core electron. Such interelectron spin polarization are found to be of importance in explaining the shifts in NMR peaks for paramagnetic molecules. Electron exchange splitting is also seen for core electrons of paramagnetic transition metal complexes.

It is clear from the above discussion that the UPS spectra of relatively large molecules are quite rich with information about ionization potentials, vibrational quanta for the ionized molecule, spin-orbit interactions, Jahn-Teller splittings, and electron exchange interactions. Unfortunately, there is frequently an overlapping of features, and broad peaks appear with no resolved vibrational structure. As an example of a small molecule with many photoionization peaks, Fig. 14.10 illustrates some of the spectral detail obtained for gaseous NO at a resolution of 10 meV for a He(I) source and at 25 meV for HE(II) source Fig. 14.10 shows that both exchange and spin-orbit splittings have been resolved.

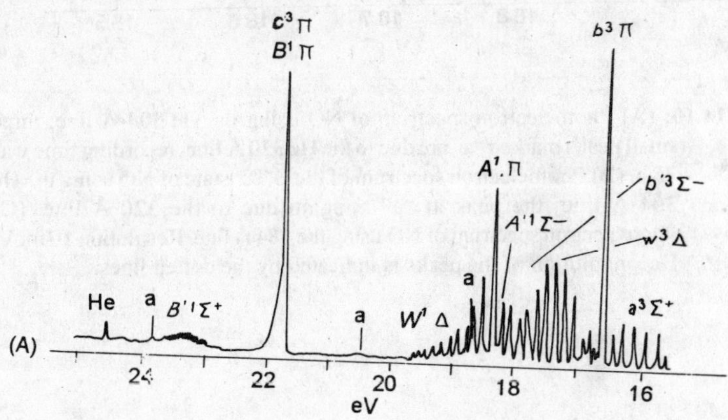

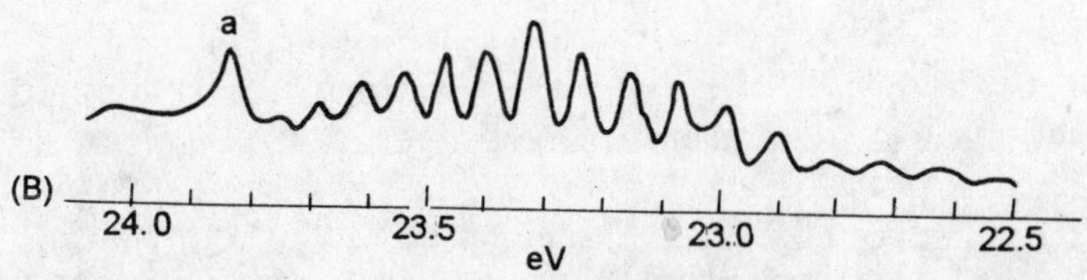

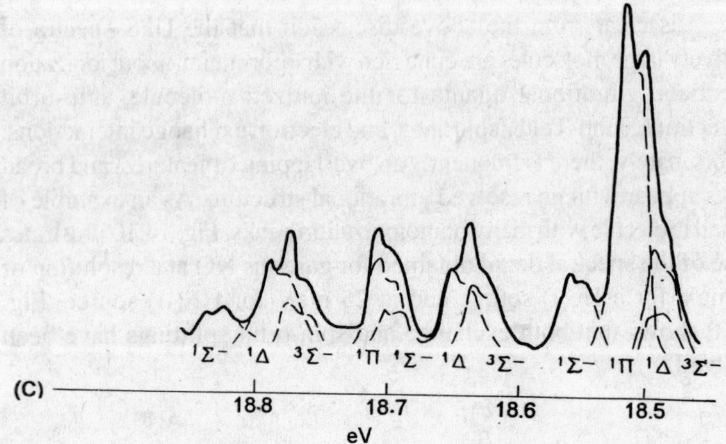

Fig. 14.10: (A) Photoelectron spectrum of NO using the He 304-Å line; three small peaks marked "a" are due to the He 320 Å line; recording time was 75 h. (B) Photoelectron spectrum of the B' $^1\Sigma^+$ state of NO using the He 304 Å line; the peak at "a" is again due to the 320 Å line. (C) Photoelectron spectrum of NO using the 584 Å line. Resolution 10 meV. Deconvolution of the peaks is indicated by the dotted lines.

15

Errors, Precision and Accuracy

Introduction

The data from physical measurements are always plagued by uncertainties or errors. An estimate of the magnitude of these uncertainties is of prime importance if experimental results are to have meaning. Unfortunately, no simple, generally applicable methods, exist that will provide a measure of reliability of experimental data with absolute certainty.

In most of the methods of instrumental analysis a singal is produced which reflects the chemical or physical property of a chemical system. The resulting signal may be used directly or transformed to one of a different nature, amplified and read on some device. The properties of the initial system are then induced from the final read out. For example in atomic absorption spectroscopy the sample is vapourised; the vapourised sample absorbs light of certain wavelengths. The degree of absorption is then correlated with the concentration of specific ion which was in the original solution.

Many, but not all, instruments contain one or more components, which bring about a change in the sample. For instance. the atomic absorption spectrophotometer requires a component to vaporize the sampler and another separate component to furnish the radiation of the correct wavelength. Electrogravimetric procedures require the input of electricity in order to bring about the necessary oxidation—reduction reaction.

Errors may arise at any of the many steps between the initial decision to perform a certain analysis and the final report. For example, any specific procedure requires standard reproducible techniques for obtaining, handling, and storing of samples. The chemical and instrumental systems must be chosen and standardized. The instruments must be calibrated. The sample must be dissolved and interferences removed, "neutralized," or corrected for in the calculations. Concentration or dilution steps may be necessary to bring the sample within the range of the instrument. Finally, the results must be calculated and their validity determined.

It is possible that by improving the calibration or by adopting a different method the reliability of a data can be improved. Moreover repeating the experiment a number of times leads to better results because of improvement in judgement in observation.

Types of Errors

The errors may be broadly classified as (i) *Determinate errors,* and (ii) *indeterminate errors.* However some times it is not possible to determine into which one of the above two classes, a given error falls.

(i) Determinate Errors

It refers to those errors which have a definite value and also have some assignable cause. It is possible to measure, account and correct for these errors. Generally these errors can be identified and corrected by calibration.

These errors can be further classified as (a) Method errors and (b) Personal errors.

(a) Method errors

The main cause of such errors is the non-ideal physical and chemical behaviour of reagents and reactions used in the process of analysis. These errors may arise due to slowness of chemical reactions, instability of reagents and chemical interferences. This type of errors are more difficult to detect and rectify than than are instrumental errors.

(b) Personal errors

It refers to those errors which are introduced into a measurement by the person performing the experiment. These may arise due to misreading the position of a meter needle, transposing numbers while making entry of the data, inadvertent spillages and some times even by bias.

(ii) Indeterminate Errors

This type of errors arise from uncertainties in measurements which are unknown and uncontrollable by experimenter. Such errors manifest themselves in the form of small random fluctuations called *noise.*

Whenever measurements are made with an instrument having a sensitivity that is not limited by its read out device, small changes among replicate measurements are observed. These small but detectable variations represent the accumulation of a very large number of uncertainties which are not detectable individually, and it is only their accumulated effect that is recognisable as noise.

This type of errors follow a random distribution and so it is possible to apply the law of probability to arrive at some conclusion about the most probable results of a series of experiments.

Distribution of Indeterminate Errors

Fig. 15.1 is the distribution curve which relates frequency of occurrence to deviations from the mean and is known as *Gaussian* or *Normal.*

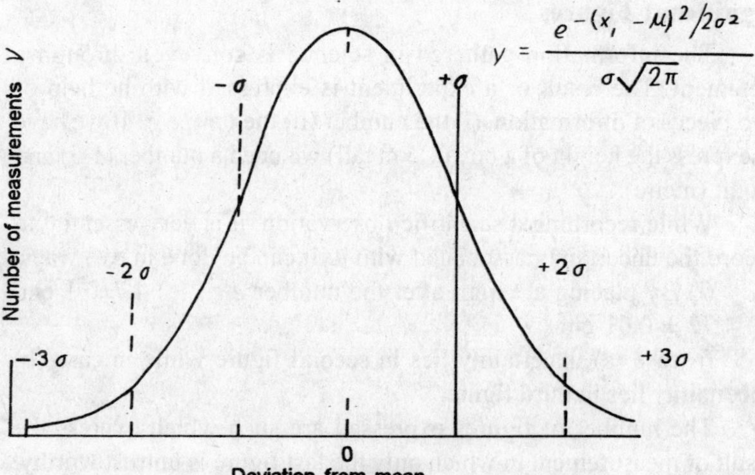

Fig.15.1: A normal or Gaussian distribution curve

The bell shaped Gaussian curve can be described by the equation

$$y = \frac{e - (x_1 - \mu)^2 / 2\sigma^2}{\sigma\sqrt{2\pi}} = \frac{e^{-z^2}2}{\sigma\sqrt{2\pi}} \qquad ....(1)$$

In this equation, x_i represents values of individual measurements, and μ is the arithmetic mean for an infinite number of such measurements. The quantity $(x_i - \mu)$ is thus the deviation from the mean; y is the frequency of occurrence for each value of $(x_i - \mu)$. The symbol π has its usual meaning, and e is the base for natural, or Napierian, logarithms. The parameter σ is called the *standard deviation* and is constant that has a unique value for any set of data comprising a large number of measurements. The *breadth* of the normal error curve, which increases as the precision of a measurement decreases, is directly related to σ.Thus, σ is widely employed to define the precision of measurements.

The exponential term in Equation (1) can be simplified by introducing the variable

$$z = \frac{x_i - \mu}{\sigma}$$

which then gives the deviation from the mean in units of standard deviations.

Significant Figures

The information gathered in science is conveyed through a statement. The result of a experiment is expressed with he help of two pieces of information. (i) the number (ii) the unit, *e.g.*, if we have to express the height of a boy (1.5 m tall) we need a number (1.5) and a unit (metre).

While recording a scientific observation, it is very essential to record the uncertainty associated with it. It can be done in two ways.

(i) By placing a ± sign after the number *e.g.*, (a) 4.7±0.1 cm, (b) 4.72 ± 0.01 cm.

In case (a) uncertainty lies in second figure while in case (b) uncertainty lies in third figure.

The number of figures expressed are such which express the result of measurement in which only the last figure is untrust worthy. This method is called the method of significant figures. This method is better as it gives an idea about the uncertainty introduced by using a device.

The following rules are used in counting the number of significant figures. These rules should be kept in mind while expressing the result of measurement:

Rule 1. All digits are significant except zeros at the beginning of the number, *e.g.*, 131 cm, 0.131 cm, 0.0131 cm, all have three significant figures.

Rule 2. The zeros to the right of the decimal point are significant cant, *e.g.*, 131 cm, 131.0 cm, 131.00 cm have three, four and five significant figures respectively.

These rules are valid if numbers are expressed in sceintific notation. In sceintific notation every number is expressed as N×10n where

N = a number with a single non-zero digit to the left of the decimal point.

and n = an integer.

While carrying about various mathematical operations it has to be kept in mind that where we add, subtract, multiply or divide the numbers obtained in different measurements the result can not be more precise than the least precise number involved in the measurement. Certain rules should be followed to arrive at the number of significant figures in any calculations.

Rules to be followed to arrive at number of significant figures in calculations involving addition and/or subtraction.

The result obtained after addition or subtraction should be reported to the same number of decimal places as that of the term with least number of decimal places. *Note that in this case the emphasis is on decimal placed and not on significant figures.* For example in case of following addition :

		32.1
	+	11.11
	+	101.001
Sum		144.211

The answer should be 144.2 and in case of following subtraction.

	25.472
−	3.312
	22.1508

The answer should be 22.151.

Rules to be followed for Rounding off the figures

In case the results obtained have more figures than are significant, *e.g.*, in above case, these are rounded off.

If the figure following the last number to be retained in less than or equal to 5. The last number is left unchanged, however, if, the figure is greater than 5, the last number to be retained is increased by 1.

Rules to be followed to arrive at number of significant figures in calculations involving multiplication and division

In multiplication and division the result should be reported to the same number of significant figures as the least precise term in the calculation, *e.g.*,

$$31.031 \times 2.21 = 68.57851$$

In this case there are five significant figures in 31.031 and three significant figures in 2.21. The answer should be in three significant figures. So the answer should be 68.6. *Note that in this case significant figures are to be counted not the figures after decimal point.*

Absolute and relative uncertainty. The uncertainty of a value may be expressed in **absolute** or **relative** terms. When expressed in absolute terms it possesses the unit of measurement. When expressed in relative terms the units of measurement cancel and the final units are in a parts per unit part relationship such as % or ppm.

Consider the weight of 16.7g. If it is assumed that the last place is in doubt this could then be written in absolute uncertainty terms as

$$\text{weight} = (16.7 \pm 0.1)\ \text{g}$$

and in relative parts per thousand, ppt, as

$$\text{weight} = 16.7\text{g} \pm (\frac{0.1\text{g}}{16.7\text{g}} \times 1000)\text{ppt}$$

or

$$\text{weight} = 16.7\text{g} \pm 6\ \text{ppt}$$

or relative parts per hundred, %, as

$$\text{weight} = 16.7\text{g} \pm (\frac{0.1\text{g}}{16.7\text{g}} \times 100)\% = 16.7\text{g} \pm 0.6\%.$$

The use of relative as opposed to absolute uncertainty yields a better comparison between values of different magnitudes or units. Consider two weights, one 1278 g and the other 18 g. They both have the same absolute uncertainty of one gram, but upon calculation of the relative uncertainties in ppt of the respective sample, as follows:

$$\frac{1}{1278\text{g}} \times 1000 = 0.8 \text{ ppt}$$

and

$$\frac{1\text{g}}{18\text{g}} \times 1000 = 56 \text{ ppt}$$

it can be seen that the relative uncertainty of the smaller is 70 limes that of the larger weight.

Consider the determination of a density of a liquid where the burrete was read to the maximum possible precision and 28.02 ml of the solution was delivered. The volume was found to weigh 31.6728g. In absolute terms the volume would be expressed as

$$(28.02 \pm 0.01) \text{ ml}$$

and the weight as

$$(31.6728 \pm 0.0001) \text{ g.}$$

Any attempt to compare the precision of the two measurements would be equivalent to a comparison of apples and organise. On the other hand, when expressed in relative terms of ppt the uncertainty of the volume measurement is

$$\frac{0.01 \text{ ml}}{28.02 \text{ ml}} \times 1000 = 0.4 \text{ ppt}$$

and that of the weight is

$$\frac{0.0001 \text{ g}}{31.6728 \text{ g}} \times 1000 = 0.003 \text{ ppt}$$

A comparison of the two uncertainty values immediately informs us that a balance which was good to the nearest 0.01 g instead of 0.0001 g would have been adequate. No loss in overall precision would have resulted. An attempt should be made to obtain relative uncertainty meaurements in the same range in any given experiment.

Introduction in Statistics

A number of statistical terms will be encountered in the

successive discussion. In order to properly understand the limitations of, and to evaluate and minimize the errors to various procedures, a limited knowledge of statistics is necessary.

The **average** is synonymous with **arithmetic mean.** It can be obtained by adding the results of a number of replications and dividing this sum the number of replications, *n.* It is designated as X. Therefore, the following triplication:

$$X_1 = 26.28 \text{ g}$$
$$X_2 = 26.30 \text{ g}$$
$$X_3 = 26.32 \text{ g}$$
$$\Sigma x = 78.90 \text{ g}$$

shows the sum of the individual values and

$$\overline{X} = \frac{\Sigma X}{n} = \frac{78.90 \text{ g}}{3} = 26.30 \text{ g}$$

is the average value, where X_1, X_2 and X_3 represent the individual values, ΣX the sum of the individual values, $\overline{x}$ the average, and *n* the number of samples.

While the average is an important, indeed necessary, component of reported results, it in itself is insufficient. The extent of the changes within any set of values should also be considered. The variation of a single value from the average is the **deviation,** *d,* and is the difference between the individual value and the mean; thus

$$d_1 = X_1 - \overline{X}$$

$$d_2 = X_2 - \overline{X}, \text{ etc.}$$

From the deviations and **average deviation,** $\overline{d}$, may be calculated. The average deviation is the arithmetic means of the *absolute* values of the individual deviations. It can be seen in the following sets of data, both with the same average (26.30).

Set ≠ 1		Set ≠ 2	
X (g)	*d* (g)	*X* (g)	*d* (g)
26.27	0.03	26.29	0.01
26.30	0.00	26.30	0.00
26.33	0.03	26.31	0.01
$\overline{X} = 26.30$ g	$\overline{d} = 0.02$ g	$\overline{x} = 26.30$ g	$\overline{d} = 0.007$ g

that the absolute average deviation of the first set is 0.02 g while that of the second set is 0.007 g.

Another means of expressing precision is with sample **standard deviations.** This is defined as the square root of the sum of the squares of the absolute deviations of the individual samples divided by one less than the number of samples. It may be represented by

$$s = \sqrt{\frac{d_1^2 + d_2^2 + d_3^2 + ...}{n-1}}$$

where d_1, d_2 and d_3 are the absolute individual deviations and n is the number of observations. An illustration follows:

Observation	%Zn	d	d^2
X_1	18.80	0.02	0.0004
X_2	18.75	0.03	0.00.9
X_3	18.80	0.02	0.0004
X_4	18.74	0.04	0.0016
X_5	18.79	0.01	0.0001
ΣX	93.88	0.12	0.0034

$$\overline{X} = \frac{93.88}{5} \qquad \overline{d} = \frac{0.12}{5} \qquad s = \sqrt{\frac{0.0034}{5-1}}$$

$$\overline{X} = 18.78\%\,Zn \quad \overline{d} = 0.02\%\,Zn \qquad s = 0.03\%\,Zn$$

While the standard deviation will not be equal to the average deviation, the two will usually be of the same order of magnitude.

In addition to average and standard dieviations, the error might be reported in terms of a **range**, *w*, which is the difference between the largest and smallest values. In the previous set of data the range.

$$w = (18{:}80 - 18.74)\%\ Zn = 0.06\%\ Zn.$$

The three parameters ($\overline{d}$, s and w) may be reported on either an *absolute* or a *relative* basis. When absolute, the units of each of the three will be the same as the units of measurement. For example, the previous determination may be reported as one of the following :

(18.78 ± 0.06)% Zn (range) or

(18.78 ± 0.02)% Zn (average deviation) or

(18.78 ± 0.03)% Zn (standard deviation).

The individual reporting the values has the responsibility of deciding which of the three method to use.

It is frequently more meaningful to express the range and deviations on a relative basis. The relative precision parameters from the example could be expressed as parts per thousand, ppt, as:

$$\text{relative error} = \frac{\text{error}}{\text{true value}} \times 1000$$

$$\text{relative range} = \frac{w}{\overline{\overline{X}}} \times 1000 = \frac{0.06\%}{18.78\%} \times 1000 = 3.2 \text{ ppt}$$

$$\text{relative average deviation} = \frac{\bar{d}}{\overline{\overline{X}}} \times 1000 = \frac{0.02\%}{18.78\%} \times 1000 = 1.1 \text{ ppt}$$

$$\text{relative standard deviation} = \frac{s}{\overline{\overline{X}}} \times 1000 = \frac{0.03\%}{18.78\%} \times 1000 = 1.6 \text{ ppt}$$

and the results reported as

18.78% ± 3.2 ppt (range) or

18.78% ± 1.1 ppt (average deviation) or

18.78% ± 1.6 ppt (standard deviation).

A relative error can also be reported on various other bases by merely changing the multiplication factors. For example, the factor would become 100 if reported in percent, and 1,000,000 if reported in parts per million, ppm. When the standard deviation is expressed as a percentage of the mean it is called the **coefficient of variance,** *v*. Thus

$$v = \frac{s}{\overline{X}} \times 100 = \frac{0.03 \text{ g}}{18.78 \text{ g}} \times 100 = 0.16\%$$

It is incumbent upon the individual to report values with explicit explanations. Thus a reported value of

18.78 ± 1.6

is incomplete and misleading. How much more explicit it is to say, "The mean of five determinations was 18.78% Zn and the relative standard deviation was 1.6 ppt".

If the errors are random and an infinite number of determinations is obtained, a plot of the frequency of occurrence (ordinate) against the observed value of measurements (abscissa) will result in a **normal**

distribution curve such as is shown in Fig. 15.2.

The total number of determinations is called the **population.** In any normal distribution curve the centre of the abscissa is the mean, μ, the small errors are more frequent than the large errors, and the positive errors are equal to the negative errors. The standard deviation of the population is called σ. Since the normal distribution curve is based upon an infinite number of samples, and since an infinite number of samples is not attainable in practice, μ and σ are theoretical values. The corresponding experimental values are $\overline{X}$ and s, respectively.

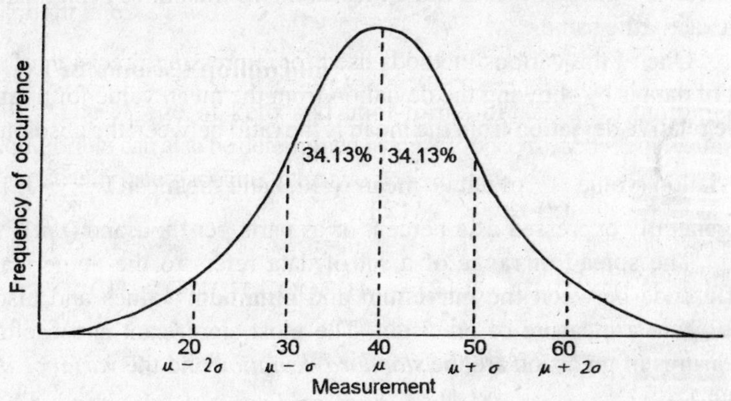

Fig. 15.2: A normal distribution curve. The mean is $\mu = 40$, and the standard deviation is $\sigma = 10$.

The percentage of all observations which fall within various multiples of the theoretical standard deviation is accurately known. Specifically, 68.26% of the values will fall into the range of $\mu \pm \sigma$, 95.46% within $\mu \pm 2.0\ \sigma$ and 99.74% within the range of $\mu \pm 3.0\ \sigma$.

Several statistical tests are available for comparing a set of experimental values with the theoretical values of a normal distribution as well as for determine what constitutes and adequate number of replicated determinations. The student is referred to any standard quantitative analysis text for these tests as well as those which indicate when a sample should be rejected on a statistical basis.

The discovery and treatment of **random** errors are a statistical problem. On the other hand, the prevention, discovery, and treatment of **systematic** errors and methods of minimizing them as they are related to specific instruments will be discussed now.

Precision and Accuracy

Two terms most widely used in discussion of reliability of data are *precision* and *accuracy.*

Precision

It refers to the variability among replicate measurements. It describes the reproduction of results *i.e.,* the agreement between the numerical values of two or more measurements that have been made in exactly the same.

One of the various methods used for expressing precision of a set of data is by showing the deviation from the mean value for a set. The relative deviation from the mean is the ratio between the absolute deviation (value (x) of data—mean value $(\overline{x})$) and the mean $\left(\frac{x-\overline{x}}{\overline{x}}\right)$. It is generally expressed as a percent or as parts per thousand (*ppt*).

The spread or range of a set of data refers to the numerical difference between the maximum and minimum values and also constitute a measure of precision. The most significant and useful measures of precision are the *standard deviation* and the *variance of data.*

Accuracy

It refers to the closeness of a measurement to its true or accepted value, x_i.. The absolute error is given by (x_i-x_t) where x_i is the true value. A positive error thus indicates that the measured quantity is greater than reference value. The relative error is expressed as $\frac{x_i-x_t}{x_i}$.

The absolute error is reported in the same units as measurement. The relative error can be expressed in parts per thousand indicated as %.

Standard Deviation

Standard deviation for a method of measurement is quite useful quantity for estimating and reporting the probable size of indeterminate errors.

For an infinite set of data, the standard deviation (σ) is given by

$$\sigma = \sqrt{\frac{\Sigma(x_i - \mu)^2}{N}}$$

The above equation holds good strictly only when N → α. The σ values can be easily found by using the above equation.

Another precision term most commonly used by stasticians is the variance which is equal to σ^2. The preference to use σ rather than σ^2 is because the units of σ are same as those of the quantity measured.

Variance has the advantage of being an additve property and several independent causes of variance exist in a system the total variance (σT^2) is the sum of individual variances.

The estimated standard deviation (s) for a small number of measurements can be defined as under

$$s = \sqrt{\frac{\Sigma(x_i - \overline{x})^2}{N-1}}$$

where (N–1) is the number of degrees of freedom. For distinguishing σ and s, σ is called *population standard deviation* and s as *sample standard deviation.*

The Coefficient of Variance

It is the standard deviation expressed as the percent of the arithmetic mean. Thus

$$CV = (\frac{\sigma}{\mu}) \times 100$$

It is useful for comparing the variability of groups of measurements having widely different mean values. For example if the results of analysis of A are reported as under

% A = 0.906 (± 0.014)

% A = 9.06 (± 0.14)

where the figures in small brackets are standard deviations. We find wide variance in the standard deviations but the coefficient of variance for the two determinations are same (1.5%).

Example

In a sample of soil the determination of iron was carried

colorimetrically. It was found as 1.67, 1.63 and 1.70 *ppm.* Calculate the standard deriation of measurement.

Solution

	x_i	$(x_i)^2$
	1.67	2.7889
	1.63	2.6569
	1.70	2.8900
$\Sigma x_i =$	5.00	$\Sigma x_i^2 = 8.3358$

$$s = \sqrt{\frac{8.3358 - (5.00)^2/3}{3-1}}$$

$= 0.0351$

$= \mathbf{0.04\ ppm.}$

Confidence Interval

The true mean value (μ) of a measurement is a constant which always remains unknown. However by making use of statistical theories we can determine the range with in which the true value might fall. The limits obtained in this way are known as *confidence limits* and the interval defined by these limits is called the *confidence interval.* The confidence limit is given by the relation

$$\text{Confidence limit} = \bar{x} \pm \frac{ts}{\sqrt{N}}$$

where t is a statistical factor which depends on the number of degrees of freedom and the confidence limit described.

Confidence Level (Z)

Referring to Gaussian distribution curve (Fig. 15.1), the breadth of the normal error curve is determined by ρ. For any given value of σ the area under a part of the normal error curve relative to the total area can be related to the parameter (Z) by equation

$$Z = \frac{x_i - \mu}{\sigma}$$

This ratio of areas (usually expressed as per cent) is known as *confidence level.* It measures the probability for the absolute deviation

$(x-\mu)$ being equal to or less than Z σ. Thus area under the curve encompassed by Z = ± 1.96σ corresponds to 95% of total area. Here the confidence level is 95% *i.e.*, 95 times out of 100 the calculated value of $(x-\mu)$ for a large number of measurements will be equal to or less than 1.96 σ. Confidence intervals for various values of Z are listed in Table 15.1.

Table 15.1

Confidence level, %	*Z*
50	0.67
68	1.00
80	1.29
90	1.64
95	1.96
99	2.58
99.7	3.00
99.9	3.29

It has been found been found that the confidence interval is decreased by $\overline{n}$ for the average of *n* replicate measurements. A more general form of equation for confidence level is

$$\mu = \overline{x} \pm \frac{Z.\sigma}{\sqrt{n}}$$

The confidence interval based on this equation apply only in the absence of determinate errors.

Confidence Limits When σ is Unknown

To account for the potential variability of *s*, a parameter μ used which is defined as

$$t = \frac{x - \mu}{s}$$

the parameter *t* is dependent not only on the desired confidence level but also upon the number of degrees of freedom. The value for *t* become equal to those of Z (Table 13.1) as the number of degrees of freedom become infinite.

The confidence limit for the mean $\overline{x}$ for *n* replicate measurements can be obtained from *t* by the following equation.

$$\text{Confidence limit for } \mu = \bar{x} \pm \frac{t.s}{\sqrt{n}}$$

Various values of t for various levels of probability are given in Table 15.2.

Table 15.2

	Factor for confidence interval %				
Degrees of Freedom	*80*	*90*	*95*	*99*	*99.9*
1	3.08	6.31	12.7	63.7	63.7
2	1.89	2.92	4.30	9.92	31.6
3	1.64	2.35	3.18	5.84	12.9
4	1.53	2.13	2.78	4.60	8.60
5	1.48	2.02	2.57	4.03	6.86
6	1.44	1.94	2.45	3.71	5.96
7	1.42	1.90	2.36	3.50	5.40
8	1.40	1.86	2.31	3.36	5.04
9	1.38	1.83	2.26	3.25	4.78
10	1.37	1.81	2.23	3.17	4.59
11	1.36	1.80	2.20	3.11	4.44
12	1.36	1.78	2.18	3.06	4.32
13	1.35	1.77	2.16	3.01	4.22
14	1.34	1.76	2.14	2.98	4.14
15	1.29	1.64	1.96	2.58	3.29

Sensitivity

The sensitivity of an instrument or a method is a measure of its ability to discriminate between small difference in analytic concentration: Two factors that limit sensitivity are (i) the slope of the calibration curve and (ii) the reproductibility or precision of measuring device.

For two methods having equal precision the one having the steeper calibration curve will be more sensitive. In case the slope of calibration curves is same in two methods then the one exhibiting better precision will be more sensitive.

The simple definition of sensitivity that has been accepted by the International Union of Pure and Applied Chemists (IUPAC) is

Calibration sensitivity, m, which is the slope of calibration curve at *the concentration of interest.*

For linear calibration curves, calibration sensitivity is independent of concentration and can be deviced from the following relationship:

$$S = mc + S_{bl}$$

Where S and S_{bl} are instrumental signals for the analytic and a blank respectively.

Mandel and Stichler proposed the following definition for analytical sensitivity, γ.

$$\gamma = \frac{m}{S_s}$$

where S_s is the standard deviation of signals and m is the slope.

Analytical sensitivity is relatively intensive to amplification factor and is independent of the measurement units for S. However it is concentration dependent since S_s often varies with concentration.

Detection Limit

It refers to the minimum concentration or weight of analyte that can be detected at a known confidence level. This limit depends upon the ratio of magnitude of the analytical signal to the size of the statistical fluctuations in the blank signal. As the limit of detection is approached the analytical signal and its standard deviation approach the blank signal S_{bl} and its standard deviation S_b. The minimum distinguishable analytical signal S_m is taken as a multiple k of the standard deviation. Thus

$$S_m = \overline{S}_{bl} + k.S_{bl}$$

The detection limit C_m is given by

$$C_m = \frac{S_m - S_{bl}}{m}$$

According to Kaiser the reasonable value for k is 3, when $k = 3$, the confidence level of detection will be 95% in most cases. Long and wine forner also recommend the user of $k = 3$.

[illegible]

[illegible] and B [illegible] respectively [illegible] for the [illegible] and [illegible] respectively [illegible]

[illegible]

Detection Limit

[illegible] the [illegible] of the [illegible] signal [illegible] the [illegible]

[illegible]

According to [illegible] theorem [illegible] signal [illegible] and [illegible]

[illegible]